BASIC THERMODYNAMICS:
Elements of Energy Systems

BASIC THERMODYNAMICS:

BERNHARDT G. A. SKROTZKI, *ME, BSEE*
Associate Editor, POWER
Member, American Society of
Mechanical Engineers and
Institute of Electrical and Electronics Engineers
Registered Professional Engineer
New York State

elements of energy systems

McGRAW-HILL BOOK COMPANY
New York
Toronto
London

BASIC THERMODYNAMICS: *Elements of Energy Systems*

Copyright © 1963 by McGraw-Hill, Inc. All Rights Reserved.
Printed in the United States of America. This book, or parts thereof,
may not be reproduced in any form without permission of the publishers.
Library of Congress Catalog Card Number: 62-18863

15 FGBP 8 5 4 3 2 1

ISBN 07-057945-8

Preface

This book first appeared as a series of articles in *Power* magazine under the title "Thermo Refresher." The series was prepared at the suggestion of the former editor of *Power*, Philip W. Swain, who recognized the need of a presentation of thermodynamics in a simple style. Too often the understanding of the basic principles of thermodynamics has been unnecessarily complicated by the use of ponderous and involved language styles. The gratifying success of this series of articles was accompanied by frequent requests to have it available in book form.

Many readers suggested that such a book would be suitable for class use as a basic text. To make it so, the original articles have been expanded to give mathematical derivations of fundamental equations wherever they were not included originally. Review topics have been placed at the end of each chapter against which the reader can test his knowledge gained. The answer to every question can be found in the text. For class use and the individual reader, problems have been added at the end of most chapters. These offer the opportunity to apply the principles covered in the chapter.

An outstanding feature of the book is the frequent use of illustrations. Especial care has been used in drawing illustrative P-V and T-S charts. Even when not graduated, all curves have been drawn in proper proportion so that the reader develops a feel for the changes to expect in the various processes. Energy flow diagrams have been used often to emphasize the simple natures of the quantities that are evaluated by rather complex equations.

Frequent reference is made to the molecular kinetic theory. In the author's teaching experience this has often helped the student's comprehension in a qualitative way. While the kinetic theory does not explain all the relations, it proves very helpful in understanding some of the limitations of purely mathematical development.

Some of the standard engine cycles have been extensively analyzed so the reader can better understand how their behavior will vary with changing controlling conditions.

The author believes that this book will fill the requirements for the basic text of a first course in engineering thermodynamics. A prerequisite for the course would be a first course in the physical sciences.

Bernhardt G. A. Skrotzki

Contents

CHAPTER

1 Energy and Work 1
2 Heat and Gas Pressure 13
3 Perfect-gas Equation 26
4 Measuring Energy 40
5 Kinetic Molecular Theory 56
6 Energy Equation of Gases 71
7 The Carnot Engine 89
8 Entropy—Index of Heat Flow 101
9 The Reversed Carnot Cycle 112
10 Process Irreversibilities 123
11 Basic Engine Cycles 133
12 Compressed-air Cycles 142
13 Internal-combustion-engine Cycles 171
14 Free-piston Gas Generator 207
15 Actual Engine Cycles 215
16 Steady Flow Energy Equation 228
17 Gas-turbine Cycles 249
18 Nozzle Gas Flow 268
19 Turbine Nozzles and Buckets 283
20 Steady-flow Compressors 298
21 Gas Properties 302
22 Properties of Mixed Gases 309
23 Energy Sources and Combustion 318
24 Vapor and Liquid Properties 340
25 Steam Processes 361
26 Steam Cycles 382
27 Steam-cycle Components 423
28 Vapor-compression Refrigeration 438
29 Gas and Vapor Mixtures 456

APPENDIX

A Work—Constant Temperature 483
B Work—Adiabatic Process 483
C Entropy—Constant Pressure 485
D Entropy—Constant Volume 485
E Entropy—Polytropic Process 486
F Perfect-gas Formulas for One Pound of Gas 487
G Optimum Intermediate Pressure 488
H Nozzle Critical Pressure Ratio 489
I Thermodynamic Properties of Sodium and Its Vapor 490
J Mollier Diagram of Thermodynamic Properties of Steam (folding insert following page 496)

INDEX 497

ABBREVIATIONS AND SYMBOLS

a = acceleration, ft per sec^2
A = area, sq ft
bdc = bottom dead center
bhphr = brake horsepower-hour
bmep = brake mean effective pressure
bsfc = brake specific fuel consumption
Btu = British thermal unit
c = per cent compressor clearance
c = specific heat, Btu per lb
c_n = polytropic specific heat, Btu per lb
c_p = specific heat at constant pressure, Btu per lb
c_v = specific heat at constant volume, Btu per lb
C = carbon atom
C_d = nozzle coefficient of discharge
C_v = nozzle velocity coefficient
CO = carbon monoxide molecule
CO_2 = carbon dioxide molecule
COP = coefficient of performance
d = dimension, ft
D = gas density, lb per cu ft
e = cycle thermal efficiency
e_b = boiler efficiency
e_c = compressor efficiency
e_e = engine efficiency
e_n = nozzle efficiency
e_r = regenerator effectiveness
e_s = stage efficiency
e_t = thermal efficiency
e_v = compressor volumetric efficiency
e_D = diffuser efficiency
e_N = nozzle efficiency
E = internal energy, Btu per lb
E_k = kinetic energy, Btu or ft-lb per lb

E_p = potential energy, ft-lb per lb
F = force, lb ($= ma$)
F = temperature, degrees Fahrenheit
g = acceleration of gravity = 32.2 ft per sec^2
h_f = enthalpy of liquid, Btu per lb
h_{fg} = enthalpy of vaporization, Btu per lb
h_g = enthalpy of vapor, Btu per lb
h-p = high pressure
H = enthalpy, Btu or ft-lb per lb
H = hydrogen atom
H_0 = stagnation enthalpy, Btu per lb = $H_1 + E_{k1}$
H_2O = water-vapor molecule
HHV = higher heating value, Btu per lb
HR = heat rate, Btu per kwhr
Ihp = indicated horsepower
J = 778.26 ft-lb per Btu
in. Hg abs = inches of mercury absolute
k = ratio of specific heats = c_p/c_v
KE = kinetic energy, ft-lb or Btu per lb
kwhr = kilowatthour = 3412.75 Btu
$\log_e$ = logarithm to the base e
l-p = low pressure
L = piston stroke, ft
m = mass = w/g
m = bleed flow, lb per lb throttle steam
M = mass
M = Mach number
M = molecular weight
n = polytropic-process constant
N = total number of items

ABBREVIATIONS AND SYMBOLS

N = nitrogen atom
O = oxygen atom
psf = lb per sq ft pressure
psfa = lb per sq ft, absolute
psfg = lb per sq ft, gage
psi = lb per sq in.
psia = lb per sq in., absolute
psig = lb per sq in., gage
P = pressure, psi, psf, psia, psig, psfa, psfg
P_c = nozzle critical pressure, psia, psfa
P_m = mean effective pressure, psi, psf
P_0 = stagnation pressure, psia, psfa
P_r = pressure ratio
P_r = reduced pressure
Q = heat transferred, Btu per lb
Q_a = heat added to cycle, Btu per lb
Q_r = heat rejected from cycle, Btu per lb
R = gas constant
R = absolute temperature, degrees Rankine
R_u = universal gas constant = 1545
RF = reheat factor
$\overline{RH}$ = relative humidity, per cent
s = distance or length, ft
s_f = entropy of liquid, Btu per lb per F
s_{fg} = entropy increase of saturated liquid to vapor, Btu per lb per F
s_g = entropy of vapor, Btu per lb per F
S = entropy, Btu per lb per F
S = sulfur atom
$\overline{SH}$ = specific humidity, lb vapor per lb dry gas
SR = steam rate, lb per kwhr
t = temperature, F
t = time, sec
tdc = top dead center

T = absolute temperature, R
T_c = nozzle critical temperature, R
T_0 = stagnation temperature, R
T_r = receiver temperature, R
T_r = reduced temperature
T_s = source temperature
$\overline{TH}$ = total heat, Btu per lb mixture
u = velocity, fps (feet per second)
v = velocity, fps
v_f = specific volume of saturated liquid, cu ft per lb
v_{fg} = specific volume increase of saturated liquid to vapor, cu ft per lb
v_g = specific volume of vapor, cu ft per lb
V = specific volume, cu ft per lb
V_c = clearance volume, cu ft
V_d = displacement volume, cu ft
V_M = mole volume, cu ft
V_r = compression ratio = V_1/V_3
V_r = reduced volume
v_t = total volume, cu ft
w = weight of a mass, lb
W = mechanical work, ft-lb or Btu per lb
W_e = engine work output, Btu or ft-lb per lb
W_f = flow work, ft-lb or Btu per lb
W_H = work in h-p cylinder, ft-lb or Btu per lb
W_L = work in l-p cylinder, ft-lb or Btu per lb
W_O = lb oxygen per lb fuel
W_p = pump work input, Btu or ft-lb per lb
x = per cent quality
y = per cent moisture
Z = compressibility factor

CHAPTER 1

Energy and Work

Before getting into the principles of thermodynamics we will study some of the basic physics on which the science is built. Thermodynamics primarily deals with energy, which can take on many forms. We find much of the machinery we use runs on the principles of thermodynamics. This is true for boilers; turbines; heat exchangers; compressors; internal-combustion engines; refrigerating systems; heating, ventilating and air-conditioning systems; processing equipment; and other apparatus. The interesting fact about these machines is that those with diametrically opposite products actually work on identical principles. Now, to work.

1·1 Energy—what is it? That is a hard question to answer in a few words. We live with energy every moment and can sense it when we move or get hot or get cold or fall and in other ways. Sensing it is not enough; we must be able to measure it and in this way perhaps define it.

Scientists usually define *energy as the ability to do mechanical work*. But in addition to this, Einstein postulated, some 50 years ago, that energy is also just another form of matter. This has been dramatically proved with the creation of the nuclear-energy industry since 1945. His simple equation shows that 1 lb of matter equals 11.3 billion kwhr of energy. As interesting as this relationship is, we will not usually be concerned with it in the engineering thermodynamics that we will talk about here.

1·2 Mechanical work. Since we hang our definition of energy on it, this needs closer examination. We naturally think of work as the daily jobs we have to do. But mechanical work has a specialized meaning: *A force moving a body through a distance does mechanical work*. Note that the force must move to do mechanical work.

Look at it this way. When we push against a brick column, Fig. 1·1, we exert considerable force. But the brick column does not budge. So we are not doing any mechanical work because our force is not moving. If we

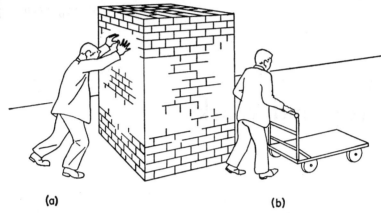

Fig. 1·1 Force does work only when it moves object it acts upon. In (a) man does no work; in (b) man does mechanical work.

push a handcart, we can move it. Now we are doing mechanical work. We calculate mechanical work as

$$W = Fs \tag{1·1}$$

where W = mechanical work, ft-lb
F = force applied, lb
s = distance moved while force acts, ft

Example: By putting a spring scale between our hands and the cart handrail, Fig. 1·1, we find we are exerting a force of 15 lb while moving the cart a distance of 12 ft. How much mechanical work have we done?

Solution: $W = 15 \times 12 = 180$ ft-lb.

1·3 Convertibility. The ability of internal energy of matter to do mechanical work depends on proper handling. Basically, internal energy is a *condition* of a substance, but it usually does not affect the composition or weight of a substance except in chemical changes. Actually, the mass does change (Einstein's relation), but so slightly that we can ignore it for engineering work. For example, the condition of a body changes when it gives up internal energy and this internal energy may do mechanical work on the body or some other body.

Internal energy can be changed into work, and work into internal energy. Since one can convert to the other, they are measured in the same units, foot-pounds. Later on we shall find that work is really one way of transferring internal energy from one substance or body to another.

ENERGY AND WORK

1·4 Energy forms. These fall into several types: mechanical, electrical, thermal, chemical, nuclear, light, radiation. Let us look at two mechanical forms that will be important to us: (1) potential and (2) kinetic.

Potential Energy. This type of energy concerns the relative position of bodies in a system. In Fig. 1·2, the weight M stood on the floor originally. By applying a force F on the rope we did work on the weight and raised it through s ft. For a frictionless pulley and weightless rope, $F = M$. Work done $= Ms$.

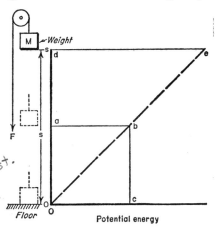

In the upper position the weight has potential energy with reference to the floor. Letting the weight fall allows it to do mechanical work equal to that done in raising it. Figure potential energy as

$$E_p = Ms \qquad (1·2)$$

$s = $ dist.

Fig. 1·2 Suspended weight gains potential energy as it rises above floor.

The units are foot-pounds. The graph of Fig. 1·2 shows the increase in potential energy with rise in height. Raising the weight halfway to a gives the weight a potential energy of ab, which equals the work done. At full height, potential energy and work done equal de.

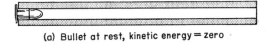

(a) Bullet at rest, kinetic energy = zero

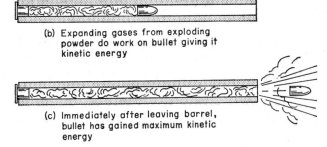

(b) Expanding gases from exploding powder do work on bullet giving it kinetic energy

(c) Immediately after leaving barrel, bullet has gained maximum kinetic energy

Fig. 1·3 Exploding powder in cartridge produces gas that does mechanical work on bullet in gun, giving it kinetic energy.

Kinetic Energy. This type of energy concerns the relative movement of bodies. A body moving relative to the earth has kinetic energy. A bullet fired from a gun acquires kinetic energy, Fig. 1·3. This energy comes from the work that the expanding gases of the exploded powder do in pushing the bullet out of the gun barrel.

Once free of the barrel, the bullet travels at constantly decreasing speed; its kinetic energy at a given velocity is figured as

$$E_k = \frac{Mv^2}{2g} = \frac{Mv^2}{64.34} \tag{1.3}$$

where E_k = kinetic energy, ft-lb
M = weight of moving body, lb
v = velocity of moving body, fps

For derivation of Eq. (1·3) see Art. 5·7.

Example: At the instant a bullet leaves a rifle it has a speed of 1,200 fps. If the bullet has a weight of 2 oz, what is its kinetic energy?

Solution:

$$M = \frac{2}{16} = 0.125 \text{ lb}$$
$$E_k = 0.125 \times \frac{1{,}200^2}{64.34} = 2{,}797 \text{ ft-lb}$$

By the time the bullet comes to a standstill, it will do work equal to this amount. Before hitting the target, the bullet does much mechanical work in pushing aside the air (wind resistance); this takes away some of the kinetic energy of the bullet and slows it down.

1·5 Changing energy form. Let us study the weight suspended in Fig. 1·4. Arrangement of weight and floor can be called a system; it has potential energy, and for our purpose we can call it the total energy of the system. This is shown by the dash-dot line at right in graph, Fig. 1·4. The dashed line shows the potential energy of the system for different heights of the weight.

When we cut the rope holding the weight, what happens to the potential energy? It decreases in proportion to the loss in height. But energy can not just disappear. What happens to the lost potential energy? As soon as the weight is loose, it starts falling with increasing velocity as shown by the dotted curve. Increasing motion means that the weight gains kinetic energy. There is the answer. Disappearing potential energy reappears as kinetic energy; the conversion is exact.

The solid straight-line curve shows the gain in kinetic energy with increasing fall and velocity gain. At any level the sum of the potential and kinetic energies equals the total energy: At a the potential energy has shrunk to value ab, kinetic energy has grown to ac, and their sum equals ad, total energy of the system.

The vertical time scale at the right of Fig. 1·4 shows the relative time taken for the weight to fall to the given height. If we are interested in changes in this system in terms of uniform intervals of time, we can replot Fig. 1·4, as in Fig. 1·5. This shows that the velocity increases uniformly with time while the kinetic energy at first increases slowly and

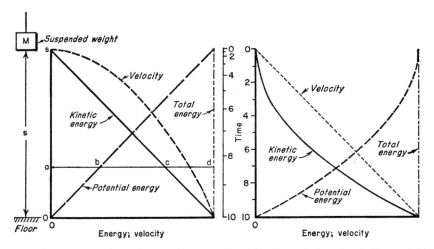

Fig. 1·4 As weight drops from position s, it exchanges potential energy for kinetic energy.

Fig. 1·5 The velocity of a falling weight increases uniformly with time, but not its kinetic energy.

then with growing rapidity as the velocity increases. The potential energy decreases in inverse fashion to the kinetic energy as we would expect.

1·6 Internal energy. The instant the weight hits the floor, it comes to a dead stop. Potential energy has dropped to zero; kinetic energy has just reached the same value as total energy. But if the speed drops to zero, the kinetic energy must drop to zero, too, according to formula for E_k, since $v = 0$.

Where did the vanishing kinetic energy go? Touching the bottom of the weight and the floor at the point of contact, we will find that they are both warmer than before impact. The kinetic energy has changed to *internal energy*, loosely called heat.

Here we see a whole series of energy transformations. Originally, work was supplied from somewhere to lift the weight above the floor. This transformed work to potential energy. During the fall, potential energy changed to kinetic energy; finally kinetic energy converted to internal energy.

Now we come to the prime consideration in our study. What is internal energy? How can we handle it? What can we do with it? These questions will be answered later as we study the various applications of converting internal energy to mechanical shaft power, creating lower temperatures, compressing gases, and so forth.

To get some idea of what internal energy is, we shall have to study the basic construction of matter. Matter may have three different *phases*, as scientists call them: (1) solid, (2) liquid, (3) gaseous.

1·7 Solids. Let us take a solid, a piece of metal, say, and start cutting it up. We shall take a small piece and cut it in half, take one of the halves and cut that in half, take one of the quarters and cut that in half, and so on and on. Pretty soon we shall find ourselves working with a microscope, halving the piece of metal dust remaining. Eventually we will find that a microscope cannot help us see the remaining particle. But all along we will notice that the nature of the metal has not changed; the pieces of all sizes are alike.

To get down to submicroscopic size we have to use imagination. On the basis of their experiments, scientists have drawn imaginary pictures to help explain the behavior of matter. These pictures are not perfect in that they reveal everything, but they go a long way in explaining actions that otherwise appear mysterious.

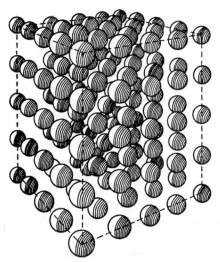

Fig. 1·6 Molecules in a solid stand in regular order with space between each two.

1·8 Atoms. Our speck of invisible metal dust might be made up of little round balls arranged in regular rows and layers, Fig. 1·6. Each little ball is an atom of metal, the smallest package that we would still recognize as a piece of the metal we started cutting up. This is true if the metal we deal with is an element, like iron, tin, or silver.

Looking closer, we find this little ball a rather fuzzy thing, the apparent surface being made up of "clouds" of electrons circling around a tiny nucleus at very high speed. The incredibly tiny nucleus, in turn, is made up of particles called neutrons and protons. The positive electrical charge carried by the protons holds the negatively charged electrons in their orbits while circling the nucleus. For the purpose of thermodynamics we need be interested only in the atom as a whole. We regard it as an equivalent solid ball, unbreakable, and always acting as a single body.

Believe it or not, these little atoms, Fig. 1·6, stand in space some distance from one another. This means that all solids include a lot of completely empty space. This is rather hard to believe on first thought, but if this space is not there, some ways that solids behave would be impossible to explain. If the matter we study should be made up of several elements, the atoms would be arranged in regular patterns making up molecules, which act like larger balls.

1·9 Forces. How can atoms stand in empty space without visible support? They do so because atoms exert forces of attraction and repulsion on one another across the empty space. These forces are similar to gravitational and electrostatic forces. Let us break off a small cube of

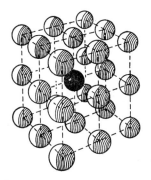

Fig. 1·7 Each atom or molecule in a solid has 26 near neighbors. Study the center one in this picture.

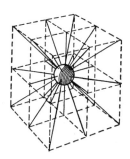

Fig. 1·8 Center lines between a molecule and its neighbors show the directions of mutual forces of attraction and repulsion.

matter, Fig. 1·7, with dimensions of three atoms along each edge. This cube has one central atom. Figure 1·8 shows the lines along which this atom interacts with each of its 26 neighbors.

The *molecular forces of attraction* are like the gravitational forces of attraction described by Newton's famous law of gravitation:

$$F = M_1 M_2 / d^2$$

Gravitational forces are proportional to the masses of the two bodies involved and inversely proportional to the square of the distance d separating them. In solids, atoms behave similarly except that the exponent for d varies from about 5 to 6 instead of being 2 as for gravitation. *Forces of repulsion* among atoms might be caused by the electrical repulsion among the negative charges of their electrons. Repulsion forces vary inversely as the 9th to 14th power of d. Figure 1·9 shows the effect of these exponents on the magnitude of the forces.

Figure 1·10 shows the net effect of the two types of forces in maintaining distance between atoms. At one given distance these forces are in balance and the net force acting on a molecule is zero. If the distance between atoms is reduced, the net force of repulsion builds up rapidly. On the other hand, increasing the distance between molecules causes the net force of attraction to increase for a short range and then fall off. When

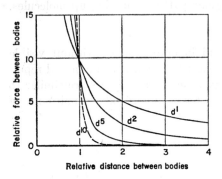

Fig. 1·9 Forces between bodies vary with the distance between them and with the value of d^n.

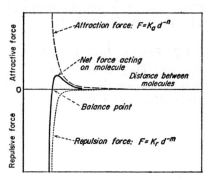

Fig. 1·10 Net forces acting on molecules in a solid depend on the sum of the opposing push and pull forces.

the distance has been doubled, the net force of attraction is very weak and practically disappears for greater distances.

Figure 1·11 gives a more concrete picture regarding forces of attraction between bodies obeying Newton's law of gravitation. If the attraction is 1 lb at distance 1 (shown by width of connecting bar), the force will be only $\frac{1}{4}$ lb at a distance of 2 and only $\frac{1}{9}$ lb at distance 3. For atoms with the force varying with $1/d^5$, doubling the distance reduces the force to 0.03 lb while tripling the distance reduces the force to a mere 0.004 lb, entirely negligible. This variation will be important to us later in understanding change of phase during boiling and condensation.

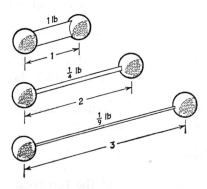

Fig. 1·11 Newton's law of gravitation shows that the pulling force diminishes with increasing distance between two bodies.

1·10 Motion. As we look at Figs. 1·6 and 1·7, the atoms appear very calm. They would look this way only if they had no internal energy; that is, their temperature would be at absolute zero. At room temperature the atoms buzz with action. Each individual atom bounces around rapidly but stays in one general location in the solid tied there by molecular forces. The hotter the solid, the more active the atoms become, Fig. 1·12a; this shows as a temperature rise.

In the more general case where a solid is made of molecules, the molecules may consist of two or more atoms each. Molecules as separate units may then be represented by the single balls in Figs. 1·6 to 1·8. Molecules may absorb energy in more than the kinetic vibration form of Fig. 1·12a.

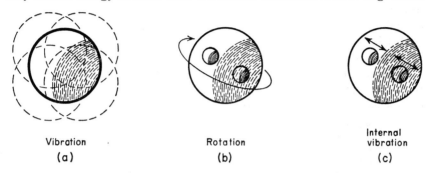

Vibration　　　　Rotation　　　Internal vibration
(a)　　　　　　　(b)　　　　　　(c)

Fig. 1·12 Molecules absorb and hold energy several ways. Three are by (a) molecules vibrating in space, (b) atoms rotating about each other within the molecule, (c) atoms vibrating inside the molecule.

Within the molecule the atoms may revolve about the center of the molecule, Fig. 1·12b. The higher the rotation speed, the more energy the molecule has. The atoms may also vibrate in respect to one another within the molecule, Fig. 1·12c. These last two forms of energy may have no effect on the measured temperature, though they may vary with it.

All these forms of motion show the ways in which a solid may absorb or give up internal energy. At a given temperature level, individual molecules in a solid have varying amounts of energy. But considering all the molecules of a mass as a group, their internal energies (in all three forms) add up to a fixed amount per unit of weight for each temperature level.

1·11 Liquids. When heating a solid we find that it melts at a certain temperature. Certain solids, like ice, change from solid to liquid with addition of heat, but

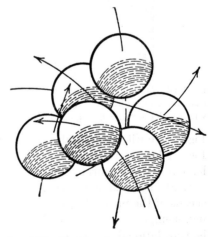

Fig. 1·13 Liquid molecules move past one another at high speed in curved paths in response to varying pulls upon one another.

the temperature stays constant during this heating. How do we explain this with our molecular picture?

As the molecules absorb more energy in the solid phase, they bounce through greater and greater ranges, with increasing speed and frequency.

The greater range of motion makes the solid expand. At a certain temperature level some molecules bounce around so hard they break away from the attractive forces of their neighbors and begin to wander off among the other molecules. They follow a curved, winding path under the pull of the forces from the molecules they pass. But their energy (speed) is high enough so they can break away each time, Fig. 1·13.

Heating the solid does mechanical work on the individual molecules and increases their motions. It takes additional work to break the molecules away from the restraining forces holding them in place. Molecules absorb this type of work as internal potential energy. In the melting process molecules successively break away from the ordered pattern in the solid phase,

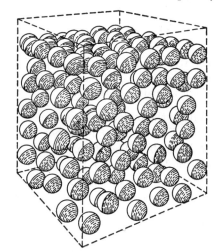

Fig. 1·14 Molecules of liquid in the cylinder bottom wander about in close order, while vapor molecules soar through space above.

Fig. 1·15 Molecules of liquid occupy bottom of the container and form a nearly level surface.

Fig. 1·6, until they are all wandering about one another in curved paths, Figs. 1·14 and 1·15. They fill the bottom of any container holding them, have a level top surface, and are closer together at the top.

Heat absorbed as internal potential energy does not affect the *average* speeds of the molecules, so the temperature does not change during melting. Temperature indicates internal kinetic energy of the molecules, that is, translational motion only.

1·12 Vapors and gases. Continued heating of a liquid raises its temperature, meaning the speed of the molecules. Again, individual molecules have different velocities, but the average speed of all the molecules corresponds to a definite temperature level.

When the liquid reaches boiling temperature, the molecules with higher speeds (energy) break away completely from their neighbors and zoom up into the space above the liquid. These molecules do additional work in breaking away from the force of their neighbors, which they absorb as potential energy and which slows them down to the average speed (temperature) of the liquid molecules. This work they have absorbed as internal potential energy.

Thus we find that temperature of a boiling liquid stays constant until all the liquid changes to vapor. Once all the liquid has evaporated, the molecules soar through space at high speeds, in practically straight-line paths, Fig. 1·16. They travel at speeds of 15 to 70 miles *per minute* at ordinary room temperatures. The molecules collide with one another and the container walls billions of times in 1 sec. Speeds increase with heating,

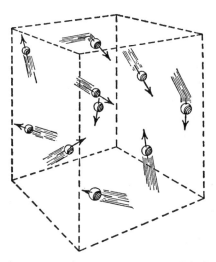

Fig. 1·16 Molecules of vapor speed through space in straight lines and collide millions of times a second with one another and the wall of confining chamber.

and of course, the temperature goes up. The vapor or gas (the difference is arbitrary) occupies much more space than the liquid for the same weight or number of molecules.

1·13 Giving up energy. We have traced the molecular actions of a solid in passing through three phases: solid, liquid, and vapor. This implies that energy was received from some source by the molecules and absorbed by them, the net effect being to speed up their motions and move them farther apart.

The reverse series of phase changes takes place when the molecules of a vapor give up their internal energy to some outside body. This slows down the molecules and allows the slower ones to agglomerate under the influence of the intermolecular forces of attraction and form drops of liquid. This continues until nearly all the vapor has condensed to a liquid

REVIEW TOPICS

1. Define energy in two ways.
2. What is the energy equivalent of 1 lb of matter?
3. Define mechanical work.
4. Can energy be changed in form?
5. Name seven forms of energy.
6. Write the equations for potential and kinetic energy of a body.
7. What are the three phases of matter?
8. Describe the internal atomic arrangement of a solid.
9. What types of forces act on molecules in a solid?
10. What does the temperature of a body indicate in regard to its molecules?
11. How do molecules behave during melting of a solid?
12. How do molecules behave during vaporizing of a liquid?

PROBLEMS

1. Burning 1 lb of coal in air will produce about 14,000 Btu of energy or 4.10 kwhr. How does this compare with the total energy contained in the matter of the 1 lb of coal? $E_T = 11.3(10)^9$ kwhr/lbm (pg 1)

2. A man walks up the stairs of a 10-story building to its roof. If the man weighs 180 lb and each story is 9 ft high, how much energy does he expend in his upward climb?

$9 \times 10 \times 180$
$16.2 (10)^3$ ft lbf

3. How much potential energy in relation to the ground has the man acquired in Prob. 2 when he reaches the roof?

4. An automobile weighing 3,300 lb is moving at 80 mph (miles per hour). What kinetic energy does it have?

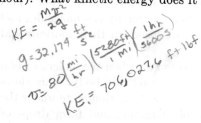

$KE = \dfrac{Mv^2}{2g}$

$g = 32.174 \ \dfrac{ft}{s^2}$

$v = 80 \left(\dfrac{mi}{hr}\right)\left(\dfrac{5280 ft}{1 mi}\right)\left(\dfrac{1 hr}{3600 s}\right)$

$KE = 706,027.6$ ft lbf

CHAPTER 2

Heat and Gas Pressure

In Chap. 1 we learned about the potential and kinetic forms of mechanical energy and how they are paralleled in molecular form to be known as internal energy. Mechanical work we defined exactly and showed it to be a form of transferring energy from one body to another.

2·1 Heat. When we mention heat in everyday living, we all understand what is meant. Perhaps the most common example would be heating a room by a steam radiator. Ordinarily we say that the steam in the radiator has heat. In thermodynamics, we are more precise and say that the steam has internal energy.

This may seem a rather fussy distinction, but there is a good reason for it. In thermodynamics, *heat* means the *action of transferring internal energy* from one body to another. Then if we say the steam heats the air in the room, that is technically correct. The internal energy of the air is being raised by the internal energy disappearing from the steam.

2·2 Heat transfer. Let us use our imaginary molecular picture to explain just how heat acts. Take an example of a closed tank divided in half by a thin wall. One half holds a hot gas, and the other a relatively cool gas, Fig. 2·1. Hot-gas temperature equals t_1, and cold-gas equals t_2.

The hot-gas molecules with their internal energy rush around at great speed, colliding with one another, the walls of the tank, and the dividing wall. Whenever a hot-gas molecule hits a wall molecule, the latter absorbs some of the kinetic energy by vibrating faster and bouncing around in all directions. The gas molecule rebounds with reduced speed because it gave up some of its kinetic energy to the wall molecule.

Increased movement or bouncing of the outer wall molecule makes it push and pull on its neighbors through the mutual attracting and repulsing forces they exert on one another. This makes the outer wall molecule do work on the other wall molecules and so give up some of its kinetic internal energy gain to the others. This action continues through the depth of the wall, so that soon the outer wall molecule exposed to the cool gas begins to vibrate faster. In this way the wall molecules pass the internal energy along from the hot side to the cold side. Each, in turn, loses some as it shares what it gains with its cooler neighbors.

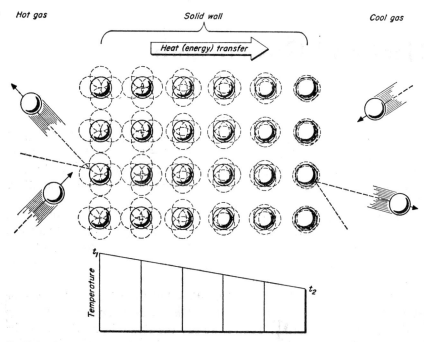

Fig. 2·1 The kinetic energy of hot-gas molecules, left, decreases as they slam into wall molecules. Wall molecules pass along energy as greater vibration to neighbors, then on to the colliding cool-gas molecules.

As the hot-gas molecules continue hitting the outer wall molecules, however, the supply of internal energy keeps coming to the wall, so we have a condition set up as shown in Fig. 2·1. The wall molecules successively have lower internal energy, which we measure as a drop in temperature through the wall from the hot side to the cool side. The rate of temperature drop through the wall depends on whether the material is a heat conductor or insulator.

Now let us study what happens between the wall and the cold-gas molecules. The outer wall molecules on the cold-gas side begin vibrating faster as they pick up energy from the interior molecules. The cold-gas molecules are rushing around colliding with one another and the wall, but at a slower rate than the hot-gas molecules. Whenever a cold-gas molecule collides with a wall molecule, the wall molecule acts like a baseball bat and sends the gas molecule (baseball) rebounding with increased speed, that is, more kinetic energy. In this way, the wall molecule gives up some of its internal energy to the gas molecule.

So there is the continuous process; the hot-gas molecules give up internal energy to the wall molecules, which share it among themselves and pass most of it on to the cold-gas molecules. As we said, this is heating—the transferring of internal energy from one body to another.

2·3 Traveling energy. Let us clarify our terms. Heat we now recognize as the transfer of internal energy, so heat might be called traveling energy. Work also falls into this class for mechanical energy; it is the transfer of potential or kinetic energy from one body to another. When this is clear to us, we realize that a substance or body does not contain heat or work in the sense that it has internal energy or mechanical energy.

In thermodynamics, we consider work and heat as the main two methods of adding or subtracting energy from a system or body. You would rightly add heat and nuclear radiation, fission, and fusion as other ways of transferring energy, but they need be considered only for special cases. Work concerns movement and position of a body in relation to a system of bodies. Heat deals with changes in internal condition of the body's molecular energy, that is, position and movement of its molecules.

2·4 Energy changes. Figure 2·2 shows the basic processes in a simplified power plant. Here a waste-heat boiler passes waste gas through its fire tubes. Some internal energy of the gas transfers as heat through the tube walls to the water and steam in the drum. Steam carrying internal energy flows into the engine. Here the steam pushes out the piston. The steam exerting a force on the piston and moving it does mechanical work by giving up part of its internal energy. At the end of the piston stroke the steam leaves the cylinder, carrying along its remaining internal energy.

Let us review what happened here. The waste gas gives up some internal energy in the boiler. This energy passes as heat into the water and steam. This heat raises the internal energy of the water to boil it off to steam. Steam flows to the engine and gives up part of its internal energy to do work on the piston. With its internal energy reduced by amount of work done, steam exhausts from the engine.

Notice that these were all *changes* in internal energy. Never do we run into a situation where a body or substance gives up *all* its energy. This boils down to always figuring heat and work from *differences* in energy, before and after.

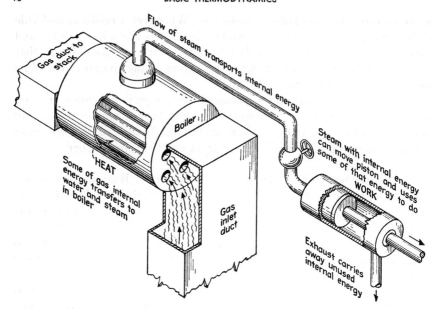

Fig. 2·2 Part of the gas internal energy flows to water and steam in the drum by heat transfer. Steam flowing from the drum enters the cylinder and gives up part of its internal energy in doing mechanical work to push the piston toward the right.

2·5 Pressure. Because of their internal kinetic energy, gas and vapor molecules fly around in space. They "fill" any container that holds them. After they collide with one another and the walls of the container, they rebound, going off in new directions. In this way they exchange energy with one another and the wall molecules. For a perfectly insulated container, the total energy in the gas stays constant.

As the gas molecules hit the walls, they exert forces that try to push the walls outward. There are billions of gas molecules in a cubic inch of gas like air, and they collide billions of times a second with a very small area of the wall. All these individual impacts add up to what we sense as a steady pressure on the container walls. If we can measure the total force F exerted on a wall with an area A, we figure the pressure P as

$$P = \frac{F}{A}$$

When F is in pounds and A in square feet, the pressure P is in pounds per square foot (psf).

First let us get acquainted with the meaning of atmospheric pressure—we are always going to be concerned with it in thermodynamics. Figure 2·3 shows the elements of a barometer. Taking a closed-end tube and completely filling it with a liquid, say mercury as in a, stop the open end with a finger. Then turning it with open end down, immerse it in a dish of mercury. When you remove your finger, some of the mercury runs out of the tube into the dish and the level falls from the closed end of the tube, Fig. 2·3b. But most of the mercury stays in the tube. The height h is about 30 in. at sea level.

This is one way of measuring atmospheric pressure. Inside the tube above the mercury level there is a perfect vacuum. (Actually there is a

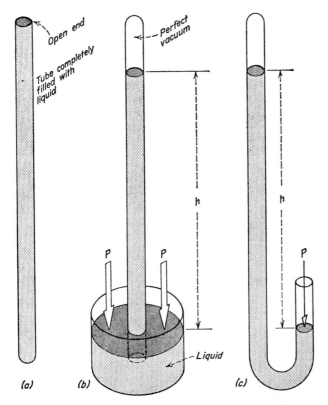

Fig. 2·3 Pressure P exerted by air molecules forces liquid up the tube that has perfect vacuum at top; h measures air pressure.

small amount of mercury vapor, but it exerts such a slight pressure that we can ignore it.) Air molecules collide with the surface of the mercury in the open dish, creating a force or pressure that pushes the liquid mercury into every unoccupied crevice and cavity. There is nothing in the upper end of the tube (no opposing pressure), so the mercury gets pushed up into the tube until the weight of the mercury column h balances the pressure of the air.

Now do not be misled. The size of the mercury surface exposed to atmospheric pressure P has no bearing on height h. We shall get the same height (reading) for a U-tube barometer as in Fig. 2·3c. Here we see more clearly that the weight of the mercury in column h is counterbalanced only by pressure P of the air on the open end.

The standard barometer or barometric pressure is 29.9213 in. Hg when the mercury is at a temperature of 32 F. We could use any liquid for our barometer—for instance, water. Our standard barometer in terms of water is 33.932 ft H_2O when the water is at 60 F. Note the great difference in height. This is easy to understand when we realize that mercury weighs 13.595 times as much as an equal volume of water.

But we were speaking of pressures in terms of pounds per square foot and then changed to heights of liquid columns. How are they related?

First, we must know that water at 60 F weighs 62.37 lb per cu ft. Since a height of 33.93 ft of water counterbalances standard atmospheric pressure, a column with cross-sectional area of 1 sq ft and height of 33.93 ft exerts a pressure at its base of

$$33.93 \times 62.37 = 2{,}116 \text{ psf}$$

and so

$$Atmospheric\ pressure = 2{,}116 \text{ psf}$$

While the pound-per-square-foot unit must be used in many thermodynamic relations, as we shall find later, it is customary to express pressures in terms of pounds per square inch, psi. All we need to remember is that there are 144 sq in. in 1 sq ft. Then

$$Atmospheric\ pressure = \frac{2{,}116}{144} = 14.7 \text{ psi}$$

The more exact value is 14.696 psi, but 14.7 is close enough for most of our calculations.

2·6 Gage versus absolute. Most of our pressure gages compare the pressure of gas or vapor against the pressure of air in our atmosphere. When measured this way, the pressure is called *gage pressure* and the units are *pounds per square inch gage* (psig) or *pounds per square foot gage* (psfg). When a gas has the same pressure as atmospheric air its gage pressure is zero, that is, 0 psig or 0 psfg.

Molecules of our air, however, bounce around rapidly and produce a pressure, even though we say that atmospheric pressure is 0 psig. Pressure can be truly zero when there are no molecules to hit a wall. This is done by removing all the molecules from a container and creating a perfect vacuum. By comparing our atmosphere with a perfect vacuum we find it has an average *absolute pressure* of 14.7 *psi absolute* (psia).

To find the true or absolute pressure of a gas or vapor we must add 14.7 psi to the pressure as read by a gage:

$$\text{psia} = 14.7 + \text{psig}$$
$$\text{psfa} = 2{,}116 + \text{psfg}$$

2·7 Pressure gages. We usually measure pressures with a bourdon tube or manometer, Fig. 2·4. The bourdon tube, which has a flattened cross section, is closed at one end. The other end connects to the region where we want to measure the pressure. The gas or vapor being measured acts on the inside tube surface and tries to force the tube into a circular cross section. The gas or vapor does this against the stresses in the tube and the pressure of the air trying to collapse the tube.

The change in cross-sectional shape makes the tube try to straighten out. This deflects the free end of the tube through the distance d, Fig. 2·4.

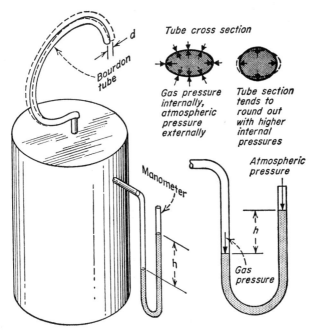

Fig. 2·4 Pressures can be measured by deflection d of a bourdon tube or by a manometer differential h. The latter is used mostly for low pressures.

Deflection d measures the pressure difference between the gas in the tank and the air of the atmosphere. Ordinarily the end of the tube links to an indicating hand sweeping over a graduated dial, and pressure can be read directly in psig.

In the manometer, Fig. 2·4, atmospheric pressure acts on the surface of the liquid in the open leg and the gas pressure acts on the surface of the liquid in the leg connected to the tank. Gas pressures higher than atmospheric push the liquid so it stands higher by distance h in the open leg. Then the sum of the liquid weight (in h) and the air pressure just balance the gas pressure.

If the gas pressure is lower than atmospheric, the liquid stands higher in the connected leg, so that h plus gas pressure just balances atmospheric.

Here h shows how much below atmospheric the gas pressure is. For this case we regard h as a negative number and add it to atmospheric to get the absolute pressure.

If the liquid is water, the height h in Fig. 2·4 would be 27.71 in. for a pressure of 1 psi. If the liquid is mercury, we find that 2.036 in. Hg equals 1 psi or, inversely, 1 in. Hg = 1/2.036 = 0.4912 psi.

Example: A manometer connected to a gas tank shows the mercury in the connected leg to be 1.5 in. higher than in the open leg. If the barometer reads 30.25 in. Hg, what is the absolute gas pressure in inches Hg abs and in psia?

Solution: Absolute gas pressure in the tank (note that 1.5 is a negative number) = 30.25 − 1.50 = 28.75 in. Hg abs.

In terms of psia, this is 28.75 × 0.4912 = 14.10 psia.

In solving this sort of problem remember that barometric pressure varies continuously. Then if the gas absolute pressure in the tank remains constant, the manometer reading h varies with the barometer reading.

2·8 Boyle's law. To understand thermodynamics we must know how gases behave. We can test one of the fundamental laws by the simple apparatus in Fig. 2·5. We need a manometer with a short sealed end and an open leg of long length. The sealed end that is graduated has an absolutely uniform bore, so we are sure of equal volumes between graduation marks.

First, we pour in mercury till equal amounts stand in both legs and we have trapped exactly four volumes of air in the closed leg, Fig. 2·5a. We keep water running over the leg with the trapped gas so its temperature stays constant at T_a.

At the time we do our experiment, barometric pressure = 30.00 in. Hg abs, so we have four volumes of air trapped in the closed leg at an absolute pressure of 30 in. Hg abs.

Next, we pour mercury into the open leg until the air volume reduces to three units. We do this slowly so the cooling water can keep the gas at exactly T_a temperature. For this volume we find the gas is under 10 in. Hg gage pressure, Fig. 2·5b. This, of course, is 30 + 10 = 40 in. Hg abs pressure.

In c we have poured in more mercury to reduce air volume to two units and find it under 30 in. Hg gage pressure or 30 + 30 = 60 in. Hg abs. Finally, in d we pour in more mercury to reduce the gas volume to one unit and find that the pressure is now 90 in. Hg gage, or 90 + 30 = 120 in. Hg abs. Before taking the measurements we have made sure each time that the trapped gas was at the temperature T_a.

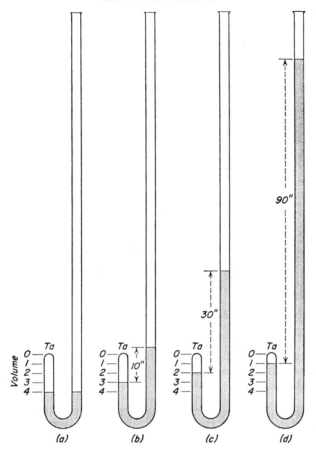

Fig. 2·5 Boyle's law can be demonstrated by trapping air in the closed end of a manometer and varying the pressure, at the same time keeping T_a steady.

Now let us list corresponding volumes and absolute pressures and then multiply them together:

Pressure, in Hg abs	Relative volume	PV
30	4	120
40	3	120
60	2	120
120	1	120

This shows that the product of absolute pressure and volume stays constant for all conditions. So we can write this relation, known as Boyle's law, as

$$PV = C$$

Measuring units for these quantities can be any convenient ones:

P = gas pressure, psfa, psia, in. Hg abs
V = gas volume, cu ft
C = constant

The constant usually differs for each gas and measuring unit used. Note that the gas temperature must stay constant for this law to hold.

Graphs are useful tools in thermodynamics, so let us plot our data for Fig. 2·5 in Fig. 2·6. This shows that, as volume shrinks, the pressure rises, gradually at first, then more and more rapidly. Starting from any point,

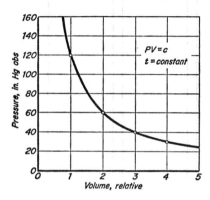

Fig. 2·6 The pressure-volume relation of Boyle's law shows that the pressure drops while the volume of gas rises.

doubling the volume halves the pressure; in reverse, halving the volume doubles the pressure.

2·9 Molecular action. We can explain the relations in Fig. 2·6 by studying what must happen to the molecules of the gas. The number of molecules in this mass of gas stays constant, but the space they move around in varies. This affects the number of collisions of the molecules *per unit area* of the enclosing container.

Actually the individual molecules move in haphazard fashion, but to simplify analysis we can assume that one portion of the total molecules bounce up and down vertically to produce pressure on the top and bottom of the container, Fig. 2·7a. The other molecules bounce back and forth horizontally to produce exactly the same pressure on the side walls of the container.

When the volume doubles, Fig. 2·7b, the molecules traveling vertically have twice the distance to go, so they now collide half as often with the top and bottom of the cylinder, dropping the pressure in half. On the other hand, those molecules traveling horizontally now cover double the wall area, so the same number of collisions produce half the pressure

($P = F/A$). Since the temperature stays constant, the molecular speeds stay constant.

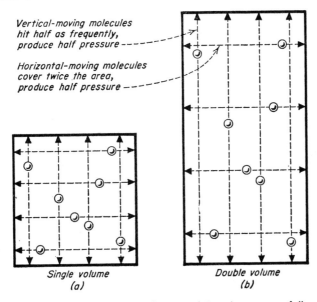

Fig. 2·7 A simplified molecular picture of gas explains why pressure falls when volume increases and temperature stays steady.

2·10 Pressure-volume relations. Whenever we deal with a *given weight* of a gas at constant temperature we can use Boyle's law to figure related pressures and volumes. We do this by the following method: When the gas is in state 1, we can write

$$P_1V_1 = C$$

For state 2,

$$P_2V_2 = C$$

Since C is the same for all states, we then write

$$P_1V_1 = P_2V_2$$

So if we know pressure and volume of a gas at one state, we do not even have to figure C to find what they would be at another state, provided the temperature does not change.

Example: We have trapped 10 cu ft of air in a cylinder with a movable leakproof piston, at atmospheric pressure of 14.7 psia. What will its pressure be when the piston is moved to reduce volume to 1 cu ft and the temperature remains the same?

24 BASIC THERMODYNAMICS

Let us substitute these quantities in our last equation:

$$14.7 \times 10 = P_2 \times 1$$

$P_2 = 14.7 \times 10/1 = 147$ psia when the volume shrinks to 1 cu ft.

Example: What volume must be provided in the above cylinder to reduce the air pressure to 3 psia?

Again, let us use our equation and substitute

$$14.7 \times 10 = 3 \times V_2$$

$V_2 = 14.7 \times 10/3 = 49$ cu ft to reduce the pressure to 3 psia.

Most gases follow Boyle's law quite closely over fairly wide ranges of pressure. Very careful measurements show that they all deviate some small amount. At very high pressures and temperatures they may disagree quite a bit. This is probably caused by the larger mutual attractive forces that the gas molecules exert on one another when they are crowded together. For many engineering purposes we can use Boyle's law with confidence. It is accurate enough to give the answers we need to solve many practical problems.

REVIEW TOPICS

1. What is the precise meaning of heat in thermodynamics?
2. Describe the molecular mechanics of transferring heat through a solid wall.
3. Name five ways of transferring energy into or out of a system of bodies.
4. Describe the energy changes taking place in a simple hot-water heating system from the furnace to the room being warmed.
5. How is pressure of a gas calculated?
6. What is the standard mercury barometer pressure?
7. What is the standard water barometric pressure?
8. What is atmospheric pressure in psfa and psia?
9. What is the relation of absolute and gage pressures?
10. Describe the action of a manometer for measuring pressure.
11. Describe the action of a bourdon tube for measuring pressure.
12. What is Boyle's law?

PROBLEMS

1. The wall of a room measures 20 by 8 ft. What total force does the atmosphere exert on the wall for a standard barometric pressure?

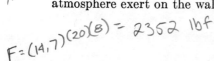

HEAT AND GAS PRESSURE   3077.74 psia

2. Gas confined in a tank has a pressure of 3,047 psig with the atmosphere at a barometric pressure of 30.74 in. Hg. What is the absolute pressure of the gas?

3. A manometer tapped into a tank full of gas shows a 1.6 in. Hg vacuum (mercury in connected leg 1.6 in. higher than in open leg) with a barometric pressure of 29.75 in. Hg. What is the absolute pressure of the gas?

4. Gas in a cylinder with a leakproof piston fills a volume of 8 cu ft and exerts a pressure of 300 psia. What will the pressure be when the volume is reduced to 1 cu ft and the temperature remains the same?

5. Gas in a cylinder with a leakproof piston occupies a volume of 1 cu ft and exerts atmospheric pressure of 30.00 in. Hg abs. To what volume must the gas be expanded to reduce the pressure to 1.00 in. Hg abs with the temperature remaining constant?

(3) $29.75 - 1.6 = 28.15$ in Hg

$P = 28.15$ in Hg $\left(.4912 \frac{psi}{in Hg}\right) =$
13.83 psia

(4) $Pv = \text{Const}$
$(300)(8) = 2400 = P(1)$
$P = 2400$ psia

(5) $(30.00)(1) = 30 = 1(v)$
$v = 30$ ft^3

CHAPTER 3

Perfect-gas Equation

We finished talking about Boyle's law in Chap. 2. This says that $PV = C$ for a given mass of gas when its temperature stays constant. But temperature changes rank as the most important factor in thermodynamic processes, so we must learn how they affect gas behavior. Effects of temperature were discovered independently by two scientists, Charles and Gay-Lussac, about the end of the eighteenth century. This was about 100 years after Boyle had discovered the $PV = C$ relation. We usually refer to the temperature relations as Charles' law.

3·1 Charles' law. To study the temperature effects, let us do some experimenting. In Fig. 3·1 we have a fixed weight of gas forced into a closed tank. A pressure gage and a thermometer allow us to take simultaneous readings of the gas pressure and temperature.

A mercury-in-glass thermometer, which we use here, is just an arbitrary indicator of temperature. Experience has shown that, when a fixed mass of mercury or other fluid gets hot, it expands; when it cools, it contracts. By letting the expansion and contraction be magnified in a narrow capillary passage we have a sensitive temperature indicator that can be graduated as we find convenient.

With these instruments let us see what happens to the constant volume of gas in the tank. In Fig. 3·1a we cover the tank completely with melting ice and cool the gas to a uniform temperature of 32 F (Fahrenheit—an arbitrary temperature scale devised by a scientist of that name). We find that the gas pressure gage reads 10 psig. In Fig. 3·1b we submerge the tank in boiling water to bring the gas temperature to 212 F. Now the pressure gage shows 19.0 psig. Finally we douse the tank in a steam bath, Fig. 3·1c, raise the temperature to 400 F, and find that the pressure rises to 28.5 psig.

Next, let us convert the gage readings to absolute pressures:

Temp, F	Psig	Psia
32	10.0	+ 14.7 = 24.7
212	19.0	+ 14.7 = 33.7
400	28.5	+ 14.7 = 43.2

When we plot the absolute pressure against the corresponding temperature readings as in Fig. 3·1, we find that the points lie on a straight line.

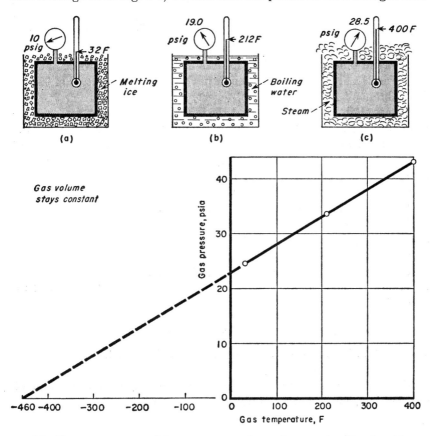

Fig. 3·1 The pressure exerted by a constant volume of gas varies directly with gas temperature. Gas pressure theoretically drops to zero at −460 F (zero internal energy).

This graph illustrates Charles' law: In a *constant* volume of gas the pressure varies directly with the temperature.

3·2 Absolute temperature. This simple straight-line relation of pressure and temperature suggests that there must be some temperature at which the pressure drops to zero. To find this we simply extrapolate the curve (dotted line) until it interesects the zero-pressure line. We find that this intersection corresponds very closely to -460 F (more accurately -459.69 F).

Let us think for a moment what zero pressure must mean on the basis of the kinetic molecular theory. No pressure means no molecular impact against the wall confining the gas—hence the molecules must be completely stripped of their kinetic energy, and they must be lying inert in the bottom of the tank.

But what does 0 F mean on the Fahrenheit scale? Nothing overly important. Fahrenheit *thought* that zero on his scale (a mixture of salt and melting ice) was the lowest possible temperature; this was back in 1708. He took his oral temperature and labeled that 100 F on his thermometer; this worked out so that melting ice was 32 F and boiling water (at 14.7 psia) was 212 F. (Since healthy people have an oral temperature of 98.6 F, did Fahrenheit have a permanent fever?)

The graph in Fig. 3·1 shows that the bottom of the "true" temperature scale stands at -460 F. Here, theoretically, the molecules of a gas lay completely inert. In actual experiments -460 F has been approached within about 0.4 F. At this level, matter exhibits many quirks; for instance, liquid helium will not stay in the bottom of a container. It spreads itself out into as thin a layer as possible, climbing up walls of the container and down the outside in defiance of gravity.

Since we are primarily interested in energy, and it appears that the internal energy of a gas is zero at -460 F, Rankine proposed that this be taken as zero on an *absolute temperature scale*. To differentiate between Fahrenheit and absolute temperatures, the latter is measured in degrees Rankine, °R. Then at 32 F, the absolute temperature is 492°R; at 212 F it is 672°R. In general, we can write

$$°R = F + 460$$

This relation is shown on the adjacent scales of Fig. 3·4.

3.3 Pressure and temperature. From the straight-line relation of Charles' law, Fig. 3·1, we can write

$$P = CT \quad \text{when } V = \text{constant}$$

where P = absolute pressure of the gas, psf
T = absolute temperature, °R
C = constant

This relation holds for any kind of gas, C being different for each.

We can also write this part of Charles' law in the form

$$\frac{P_1}{P_2} = \frac{T_1}{T_2}$$

The dotted line of Fig. 3·1 implies that the gas would stay in the form of a gas until it was cooled to 0°R. Actually, most gases would first condense to a liquid and then freeze to a solid long before temperature dropped to zero. But this does not invalidate our theory, because temperature reflects only molecular motion, that is, molecular kinetic energy and not the intermolecular forces of attraction and repulsion that control the molecular potential energy. Potential energy comes into play during the change of phases from gas to liquid to solid or the other way round.

Charles' law shows that, as absolute temperature rises, molecular speeds increase. This increases the number of collisions per unit time between wall and gas molecules, which register as a rise in pressure.

3·4 Volume and temperature. Now let us study another aspect of Charles' law by keeping the pressure constant and letting the volume vary as the temperature changes. In Fig. 3·2 we enclose a constant weight

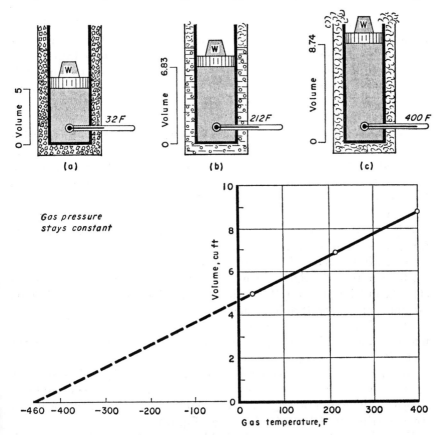

Fig. 3·2 Gas held under constant pressure varies its volume directly with changes in its temperature. Projection of curve shows that its volume drops to about zero at −460 F.

of gas in a cylinder and piston. The piston carries a constant weight W that keeps a constant pressure on the trapped gas. We bring the gas to the same temperatures as in Fig. 3·1 and note the changes in volume. The graph of Fig. 3·2 shows the plot of V, cu ft per lb, versus T. The points lie on a straight line. Extrapolating the line shows that it passes through zero volume when the temperature falls to -460 F or $0°R$. This gives us the other aspect of Charles' law: When pressure of a gas is held constant, its specific volume varies directly with its absolute temperature.

We can write equations for this part of Charles' law in the forms

$$V = CT$$

or

$$\frac{V_1}{V_2} = \frac{T_1}{T_2}$$

We can also explain this behavior by our molecular theory. As the gas molecules absorb energy, their speeds increase. This raises the number of collisions per second with the cylinder wall and piston. A greater number of collisions on the piston raises the gas pressure and moves the weighted piston upward. This increase in volume causes the vertically moving molecules to travel a longer distance between collisions and so to drop the gas pressure to the level where it again balances the weighted piston. Horizontally moving molecules cover a greater wall area and so also drop the gas pressure.

3·5 Perfect-gas equation. Actual gases do not comply strictly with Charles' law, but in many working ranges, the law can be applied to practical problems with sufficiently close results. The two parts of Charles' law, Figs. 3·1 and 3·2, can be combined into a single expression known as the perfect-gas equation. Figure 3·3 shows how they combine with the following reasoning:

P is pressure in pounds per square foot, V is specific volume in cubic feet per pound of gas, and T is the gas absolute temperature in degrees Rankine. In Fig. 3·3 we have a fixed weight of gas in state 1, with pressure P_1, temperature T_1, and volume V_1. We then heat the gas till it reaches state 2, but its pressure P remains constant, so that $P_1 = P_2$, but its volume and temperature have increased to V_2 and T_2.

The P-V graph shows the change in state. Next, we hold the gas volume constant and heat the gas some more so that its pressure rises from P_2 to P_3, but $V_2 = V_3$ and T_2 rises to T_3.

Now let us explain the equations. Formula (1) is the simple statement of the constant-pressure Charles' law. But the two changes of state as shown in the graph of Fig. 3·3 show that $V_3 = V_2$. Substituting in (1) we can write (2). Solving for T_2 we get Eq. (3).

From the constant-volume Charles' law we get Eq. (4). But the graph shows us that $P_2 = P_1$, so substituting for P_2 we get Eq. (5). Solving for T_2 we get Eq. (6). Equations (3) and (6) give us two expressions for T_2.

PERFECT-GAS EQUATION

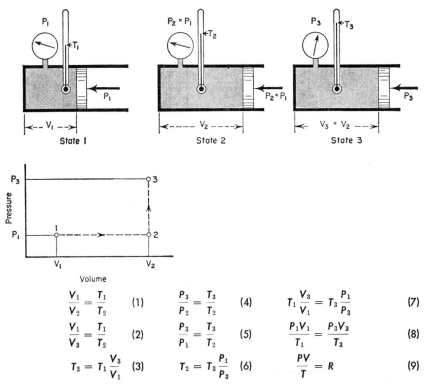

$$\frac{V_1}{V_2} = \frac{T_1}{T_2} \quad (1) \qquad \frac{P_3}{P_2} = \frac{T_3}{T_2} \quad (4) \qquad T_1\frac{V_3}{V_1} = T_3\frac{P_1}{P_3} \quad (7)$$

$$\frac{V_1}{V_3} = \frac{T_1}{T_2} \quad (2) \qquad \frac{P_3}{P_1} = \frac{T_3}{T_2} \quad (5) \qquad \frac{P_1 V_1}{T_1} = \frac{P_3 V_3}{T_3} \quad (8)$$

$$T_2 = T_1\frac{V_3}{V_1} \quad (3) \qquad T_2 = T_3\frac{P_1}{P_3} \quad (6) \qquad \frac{PV}{T} = R \quad (9)$$

Fig. 3·3 The general gas equation $PV/T = R$ can be derived by using constant-volume and constant-pressure versions of Charles' law on three states of a perfect gas. Two states must have the same pressure; one of these and the third state the same volume.

By simply equating the right-hand sides we get Equation (7). Rearranging the terms gives us (8). This tells us that dividing the product of pressure and volume at any state by the absolute temperature at that state will be the same as the product of pressure and volume divided by absolute temperature at any other state for the same mass of gas. We can generalize this by Eq. (9), which says that for any state $PV/T = R$ where R is the gas constant, it differs for each gas. Table 3·1 lists the gas con-

Table 3·1 Gas constants for perfect-gas equation

Gas	Gas constant R, ft-lb per lb, F
Dry air	53.3
Carbon monoxide, CO	55.2
Hydrogen, H_2	767
Nitrogen, N_2	55.2
Oxygen, O_2	48.3
Helium, He	386
Carbon dioxide, CO_2	35.1
Water vapor, H_2O	85.8

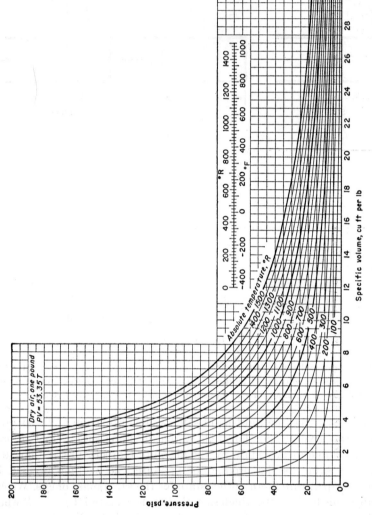

Fig. 3·4 Pressure-volume relations for 1 lb of air tie together Boyle's and Charles' laws. The pair of adjacent scales shows how Fahrenheit and Rankine temperatures are numerically related.

stants for eight different gases. (Do not confuse R with degrees Rankine, °R.)

The beauty of the perfect-gas equation is that it gives us both Boyle's and Charles' laws. By keeping T constant we get $PV = C$, keeping P constant we get $V = CT$, or keeping V constant we have $P = CT$, the C's being all different constants.

In studying thermodynamics, we shall find that plotting gas changes on P-V coordinates simplifies understanding of the processes. So let us plot the perfect-gas equation for 1 lb of dry air, Fig. 3·4. Here we have a series of Boyle's law curves for even values of absolute temperatures. Study the curves for future reference.

3·6 Properties. Temperature, pressure, and volume are called properties of a gas. If we know two of them, we automatically can find the third by the perfect-gas equation. This assumes, of course, that we know the kind of gas. These properties also measure the relative amount of internal energy that a gas contains; more will be said on this later.

3·7 Problems. To check our understanding of the perfect-gas equation let us use it in several examples.

Example 1: Two tanks, each 4 ft in diameter and 8 ft long internally, are connected by a pipe and valve. With the valve closed, tank A contains ½ lb CO_2 at 60 F and tank B, 3 lb CO_2 at 60 F. (a) Find the gas pressure in each tank with the valve closed. (b) What is the common pressure after opening the valve, assuming the temperature stays at 60 F?

Solution: First let us find the volume of each tank:

$$V = \pi \left(\frac{d}{2}\right)^2 L$$
$$= 3.14(4/2)^2 \times 8$$
$$= 100.5 \text{ cu ft}$$

Tank A. Next, let us find the pressure of CO_2 in tank A. From the Table 3·1 we find that for CO_2, $R = 35.1$. The temperature of CO_2 is $60 + 460 = 520°R$. The specific volume of the gas is

$$V = \frac{100.5}{0.5} = 201.0 \text{ cu ft per lb}$$

From the perfect-gas equation $PV = RT$ we can find

$$P = \frac{RT}{V} = 35.1 \times \frac{520}{201.0}$$
$$= 90.8 \text{ psfa}$$

Tank B. Now let us find the specific volume of CO_2 in tank B:

$$V = \frac{100.5}{3}$$
$$= 33.5 \text{ cu ft per lb}$$

Now we are ready to compute its pressure:

$$P = \frac{RT}{V} = 35.1 \times \frac{520}{33.5}$$
$$= 545 \text{ psfa}$$

After Opening Valve. Gas from B flows into A until both have same pressure. With the valve open, the total volume equals the sum of both tanks. Then the specific volume of mixed gases is

$$V = \frac{100.5 + 100.5}{3.0 + 0.5}$$
$$= 57.4 \text{ cu ft per lb}$$

The pressure in both tanks

$$P = \frac{RT}{V}$$
$$= 35.1 \times \frac{520}{57.4}$$
$$= 318 \text{ psfa}$$

Summary: (a) Tank A pressure = 90.8/144 = 0.631 psia. Tank B pressure = $545/144$ = 3.78 psia. (b) Common pressure = $318/144$ = 2.21 psia.

Example 2: An unlabeled gas bottle holds 5 cu ft of gas that weighs 20 lb and exerts a pressure of 763 psig at 120 F. What gas does it contain?

Solution:

$$V = 5/20$$
$$= 0.25 \text{ cu ft per lb}$$
$$P = (763 + 14.7)144$$
$$= 112{,}000 \text{ psfa}$$
$$T = 120 + 460$$
$$= 580°R$$
$$R = \frac{PV}{T}$$
$$= 112{,}000 \times \frac{0.25}{580}$$
$$= 48.3$$

Comparing R with the constants listed in Table 3·1 the gas must be oxygen.

Example 3: An inflated automobile tire has a volume of about 1,300 cu in. With an air temperature of 60 F it has an air pressure of 25 psig. (a) What will be the air pressure when the temperature rises to 120 F? (b) How many pounds of air does the tire hold?

Solution: First let us convert these units so we can use them in the perfect gas equation.

$$P_1 = (25 + 14.7)144$$
$$= 5{,}720 \text{ psfa}$$
$$T_1 = 60 + 460$$
$$= 520°R$$
$$T_2 = 120 + 460$$
$$= 580°R$$
$$\frac{P_2}{P_1} = \frac{T_2}{T_1}$$
$$P_2 = \frac{P_1 T_2}{T_1}$$
$$= 5{,}720 \times {}^{580}\!/_{520}$$
$$= 6{,}380 \text{ psfa}$$

or $(6{,}380/144) - 14.7 = 29.6$ psig at 120 F—the answer to (a).

To find the air weight we first look up air in Table 3·1 and find $R = 53.3$. Then at 25 psig and 60 F, the specific volume for air is

$$V = \frac{RT}{P}$$
$$= 53.3 \times \frac{520}{5{,}720}$$
$$= 4.85 \text{ cu ft per lb}$$

But volume for the tire is

$$v_t = \frac{1{,}300}{1{,}728} = 0.752 \text{ cu ft}$$

$$\text{Lb air in tire} = \frac{v_t}{V}$$
$$= \frac{0.752}{4.85} = 0.155 \text{ lb}$$

3·8 Molecular weights. The perfect-gas equation covers a more general relationship among all gases than we might suspect on first introduction. To understand this we must remember that the atoms and molecules of a gas have a definite mass or weight per particle, Table 3·2. For instance, a hydrogen atom has a mass of 1.008 units and an oxygen atom 16.000 units. But in the usual forms of hydrogen and oxygen gases, we find that the atoms pair off to form molecules, so that a hydrogen molecule weighs 2.016 units and an oxygen molecule 32.000 units.

Table 3·2 Molecular weights of gases

Gas	Molecular weight
Dry air	28.96
Carbon monoxide, CO	28.00
Hydrogen, H_2	2.016
Nitrogen, N_2	28.02
Oxygen, O_2	32.00
Helium, He	4.00
Carbon dioxide, CO_2	44.00
Water vapor, H_2O	18.01

3·9 Avogadro's law. Back in 1811 an Italian, Avogadro, proposed a theory that later proved correct: *Equal volumes of perfect gases held under exactly the same conditions of temperature and pressure each have equal numbers of molecules.* Many experiments have since shown that

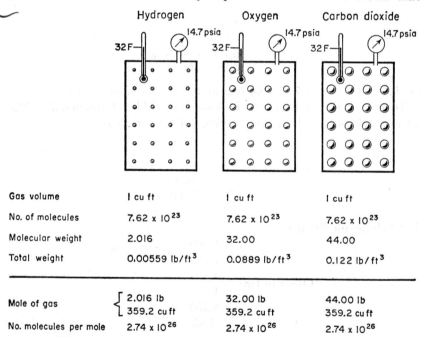

Fig. 3·5 Avogadro's law tells us that different gases with like volume, temperature, and pressure have the same number of molecules. These fix the total weight or mass of the gas.

there are 7.62×10^{23} molecules in a cubic foot of *any* perfect gas when it exerts a pressure of one atmosphere (14.696 psia or 2,116 psfa) and has a temperature of 32 F (492°R), Fig. 3·5.

For most engineering work, however, we do not need to know the number of molecules. But Avogadro's law tells us that, if a given volume of hydrogen weighs 2.016 lb, then an equal volume of oxygen at the same pressure and temperature will weigh 32.00 lb, the same as the ratio of their molecular weights, since there are an equal number of molecules in the two volumes.

With this basic idea let us look again at the perfect-gas equation and write separate equations for hydrogen and oxygen:

$$PV_H = R_H T$$
$$PV_O = R_O T$$

Assuming that we have both gases under the same P and T, we see immediately that the gas constants must have the same ratio as the specific volumes. Dividing the hydrogen equation by the oxygen equation we get

$$\frac{V_H}{V_O} = \frac{R_H}{R_O}$$

But the gas density $D = 1/V$; then

$$\frac{D_O}{D_H} = \frac{R_H}{R_O}$$

Since densities of gases have the same ratio as their molecular weights,

$$\frac{32.000}{2.016} = \frac{R_H}{R_O}$$
$$32.000 R_O = 2.016 R_H$$
$$32.000 \times 48.29 = 2.016 \times 766.5 = 1,545$$

This tells us that multiplying the gas constant of any perfect gas by its molecular weight gives us a *universal gas constant* of 1,545. If we call this univeral gas constant R_u and the molecular weight M, then

$$M_H R_H = M_O R_O = MR = R_u = 1,545$$

While this holds for theoretically perfect gases and many actual gases, some gases deviate slightly from this value.

Now let us see how we can use R_u. For any perfect gas

$$PV = RT$$

Multiply both sides by M:

$$PMV = MRT$$
$$PV_M = R_u T$$

We could call this the general perfect-gas equation. The specific volume of any gas multiplied by its molecular weight M is written as V_M, which has the special name of *mole volume*. By rearranging the general equation we can solve for V_M to get

$$V_M = \frac{R_u T}{P}$$

Let us study this carefully. Since R_u is a constant for all perfect gases, and if we set a standard P and T for all gases, then the mole volume is a constant for all gases. Choosing P as 2,116 psf and T as 492°R, we then find that

$$V_M = \frac{1{,}545 \times 492}{2{,}116} = 359.2 \text{ cu ft}$$

This means that the mole volume of *any* perfect gas is 359.2 cu ft at standard P and T. Note that V_M is not a *specific* volume but the *total* volume of M lb of a gas. In fact we call this a *mole* of gas. A mole of gas always weighs M lb, has a gas constant of 1,545, and a volume of 359.2 cu ft at 2,116 psf and 492°R. This mole idea becomes very useful when dealing with gas mixtures—more of this later. Let us try out this general perfect-gas equation and see what we can do with it.

Example 1: A gas bottle holds 3 cu ft of CO_2 under a pressure of 300 psia at a temperature of 140 F. How many pounds of CO_2 are in the bottle?

First let us find the molecular weight of CO_2. Carbon, C, has an atomic weight of 12 and oxygen, O, an atomic weight of 16. Then for CO_2 the molecular weight $= 12 + 2 \times 16 = 44$.

Substituting in the general perfect-gas equation and remembering that P must be in pounds per square foot and T in degrees Rankine, we get

$$PV_M = R_u T$$
$$PMV = R_u T$$
$$(300 \times 144)\, 44V = 1{,}545(140 + 460)$$
$$V = 0.4877 \text{ cu ft per lb}$$

Since there are 3 cu ft in the bottle, the weight of CO_2 is

$$\text{Weight} = \frac{3}{0.4877} = 6.15 \text{ lb}$$

REVIEW TOPICS

1. Define Charles' law by pressure variation and volume variation.
2. What is meant by absolute temperature?
3. How is absolute zero temperature related to Fahrenheit scale?
4. What scales are used to measure absolute temperatures?
5. Define the perfect-gas equation.
6. Show how Boyle's law and Charles' two laws can be obtained from the perfect-gas equation.
7. What are properties of a gas?
8. State Avogadro's law.

PERFECT-GAS EQUATION

9. What is the universal gas constant?
10. What is a mole volume?
11. What are the standard pressure and temperature for a mole of gas?

PROBLEMS

1. A perfect gas has a pressure of 450 psia when its temperature is 45 F. What will its pressure be at a temperature of 550 F if its volume remains constant?

2. A perfect gas has a temperature of 1200 F at a pressure of 980 psig. To what temperature must it be cooled to reduce the pressure to 0 psig while the volume remains constant?

3. A perfect gas held in a cylinder behind a leakproof piston has a volume of 2 cu ft at a temperature of 45 F. What will be its volume if heated to 1000 F at a constant pressure?

4. A gas tank with internal dimensions of 4 ft height and 9 in. diameter holds hydrogen at a pressure of 3,000 psig and temperature of 65 F. Some of the hydrogen is withdrawn, so that the pressure of the gas remaining in the tank is 1,850 psig at -40 F. What is the mass of the hydrogen withdrawn from the tank?

5. Carbon dioxide held in a cylinder with movable piston has an initial pressure of 2,750 psig and temperature of 460 F. (a) What is the specific volume of the CO_2? (b) If the pressure is reduced to 1,565 psig and the gas cooled to 40 F, what is the specific volume? (c) With 3 cu ft in the cylinder what is the mass of the CO_2 in condition a? (d) What is the mass of the CO_2 in condition b?

6. An automobile tire has an internal volume of 1,800 cu in. when inflated to a pressure of 28 psig at a temperature of 80 F with air. (a) What is the mass of the air in the tire? (b) How much must the tire volume expand to hold the air pressure constant when its temperature rises to 180 F?

7. A gas bottle has a volume of 5 cu ft. Using the universal gas constant and molecular weights in Table 3·2, find the weight of each gas the bottle can hold at a pressure of 1,265 psig and 140 F.

CHAPTER 4

Measuring Energy

In Chap. 1, we discussed the mechanical forms of energy and how they were measured in foot-pound units. We learned that there were parallel forms in the internal energy of substances that we loosely call heat. Such internal energy is the mechanical energy of the molecules and atoms making up the substance. Since it is impossible to see these particles and there are so many of them, it becomes very difficult to measure internal energy in foot-pound dimensions; we have to use other methods.

4·1 Internal energy. Water, being plentiful and fairly uniform in its behavior and composition, makes a convenient substance with which to measure internal energy. We can increase the internal energy of the water in a can by heating it with a flame, standing it in the sunshine, or stirring it violently. You will be able to think of many more ways of raising the internal energy of the water so that its temperature will rise.

To measure internal energy absorbed by water we must know two things: (1) the mass or weight of the water, (2) the change in temperature. So let us take 1 lb of distilled water in a can at a temperature of 63 F, then carefully heat it over a flame until its temperature rises to 64 F, Fig. 4·1. The amount of internal energy gained by the water during this heating is the basic energy unit called one *British thermal unit, Btu.* The same amount of energy would raise the temperature of 2 lb of water by only $\frac{1}{2}$ F or raise the temperature of $\frac{1}{2}$ lb of water by 2 F. When we use this unit, we in effect are saying that X Btu is the amount of energy that will heat X lb of water by 1 F.

An equivalent definition of a Btu states that it is $\frac{1}{180}$th of the energy needed to heat 1 lb of water from 32 to 212 F (freezing to boiling temperatures). This exactly equals the amount of heat needed to raise water from 63 to 64 F. At other temperature levels it takes slightly different amounts of energy to raise the temperature by 1 F. For instance, at 41 F it takes about 1.005 Btu, at 86 F it takes about 0.998 Btu. But the average energy

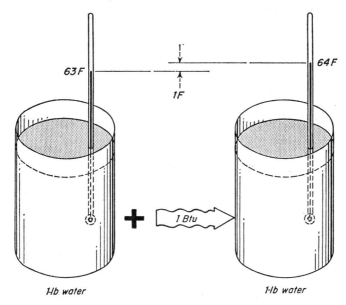

Fig. 4·1 The British thermal unit, Btu, is the amount of energy that warms 1 lb of water at 63 F temperature by 1 degree to 64 F.

per degree Fahrenheit from 32 to 212 F is the same as 1 Btu from 63 to 64 F, Fig. 4·1.

4·2 Mechanical equivalent of heat. If you think about this a bit, you will see that this method of measuring internal energy is simple, but arbitrary. We can take a common substance of an amount chosen at random and measure temperature rise on a scale of somewhat arbitrary units. How do we relate this to the more scientific mechanical energy unit—foot-pounds?

This was done back in 1843 by an experiment performed by an amateur scientist James Joule (he was a brewer). The experiment has since been performed many different ways by others, and the relation between foot-pounds and Btu has been considerably refined. Figure 4·2 shows schematically the elements of his experimental equipment. The tank holds a known amount of water. A set of fixed vanes in the tank are arranged so paddles carried on the vertical rotating shaft can pass between them.

The shaft is rotated by a cord wrapped around the pulley at the upper end. The cord is pulled by known weights falling through a measured distance. You can see that the water will be churned by the paddles every time the weights fall. This heats the water by frictional effect and raises the water temperature, which in turn is measured by the thermometer. From the weight of water and temperature rise, the Btu's of internal energy gained can be figured and compared with the foot-pounds of mechanical work done by the falling weights.

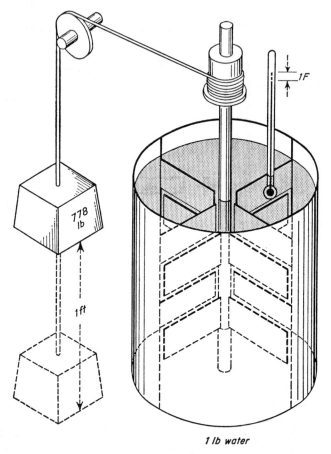

Fig. 4·2 Joule's experiment showed that using 778 ft-lb of work to churn 1 lb of water warmed it by 1 F, or by 1 Btu.

Many experiments have shown that 1 Btu = 778.26 ft-lb. We shall use 778 ft-lb per Btu as being close enough for our work. In honor of Joule, the figure 778 is given the symbol J. The mechanical equivalent of heat can be checked several different ways, also by theoretical calculations. Because thermodynamics deals with the changes of energy between thermal and mechanical forms, we shall frequently use J in our figuring.

4·3 Electrical equivalent of heat. Today we make much use of electrical energy and we obviously have to know how the kilowatt-hour unit of energy is related to the Btu. The form of experiment suggests itself. Simply immerse an electric heater in a known amount of water. Measure the kilowatt-hours of electric energy expended by the heater while it raises the temperature of the water. This type of investigation shows that 1 kwhr = 3412.75 Btu. We round this out to 3413 Btu per kwhr.

4.4 Energy. From what we have talked about so far, we see that energy can be measured in many different units. We have not met all of them yet. In addition to the English system of units that we use, there is the centimeter-gram-second system used by scientists. We are not going to talk about the latter in this book.

Ordinarily we use the energy units that are most convenient to the problem we are handling. There is nothing, however, to bar us from solving a mechanical problem in terms of kilowatt-hours or an electrical problem in terms of foot-pounds. We just have to keep in mind the ratio of the different energy units:

$$1 \text{ Btu} = 778 \text{ ft-lb} = 0.000293 \text{ kwhr}$$
$$0.00128 \text{ Btu} = 1 \text{ ft-lb} = 0.000000377 \text{ kwhr}$$
$$3413 \text{ Btu} = 2{,}656{,}000 \text{ ft-lb} = 1 \text{ kwhr}$$

You can use this little table or simply remember that

$$1 \text{ Btu} = 778 \text{ ft-lb}$$

and

$$1 \text{ kwhr} = 3413 \text{ Btu}$$

The last two equalities will usually be enough to solve most problems quickly.

4.5 Calorimetery. In the everyday operation of power plants we wish to know the heating value of fuels burned in the furnaces of the plant. These heating values are given as Btu per pound of fuel. Figure 4·3 shows basic elements of a bomb-type calorimeter used to find the heating value.

A very small sample of the fuel to be tested, powdered coal for example, is carefully weighed and placed in the fuel pan of the bomb. The bomb is assembled and tightly shut. Through a valve, not shown, compressed oxygen is forced into the bomb up to a pressure of 600 psig. The oxygen is about three times that needed to ensure completely burning the fuel.

The bomb is immersed in a known amount of water. The fuel starts burning after being ignited by a fuse wire in contact with it. An electric current heats the fuse wire, not shown in Fig. 4·3.

All the heat developed by the burning fuel passes into the water surrounding the bomb. The fuel heat raises the water temperature which we measure by the high-accuracy thermometer which can be read to $\frac{1}{100}$ F.

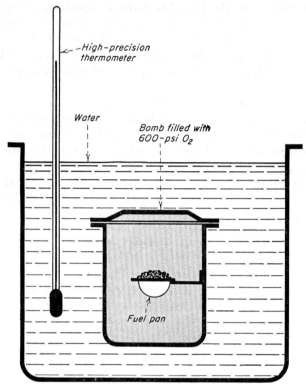

Fig. 4·3 A bomb calorimeter measures the heating value of a fuel by absorbing combustion energy in a known amount of water and noting the temperature rise. All losses must be accounted for.

Example 1: Let us find the heating value of a bituminous coal in Btu per pound. We placed 0.004 lb of the coal in the fuel pan of the calorimeter bomb. The calorimeter contains 6.00 lb of water. Before the fuel was ignited, the water temperature was 72.55 F; after burning, the thermometer read 81.97 F. Assuming that there were no heat losses from the calorimeter and that the fuel was completely burned, what is the coal heating value?

The basic principle of the calorimeter is:

$$\text{Heat absorbed by water} = \text{heat developed by burning fuel}$$

$$\text{Water temperature rise} = 81.97 - 72.55 = 9.42 \text{ F}$$

$$\text{Heat absorbed by water} = 9.42 \times 6.00 = 56.52 \text{ Btu}$$

$$\text{Heat developed by burning 0.004 lb fuel} = 56.52 \text{ Btu}$$

$$\text{Heating value of coal} = \frac{56.52}{0.004} = 14{,}130 \text{ Btu per lb coal}$$

As you might suspect, this example has been highly simplified to emphasize the basic principles. In an actual calorimeter, many corrections must be made to find the true temperature rise, the principal one being to correct for heat gain or loss of the water by radiation. The calorimeter usually has a stirrer to mix the water for even temperature distribution; it also has a water jacket to reduce radiation.

4·6 First law of thermodynamics. We found that energy exists in many forms and that it can change from one form to another quite easily. While talking about the calorimeter, we made a significant statement: Heat absorbed by water equals heat developed by burning fuel. This is a practical application of the *First Law of Thermodynamics* which simply says that energy cannot be created or destroyed; this is also called the *Law of Conservation of Energy.*

Since Einstein's premise on the interchangeability of matter and energy, it has been shown that this law is not absolutely correct. But since a minute amount of matter equals a large amount of energy, we can safely ignore the changes in matter involved in the transfer of energy. As a working tool our first law is good enough for our needs. It allows us to make energy balances around any machine or system of machines.

4·7 Specific heats. In many cases we find it impossible to measure energy directly as so many Btu or foot-pounds or kilowatt-hours; this is

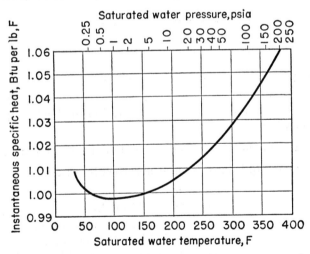

Fig. 4·4 Specific heat of saturated water varies with temperature and pressure.

particularly true for thermal forms of energy. We actually calculate thermal energy from other indicators that we call properties, such as pressure, temperature, volume, and mass. To calculate thermal energy from these data we need constants called specific heats.

BASIC THERMODYNAMICS

These constants have been determined by many experimenters. The most basic one is the specific heat of water—we define our thermal energy unit, Btu, in terms of what happens to water. Specific heat of a substance is the number of Btu's required to warm one pound of it by one degree Fahrenheit. For high-accuracy work we must realize that the specific heat for most substances varies with the temperature. Table 4·1 shows how specific heats vary for water with temperature (at atmospheric pressure) as well as for some solids at given temperatures. Figure 4·4 shows how specific heat of saturated water varies with temperature and pressure. Whenever energy changes in a substance are calculated over a wide temperature variation, the proper average specific heat for the range must be used.

Table 4·1 Specific heats of some common substances, Btu per lb, F

Substance	Btu per Lb, F
Ice at 32 F	0.492
Water at 32 F	1.0087
Water at 40 F	1.0048
Water at 63 F	1.0000
Water at 100 F	0.9975
Water at 160 F	1.0000
Water at 212 F	1.0065
Steam at 212 F	0.482
Lead at 70 F	0.031
Silver at 70 F	0.056
Copper at 70 F	0.092
Iron at 70 F	0.107
Aluminum at 70 F	0.214

Example 2: During a forging operation a ½-lb iron piece at 1600 F is plunged into 10 lb of water at 50 F. Assuming an average specific heat for water of $c_w = 1.00$ and for iron of $c_i = 0.17$, find the final temperature of the water and iron.

We shall assume no losses of energy and apply the first law by writing

Heat given up by iron = heat absorbed by water

Let
M_w = mass of water, lb
M_i = mass of iron, lb
t_w = initial water temperature, F
t_i = initial iron temperature, F
t_f = final temperature of iron and water, F

Then we can substitute in the first equation

$$M_i c_i (t_i - t_f) = M_w c_w (t_f - t_w)$$
$$0.5 \times 0.17(1600 - t_f) = 10 \times 1.00(t_f - 50)$$
$$t_f = 63.1 \text{ F}$$

$(.5)(-.17)t_f + (.17)\frac{(1600)}{2} = 10 t_f - 500$

$c_i 800 + 500 = 1 t_f (10 + \frac{c_i}{2})$

This example demonstrates the transfer of energy between substances and the high heat capacity of water. The ½ lb of iron cools from 1600 to 63.1 F, while the 10 lb of water warms from 50 to 63.1 F. To understand more clearly what happens here, let us compare the temperature rises of 1 lb of iron and 1 lb of water when each absorbs 1 Btu of energy. The temperature of the iron rises by $1/0.107 = 9.34$ F, but the temperature of the water rises by only $1/1.00 = 1$ F. (Here we are assuming iron at room temperature, so $c_i = 0.107$.)

As a matter of interest the specific heat of ice is 0.492 Btu per lb, F. Saturated steam at atmospheric pressure has a specific heat of 0.482 Btu per lb, F. Both of these water phases have less than half the specific heat of the liquid phase; water is one of the few substances having a high heat capacity.

4·8 Gas specific heats. In dealing with specific heats of liquids and solids, we have only the complication of variation with temperature. To

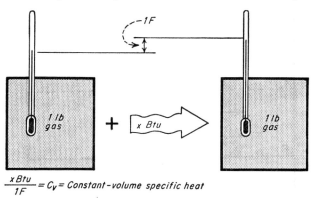

$$\frac{x\,Btu}{1\,F} = C_v = \text{Constant-volume specific heat}$$

Fig. 4·5 Specific heat of a gas at constant volume is the amount of heat that raises the temperature of 1 lb by just 1 F.

make an accurate computation of energy change we must be sure to use the proper average specific heat for the temperature change involved. When dealing with gases, we not only have a much more marked change in specific heat with temperature but must also know what changes in volume or pressure are taking place.

When we confine a gas in a closed container, Fig. 4·5, and raise its internal energy by heating, the volume and mass stay constant. For this condition a gas has a *constant-volume specific heat* c_v. The physical meaning of c_v parallels that for a liquid or solid. It is the unit energy absorbed as internal energy when the temperature (motion of the molecules) of the substance is increased. In a solid, the molecules bounce around faster in one general location; in a liquid, the molecules move faster past one another in close order; in a gas, the molecules fly through space in

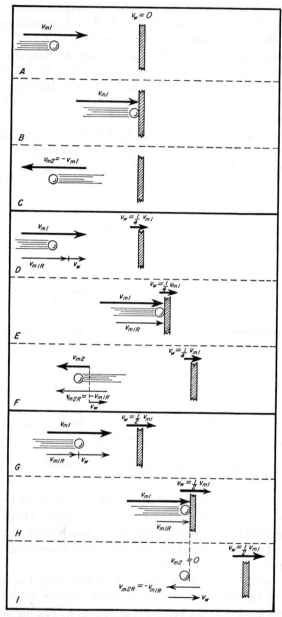

Fig. 4·6 A to C: The molecule does no work on a fixed wall, and retains its energy. D to I: The molecule does work on a moving wall and slows down on rebound.

complete chaos, slamming into one another and the walls of the container billions of times a second.

The vital difference between the gas phase and the other two, solid and liquid, is the pressure or force that the gas exerts upon its container. *This pressure means that a gas can do mechanical work on any restraining wall if the wall moves.* To do such work the gas must use part of its internal energy. When the billions of molecules slam into a stationary plane wall, they rebound with equal speed, since their kinetic energy stays constant, Fig. 4·6A, B, C. Kinetic energy of the molecule before and after hitting the wall is $KE = mv_{m1}^2/64.34$. Since the wall can not move, it does not absorb any of the kinetic energy from the speeding molecule. It does, however, resist the force exerted by the molecule while its direction is reversed.

When the wall does move (like the piston in a cylinder), we have the setup pictured in Fig. 4·6 D, E, F. Here we assume a simple case where the wall moves at a speed of v_w, just one-fourth of the molecule speed of v_{m1}. Both the wall speed and the molecule speed are measured relative to the earth. But the molecule speed relative to the wall speed is the difference of their absolute speeds $v_{m1} - v_w = v_{m1R}$.

When the molecule collides with the moving wall, it rebounds with a relative speed v_{m2R} that equals the relative approach speed v_{m1R} but is opposite in direction. During the collision, the molecule exerts a force on the wall. This force moves through a distance, so the molecule does mechanical work on the moving wall. To do this work, the molecule gives up part of its kinetic energy.

We can calculate the energy loss (the work done) of the molecule by finding the absolute speed of the molecule after collision. The molecule leaves the moving wall with a relative speed of v_{m2R}. But the wall itself is moving away with an absolute speed of v_w, so we must subtract v_w from v_{m2R} to find the speed of the molecule relative to the earth, v_{m2}. This is done graphically in Fig. 4·6F. The work done by the molecule is then $m(v_{m1}^2 - v_{m2}^2)/64.34$.

Figure 4·6G, H, I shows the conditions under which a molecule gives up all its kinetic energy to a moving wall. This happens when the wall moves at half the initial molecular speed. As the molecule hits the wall, it comes to a dead stop, so that $v_{m2} = 0$. Note in Fig. 4·6I that the relative speed of the molecule leaving the wall is v_{m2R}, but the motion lies entirely in the receding wall; the molecule remains at the point of impact.

4·9 Thermal-mechanical conversion. If you grasp the meaning of the story told in Fig. 4·6 you will understand the basic process of changing heat energy to mechanical energy—one of the principal concerns of thermodynamics. Since internal energy is simply a form of molecular motion, we have to set up an arrangement where these energized molecules can impinge on moving walls and do work on a practical scale.

Specific heats play an important part in giving us information on how much energy we can expect to convert in given processes. As we said before, the specific heat of a gas depends on how it is handled during the heating. We have already discussed the constant-volume specific heat c_v. Instead of holding a gas in a closed container, we can put it in a cylinder plugged with a movable piston, Fig. 4·7. Now we have a choice of rates of volume expansion or contraction that we can allow the gas to undergo while heating or cooling it.

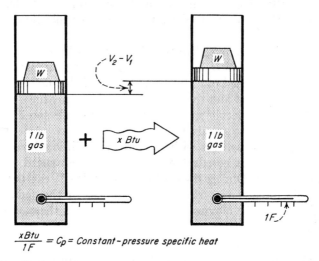

$\dfrac{xBtu}{1F} = C_p =$ Constant-pressure specific heat

Fig. 4·7 Specific heat of gas at constant pressure is always larger than at constant volume because it does work on the piston during expansion.

Of this infinite choice of heating processes, there are only two of practical significance (1) the constant-volume heating we already talked about and (2) the constant-pressure heating. In the latter, we allow the gas to expand so that its pressure stays exactly constant while it is heated, Fig. 4·7. But here we have the piston as a moving wall as in Fig. 4·6, so that the confined gas does mechanical work at the same time that its internal energy is being raised. Charles' law tells us that constant-volume heating raises the pressure, since $P = CT$. Letting the gas expand against the piston makes it do mechanical work and give up some of its internal energy. So when we heat a gas in a constant-pressure process, we must supply energy not only to raise its internal energy but also to do work in expanding against a constant pressure. In that case c_p (constant-pressure specific heat) must be larger than c_v.

Figure 4·8 shows variation of c_v for carbon dioxide, nitrogen, and oxygen with temperature. The specific heats show a marked increase with

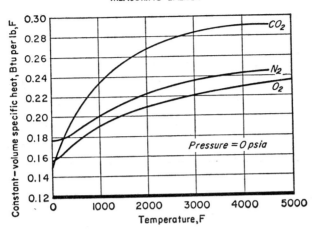

Fig. 4·8 Constant-volume specific heats for gases depend on temperature level and number of atoms in the gas molecule.

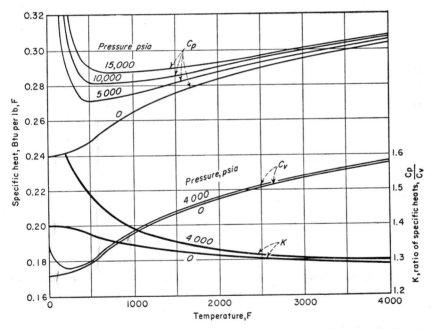

Fig. 4·9 Specific heats of air at constant volume and constant pressure vary with both temperature and pressure. The ratio of specific heats also varies with these properties.

rising temperature. For this reason we must always know what the temperature range is when we make any energy calculations.

Figure 4·9 shows the variation of both c_v and c_p for dry air with temperature at different pressure levels. At temperatures above 200 F, pressure has only a negligible effect on c_v. At room temperatures, the effect is more noticeable, but even at 0 F raising the pressure to 4,000 psia increases c_v by less than 10%. Pressure, however, has a marked effect on c_p, especially at the lower range of temperatures.

Later on, we shall find that the *ratio of specific heats* $k = c_p/c_v$ will be an important factor in our figuring. Figure 4·9 shows how k varies with temperature and pressure. Since c_p is sensitive to pressure, we find that k

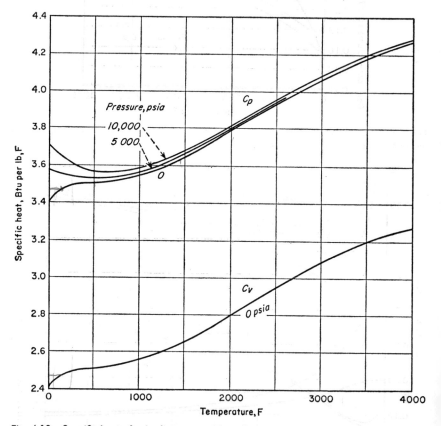

Fig. 4·10 Specific heats for hydrogen run about fourteen times higher than those for air, but the ratio of specific heats k for both gases varies about the same with temperature.

is too. Increasing temperature reduces k, but increasing pressure raises k, especially at the lower temperatures.

Figure 4·10 shows specific-heat variations for the light gas hydrogen. These run more than fourteen times the specific heats for the heavier gases in Figs. 4·8 and 4·9.

4·10 Mechanical work. The constant-pressure heating process of a gas gives us a clue on how to get mechanical work out of thermal energy. Next, we must find a way of figuring the relation between the two.

In Fig. 4·11, suppose we have 1 lb of gas trapped behind the piston in the cylinder with a volume of V_1 and a pressure of P_1. If we add energy to the gas by heating and let the gas expand so that P_1 stays constant, the volume grows to V_2. How much mechanical work has the gas done on the piston in moving it through the length L?

Mechanical work is measured by the product of a force working through a distance. Let us first find the force. The total force acting on the piston

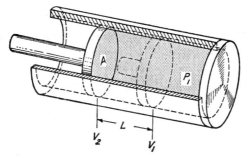

Fig. 4·11 The piston and cylinder are among the most important thermodynamic machines; heat added to gas, pressure being kept constant, makes gas do work on the piston while it expands from V_1 to V_2.

is simply the product of the pressure P in pounds per square foot and the area of the piston A in square feet, so that

$$F = PA$$

The piston is moved through the distance L by this force, so that the work done

$$W = LF$$

Substituting,

$$W = LPA$$

But studying Fig. 4·11, we find that LA is the volume swept out by the expanding gas, and

$$LA = V_2 - V_1$$

Substituting,

$$W = P(V_2 - V_1) \tag{4·1}$$

This last equation shows that we need know only the changes in the measurable properties of the gas to calculate the work done by 1 lb of gas.

4·11 Relation of c_p and c_v. This last equation for a constant-pressure process can be used to find the relation between c_p and c_v. Remember that for 1 lb of gas $PV = RT$; we can substitute in the last equation and get

$$W = R(T_2 - T_1) \quad \text{ft-lb}$$
$$= \frac{R}{J}(T_2 - T_1) \quad \text{Btu}$$

This evaluates the mechanical work done in a constant-pressure heating process. But this process has two parts so we can write:

Heat input = work done + gas internal-energy rise

We can figure the internal-energy rise from the constant-volume heating process. So knowing the gas and the c_v for this gas, the internal energy rise

$$E_2 - E_1 = c_v(T_2 - T_1) \tag{4·2}$$

On the other hand, the total heat input during a constant-pressure process

$$Q = c_p(T_2 - T_1) \tag{4·3}$$

Bringing these factors together in the basic equation,

$$Q = P(V_2 - V_1) + (E_2 - E_1)$$

Substituting,

$$c_p(T_2 - T_1) = \frac{R}{J}(T_2 - T_1) + c_v(T_2 - T_1)$$

Dividing through by $(T_2 - T_1)$ we get

$$c_p = \frac{R}{J} + c_v$$

Rearranging,

$$c_p - c_v = \frac{R}{J} \tag{4·4}$$

Remembering that $k = c_p/c_v$ and substituting, we get relations

$$kc_v - c_v = \frac{R}{J} \tag{4·5}$$

$$c_v = \frac{R/J}{k-1} \tag{4·6}$$

$$c_p = \frac{(R/J)k}{k-1} \tag{4·7}$$

The important relation to note here is the difference of the two specific heats equaling the gas constant over J, which remains constant for most processes. Always make sure that c_p, c_v, and k apply to the pressure and temperature ranges you deal with in future calculations.

REVIEW TOPICS

1. Why is water chosen as a standard for measuring internal energy?
2. What two factors must be known to measure internal energy of a substance?

3. What is a British thermal unit? Define two ways.
4. How can the mechanical equivalent of heat be evaluated?
5. What is the foot-pound equivalent of 1 Btu?
6. What is the Btu equivalent of 1 kwhr?
7. Describe the use of the bomb-type calorimeter in measuring fuel heating value?
8. State the First Law of Thermodynamics.
9. What is the prime use of the First Law of Thermodynamics?
10. What are properties of matter?
11. Define the specific heat of a substance.
12. Do specific heats of a substance vary? How?
13. Define the various specific heats of a gas.
14. Describe the energy interchange between a moving wall and an impacting gas molecule.
15. What is the formula for the ratio of specific heats k?
16. Describe how an expanding gas converts internal energy to mechanical work.

PROBLEMS

1. Many households in the United States use about 4,500 kwhr of electrical energy per year. How high would this energy raise a weight of 1 short ton?
2. How many foot-pounds of energy must be used to heat 1 lb of water from freezing (32 F) to boiling (212 F) temperature?
3. Find the heating value of a 0.005-lb sample of fuel oil placed in a bomb calorimeter holding 7 lb of water when the water temperature rises from 62.5 to 76.2 F.
4. Find the final temperature of 5 lb of water at 70 F when the following metals are plunged into it:
    (a) 1 lb of lead at 500 F
    (b) 1 lb of silver at 500 F
    (c) 1 lb of copper at 500 F
    (d) 1 lb of iron at 500 F
    (e) 1 lb of aluminum at 500 F
5. An automobile moving at 80 mph weighs 3½ tons. What kinetic energy does it have?
6. In Fig. 4·9 check the value of k from c_v and c_p at temperatures of 500, 1000, 2000, 3000, and 4000 F.
7. How much mechanical work is done by 3 lb of air at atmospheric pressure when it expands from 13.4 to 14.0 cu ft per lb?
8. Find the value of R for air and hydrogen from Figs. 4·9 and 4·10 at 1000, 2000, 3000, and 4000 F.

CHAPTER 5

Kinetic Molecular Theory

5.1 Basic assumptions. Over the past 100 years scientists have built up the kinetic molecular theory to explain the behavior of gases. This theory assumes that atoms and molecules of a gas act as hard, perfectly elastic spheres or balls. While we now know that an atom consists of neutrons, protons, and electrons inside a relatively enormous amount of empty space, this knowledge plays no part in the kinetic theory.

From experimental work and the kinetic theory we can roughly calculate the following data about oxygen:

Diameter of O_2 molecule	1×10^{-8} in.
Mass of O_2 molecule	1.2×10^{-27} lb
Average molecular velocity at 32 F, 14.7 psia	950 mph
Number of molecules at 32 F, 14.7 psia	7.62×10^{23} per cu ft
Number of collisions per molecule at 32 F, 14.7 psia	4.2×10^{9} per sec

These numbers are all very large or very small—in either case hard to visualize, since we do not "experience" them in everyday events. As indicated by the speed of the molecules and their enormous number we can see that there is a vast amount of activity and energy in oxygen even at the "freezing" temperature of 32 F.

Slamming of these molecules into the wall of the tank holding the gas produces what we call pressure. To ensure understanding the implications of the kinetic theory let us review the basic ideas of force, motion, energy, and work.

5.2 Newton's laws of motion form the basis of the kinetic theory. Here they are:

1. A stationary body stays at rest and a moving body keeps moving in a straight line at constant speed as long as no unbalanced force acts on them.

2. A force acting on a body equals the mass of the body multiplied by the acceleration of the body in the direction of motion of the force.

3. For every action of a body there is an equal and opposite reaction on another body.

5·3 First law. We have no trouble with the body at rest. We know it cannot move unless a force is applied to it by some other body. The body B in Fig. 5·1b is at rest, relative to the earth, and will stay so unless something comes along to move it.

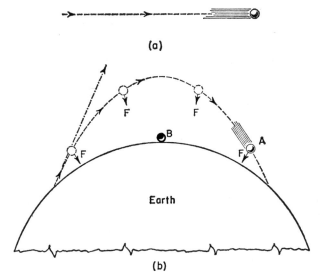

Fig. 5·1 (a) A moving body unaffected by unbalanced force moves in straight line at constant speed. (b) The gravitational pull of the earth prevents straight-line motion of A by action of F.

When we throw a stone into the air, it always falls back to earth again in an arc, like A in Fig. 5·1b. It does not travel in a straight line. If we could take this stone into outer space, far from the earth, planets, and stars, and throw it, it would move in a straight line as in Fig. 5·1a, because no unbalanced force would act on it. On earth we cannot get away from the unbalanced force F exerted by gravitational pull of the earth. If an object is thrown with enough initial speed, it can escape from the earth—after F diminishes to zero or effectively so.

The idea of a moving body keeping going at constant speed after it has been started is contrary to our experience. To accept this notion we have to recognize that frictional effects exert unbalanced forces on moving bodies that brings them to a halt.

In the world of molecules there are no frictional effects. Between collisions the molecules travel freely in comparatively wide-open empty space. The kinetic theory assumes that on the whole the molecules are so far apart that the forces of attraction between them become so small they can be ignored. The molecules travel so fast that gravitational pull of the earth does not deflect them appreciably between collisions. This lets the molecules completely fill the tank holding the gas.

5.4 Second law. This law simply tells us how we can measure a force:

$$F = ma = m\frac{(v_2 - v_1)}{t} \tag{5.1}$$

where F = force, lb
m = mass, slugs
t = time, sec
a = acceleration, ft per sec^2
v = velocity, fps

We usually measure both force and mass in pounds. To use these units we must rewrite Eq. (5·1) as

$$F = \frac{w}{g}a = \frac{w}{g}\frac{(v_2 - v_1)}{t} \tag{5.2}$$

where w = weight of mass, lb
g = gravitational acceleration = 32.2 ft per sec^2

Equations (5·1) and (5·2) show that 1 lb mass equals 32.2 slugs. We shall generally use the simpler form of (5·1) in our discussion because the w/g term is simply a conversion factor for reconciling measuring systems.

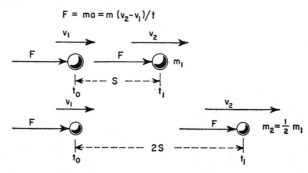

Fig. 5·2 Newton's Second Law of Motion relates the effect of force F on change of velocity of a body having a known mass m.

Figure 5·2 shows the physical meaning of the second law. The acceleration (rate of change in velocity v) acts in the direction of the force F. When the same force acts on a mass half as large $m_2 = \frac{1}{2}m_1$ the distance covered s in the time of action of the force $t_1 - t_0$ doubles and the change in velocity $v_2 - v_1$ or acceleration doubles.

In dealing with forces, velocities, and accelerations, we must keep in mind that they are vector quantities. This means that in adding and subtracting them we must take account of their direction as well as their magnitude. When they all act along one straight line, these quantities can be simply added and subtracted arithmetically.

5·5 Third law. A force cannot exist by itself. It has to be exerted by a body or substance. When such a body exerts a force on another body, it experiences a balancing reactive force in the opposite direction exerted by the body being acted upon. For instance, when you throw a ball, your hand exerts an accelerating force on the ball, giving it motion. In turn, the ball exerts a reactive force on the hand equal to the force developed by the hand.

Figure 5·3 shows what happens in a gun barrel firing a bullet. When the powder explodes, it exerts an accelerating force on the bullet and at the same time an equal reactive force on the breech, making the gun recoil.

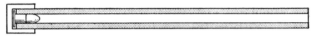

(a) No motion - no forces

(b) Equal forces act on bullet and gun breech

Fig. 5·3 Newton's Third Law of Motion shows that a force causing action on one body has an equal reactive force on the body exerting the original force.

If the gun was not fastened to a mount and free to travel, it would move in the direction opposite to the bullet, but at a slower speed because of its much greater mass.

To see how these motions are related let us rewrite Eq. (5·1) in the form

$$Ft = mv_2 - mv_1 \qquad (5·3)$$

Here Ft is the *impulse* and mv is the *momentum*. This equation says that the change of momentum of a body equals the impulse acting on it.

In Fig. 5·3 if the force F (active and reactive) stays constant during the time of action, then letting the active force on the bullet be F_1 and the reactive force on the gun barrel and breech be F_2 the third law says that $F_1 = -F_2$.

Impulse on the bullet $= F_1 t = m_1(v_1 - 0)$
Impulse on the gun $= F_2 t = m_2(u_2 - 0)$

Then
$$F_1 = -F_2 = \frac{m_1 v_1}{t} = \frac{-m_2 u_2}{t}$$

Clearing, $\qquad m_1 v_1 + m_2 u_2 = 0 \qquad (5·4)$

This says that the sum of the momenta of two interacting bodies always equals zero.

5·6 Conservation of momentum.

Knowing the masses of two interacting bodies m_1 and m_2, we can easily figure what their relative velocities could be by Eq. (5·4). This, however, applies only to masses starting from zero velocity. We can develop a more general equation by the same method used for (5·4), which expresses the law of conservation of momentum.

This is analogous to the first law of thermodynamics concerning energy. Let us derive the momentum law:

For mass m_1
$$F_1 t = m_1 v_2 - m_1 v_1$$

For mass m_2
$$F_2 t = m_2 u_2 - m_2 u_1$$

Third law says that
$$F_1 = -F_2$$

Substituting, combining, and canceling t's,

$$m_1 v_2 - m_1 v_1 = m_2 u_1 - m_2 u_2$$

Rearranging,
$$m_1 v_1 + m_2 u_1 = m_1 v_2 + m_2 u_2 \qquad (5·5)$$

This is the law of conservation of momentum. In words the sum of the momenta of two interacting bodies before collision equals the sum of the momenta after collision.

In Eq. (5·5) we usually know m_1, m_2, v_1, and u_1, but we have two unknowns, v_2 and u_2. Using this in analyzing the kinetic theory of molecules we would know velocities before impact but only an infinite number of possible pairs of values of velocities after impact. To find exact values we have to use our first law of thermodynamics.

5·7 Molecular kinetic energy.

To know what happens after impact we must remember that the perfectly elastic spheres will have the same

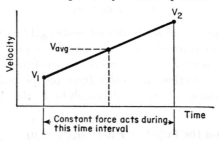

Fig. 5·4 A constant force acting on a body produces an acceleration that gives a uniform change of velocity during the period.

total kinetic energy after collision that they had before collision. Let us evaluate the kinetic energy of a molecule by using Galileo's equations of motion.

When a force acts on a body, it produces uniformly accelerated motion,

KINETIC MOLECULAR THEORY

changing velocity from v_1 to v_2, Fig. 5·4. The average velocity during the accelerating period is

$$v_a = \frac{v_1 + v_2}{2} \tag{5·6}$$

Acceleration is

$$a = \frac{v_2 - v_1}{t}$$

Rearranging,
$$at = v_2 - v_1$$
$$v_2 = v_1 + at$$

Substituting in (5·6),

$$v_a = \frac{v_1 + (v_1 + at)}{2} = v_1 + \frac{1}{2}at = \frac{s}{t}$$

where s is the distance traveled during time t. Then

$$s = v_1 t + \tfrac{1}{2}at^2 \tag{5·7}$$

If the body is started from rest, this equation becomes

$$s = \tfrac{1}{2}at^2 \tag{5·8}$$

From an energy standpoint, work done on a molecule in starting it from rest equals the kinetic energy KE acquired by the molecule, or

$$KE = Fs$$

Substituting,
$$KE = ma \times \tfrac{1}{2}at^2 = \tfrac{1}{2}ma^2 t^2$$

but the final velocity when starting from rest $v = at$; then

$$KE = \frac{1}{2}mv^2 = \frac{Wv^2}{2g} \tag{5·9}$$

Equation (5·5) gives us the momentum-conservation law. The companion energy-conservation equation gives us

$$\tfrac{1}{2}m_1 v_1^2 + \tfrac{1}{2}m_2 u_1^2 = \tfrac{1}{2}m_1 v_2^2 + \tfrac{1}{2}m_2 u_2^2 \tag{5·10}$$

The sum of the kinetic energies of two bodies or molecules before collision equals the sum of the kinetic energies after collision. Rearranging (5·10),

$$m_1(v_1^2 - v_2^2) = m_2(u_2^2 - u_1^2)$$

Factoring, we get

$$m_1(v_1 + v_2)(v_1 - v_2) = m_2(u_2 + u_1)(u_2 - u_1)$$

From (5·5) we have

$$m_1(v_2 - v_1) = m_2(u_1 - u_2) \tag{5·10a}$$

Dividing the last equation into the preceding one, we get

$$v_1 + v_2 = u_1 + u_2 \quad \text{or} \quad v_1 - u_1 = u_2 - v_2 \tag{5.11}$$

The last relation means that the relative speed of approach between two bodies before collision equals the relative speed of recession after collision. Note that this is independent of the masses of the bodies involved.

Now we are ready to find the solution for v_2 and u_2. Next let us rearrange (5·11):

$$v_2 = u_1 + u_2 - v_1$$

Substitute this in (5·10a):

$$m_1(u_1 + u_2 - 2v_1) = m_2(u_1 - u_2)$$
$$m_1 u_1 + m_1 u_2 - 2m_1 v_1 = m_2 u_1 - m_2 u_2$$
$$u_1(m_1 - m_2) + u_2(m_1 + m_2) = 2m_1 v_1$$

When $m_1 = m_2$, the case for collisions between like molecules is

$$2m_2 u_2 = 2m_1 v_1 \quad \text{or} \quad u_2 = v_1 \qquad (5·12)$$

This means that the velocity of one molecule after collision equals the velocity of the other before collision. If we substituted $v_1 = u_1 + u_2 - v_2$ from (5·11) in (5·10a), we would find that $v_2 = u_1$. In other words, like

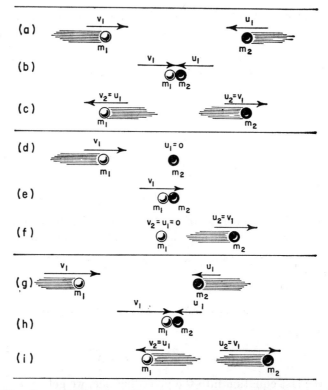

Fig. 5·5 When free molecules of equal masses moving on a common straight path collide, they simply exchange their speeds.

masses or molecules merely exchange velocities during collisions. This simply demonstrates the conservation of momentum, showing that momentum existing in a given direction still persists after collisions between bodies.

Figure 5·5 shows how molecules exchange velocities when they move along the same straight line. This assumes that $m_1 = m_2$. When the masses differ, we have the more complicated equation

$$u_2 = \frac{2m_1v_1 - u_1(m_1 - m_2)}{m_1 + m_2}$$

When m_1 and m_2 move along different lines before collision, the exchange must be figured by vectors.

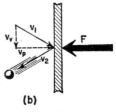

5·8 Molecular collision with walls.
In this type of impact the wall does not move but simply sets up a reactive force F when hit by the molecule, Fig. 5·6. We can explain this action by saying the wall exerts an impulse on the molecule to change its momentum:

$$F_1 t_1 = mv_2 - mv_1$$

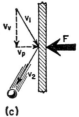

Fig. 5·6 A molecule colliding with a wall exerts a force balanced by the reactive force of the wall. Only the v_p component produces force on a wall.

But Eq. (5·11) tells us that the sum of the velocities of two colliding bodies must be equal before and after impact; then since wall velocity is zero,

$$v_1 + v_2 = 0 \quad \text{or} \quad v_1 = -v_2$$

In words, molecular speeds before and after collision with walls are equal but opposite in direction. Substituting in previous equation we get

$$F_1 t_1 = m[v_2 - (-v_2)] = 2mv_2 = 2mv_1$$

or

$$F_1 = \frac{2mv_1}{t_1}$$

A single molecule exerts a force F_1 for a very short time t_1. To get a continuous force F acting on the wall we must have a continuous stream of n molecules per second hitting in succession so that

$$t = t_1 + t_2 + t_3 + \cdots + t_n = 1 \text{ sec}$$

Then since

$$t_1 = \frac{2mv_1}{F_1} \qquad t_2 = \frac{2mv_1}{F_2} \qquad t_n = \frac{2mv_1}{F_n}$$

and
$$F = F_1 = F_2 = F_3 = \cdots = F_n$$

it follows that

$$t = \frac{2mv_1}{F_1} + \frac{2mv_1}{F_2} + \frac{2mv_1}{F_3} + \cdots + \frac{2mv_1}{F_n} = \frac{n2mv}{F}$$

or, since $t = 1$ sec,

$$F = n2mv \qquad (5 \cdot 13)$$

Actually we could have the stream of molecules come so that the force on the wall would fluctuate, and then over a period of time we could calculate an average force. In practical cases there are billions of molecules hitting a wall section continuously and at random. This causes very minute fluctuations. These are so small, however, that our most sensitive pressure gages cannot detect them.

So far we have considered motions and forces acting along a common straight line. Figure 5·6b and c shows molecules hitting a wall at an angle other than perpendicular. The angle of rebound of the molecule equals the angle of impact in all cases. The only component of velocity that contributes to producing force on the wall is the component perpendicular to the wall, v_p. The component parallel to the wall plays no part in force production but equals the parallel component of the rebound velocity in size and direction.

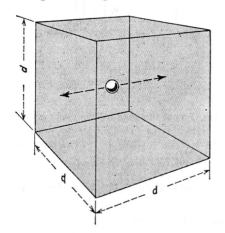

Fig. 5·7 Kinetic-theory build-up starts with one molecule bouncing between opposite parallel walls in a cubical container.

5·9 Gas pressure. We are now ready to develop the kinetic theory as it was initiated by Joule in 1847. Let us study the behavior of one molecule, Fig. 5·7, bouncing back and forth between two parallel walls in a cubical container with side dimensions d.

When the molecule moves with the velocity $v = s$ fps, it makes $s/2d$ round trips in 1 sec. This allows it to make $s/2d$ hits on each wall per second. The area of each wall is d^2. In equation form

$$\frac{v}{2d} = n \text{ impacts per second against each wall}$$

Then average force against each wall is

$$F = n2mv = \frac{v}{2d}2mv = \frac{mv^2}{d}$$

Since we figure pressure as the force per unit of area:

$$P = \frac{F}{d^2} = \frac{mv^2/d}{d^2} = \frac{mv^2}{d^3} = \frac{2}{d^3}\frac{mv^2}{2}$$

This equation shows that the pressure is directly proportional to the kinetic energy of the molecule or to the mass of the molecule and the square of its velocity.

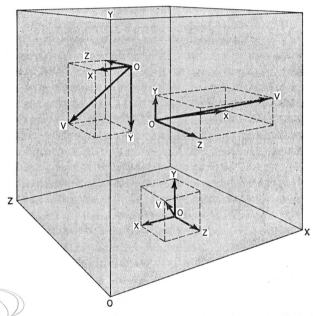

Fig. 5·8 Any velocity of a molecule can be resolved into three components that are perpendicular to the wall of a cubical tank.

We can easily recognize that the pressure will also depend on the number of molecules colliding with the walls. In fact the pressure is directly proportional to the number N of molecules in the container.

From experience we know that the pressure on all walls of a container will be equal. This shows that the molecules, while moving in complete chaos, nevertheless, distribute themselves to produce an even pressure effect on the walls.

In light of the even pressure, let us consider the direction of molecular movements in the body of the gas. Figure 5·8 shows three different velocity vectors OV of the infinite possible number of directions. While these molecules move in all possible directions, the only components that produce pressure on the walls are those perpendicular to the walls. In Fig. 5·8 each of the velocities have been resolved into components perpendicular to the six walls of the container, OX, OY, and OZ.

In an entire mass of gas of N molecules the *average* of components OX, OY, and OZ must all be equal, since the pressures on the walls are equal. Then the kinetic energy of the N molecules must be divided equally (on the average) between the *three* pairs of facing walls, since the energy depends directly on the velocity components of all N molecules perpendicular to the wall, or $P \propto \frac{1}{3} KE$.

5·10 Velocity averages. If v^2 is the average of all the squares of individual molecular velocities for N molecules,

$$v^2 = \frac{v_1^2 + v_2^2 + v_3^2 + \cdots + v_N^2}{N} \tag{5·14}$$

Then the total pressure on any wall area is

$$P = \frac{N}{3} \frac{2}{d^3} \frac{mv^2}{2}$$

Now d^3 is the volume of the cube and Nm is the total mass of the gas. Then the specific volume is

$$V = \frac{d^3}{Nm}$$

So
$$P = \frac{2}{3} \frac{1}{V} \frac{v^2}{2}$$

or
$$PV = \frac{2}{3} \frac{v^2}{2} = \frac{v^2}{3} \tag{5·15}$$

Comparing the last equation with the general gas equation $PV = RT$, we can conclude that $RT = v^2/3$. Since R is a constant depending on the gas only, we can conclude that *temperature measures the averaged square of the molecular velocities* and no other factor, see Eq. (5.14). By taking the square root of this average square we get from (5·14)

$$v = \sqrt{\frac{v_1^2 + v_2^2 + v_3^2 + \cdots + v_N^2}{N}}$$

v is called the rms (root-mean-square) velocity. It is higher than the average velocity of all the molecules.

5·11 Velocity distribution. This development is based on the rms velocity of the molecules. We might jump to the conclusion that at any given temperature all the molecules have the same velocity. This is not true—in fact, if all molecules could be brought to a common velocity, differing only in directions, they would immediately acquire a wide range of speeds all the way from zero to almost infinity.

To understand why this must be so let us study Fig. 5·9a. Here two molecules of the same mass approach each other and collide. A line connecting the centers of m_1 and m_2 is coincident with direction of travel of m_1 and exactly at right angles to line of travel of m_2.

Equation (5·12) showed that molecules in collision simply exchange velocities in their direction of travel. In Fig. 5·9a, m_2 has zero velocity in the direction of motion of m_1, so m_2 acquires all velocity of m_1, which is v_1, and gives none in return to m_1. This means that m_1 comes to a dead stop so that $v_2 = 0$ as shown in Fig. 5·9b.

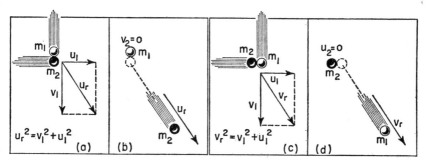

Fig. 5·9 When one molecule hits another at right angles, the first one stops and sends the second one off at an angle with more energy.

Molecule m_2 now has the vector sum of its own original velocity u_1 and v_1 which gives it a leaving velocity of u_r. Since $u_r^2 = u_1^2 + v_1^2$ we can see that m_2 acquires all the kinetic energy originally held by both m_2 and m_1. Figure 5·9b shows the molecules after collision. Figure 5·9c and d shows what happens when m_2 hits m_1. Their roles are just reversed.

If a molecule like m_2 is hit at right angles several times in succession, it will acquire very high velocities and energy and leave zero-energy molecules in its wake. So obviously there must be a wide range of velocities in any group of gas molecules. Figure 5·10 shows the more general type of collision between molecules. The left-hand collision picture gives the vector diagram for computing velocity exchange. The right-hand diagram shows conditions in place before and after collision. To keep these exchanges in proper perspective do not forget that at atmospheric pressure and temperature any one molecule suffers *billions* of collisions per *second*. On the average they travel extremely short distances, so we are not aware of all this activity.

To measure this velocity distribution Maxwell and Boltzmann in the middle 1800's made a statistical analysis of exchanges within a gas. The method is quite involved and beyond the scope of mathematics we intend to cover, but Fig. 5·11 shows the results for a study of air. At a temperature of 492°R the most probable velocity is about 900 mph. The rapid drop of the curve from this value shows that there are fewer molecules with lower and higher speeds and just a trace of molecules with zero speeds or very high speeds (three or more times the most probable).

As the gas temperature rises progressively to 984 and 1476°R, the most probable speed also increases in proportion to the square root of the rise in temperature. Doubling the temperature increases the most probable speed by 41.4%; tripling the temperature raises the most probable speed by 73.1%. The proportion of molecules at the most probable speed also decreases with rising temperature, but the proportion of molecules at

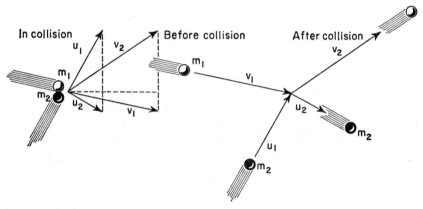

Fig. 5·10 In a general collision among molecules, they exchange velocity components in parallel with line connecting their centers but keep the components at right angles to this line.

lower speeds drops while the proportion at higher speeds increases with rising temperature.

Keep this picture in mind when we discuss reversibility later in the book. While most experts in thermodynamics say we do not need to know the kinetic theory, this physical picture will help to interpret many relations that classical thermodynamics deals with only by equations.

5·12 Temperature and velocity. We see that temperature gives us a yardstick of the speed of the molecules bouncing around, since $RT = v^2/3$. R, the gas constant, depends on the mass of the gas molecule M.

In a gas, absolute temperature T measures the rms velocity of all the molecules:

$$T \propto v = \sqrt{\frac{v_1^2 + v_2^2 + v_3^2 + \cdots + v_N^2}{N}}.$$

Actually we can give a broader meaning to T. Remembering that the universal gas constant $MR = R_u = 1{,}545$ (Art. 3·9) where M is the molecular weight of the gas, we can write $MRT = Mv^2/3$. Recognizing

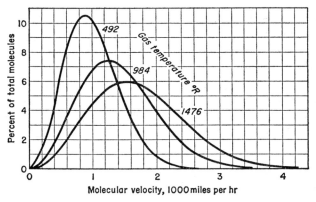

Fig. 5·11 Molecular velocity distribution ranges over a larger spread of speeds as gas temperature rises. The speed for the greatest percentage of molecules at each temperature is the most probable value.

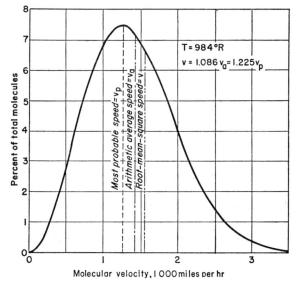

Fig. 5·12 Molecules in a gas like air have a wide range of speeds at any given temperature; different averages can be used to measure the speeds.

MR as a constant we see that T depends on Mv^2, the kinetic energy of the molecules.

What does this mean when we compare temperatures of gases with different molecular weights? Just one thing—for equal temperatures heavier gases have slower molecules, since

$$T = M_1 v_1^2 = M_2 v_2^2 = M_3 v_3^2 = etc.$$

Using this formula we shall see that, if at 32 F hydrogen molecules move at 3,800 mph, oxygen molecules, sixteen times heavier, will move at only 950 mph when they both have the same kinetic energy and temperature.

Perhaps the most interesting fact of molecular motion in a gas is the wide range of speeds molecules have at a given temperature, Fig. 5·11; Fig. 5·12 reproduces the distribution curve for air at 984°R (524 F). Remember, the absolute temperature measures one particular kind of average, the *root-mean-square* defined above, v. Figure 5·12 shows that for 984°R air, the largest single group of molecules (about 7.5%) has a speed called the *most probable*, v_p. We could also figure the *arithmetical average speed*, v_a. These three measures of the collective speed of a group of molecules are definitely related:

$$v = 1.086 v_a = 1.225 v_p$$

So much for the kinetic theory.

REVIEW TOPICS

1. What characteristics are assumed for molecules in the kinetic molecular theory?
2. State Newton's three laws of motion.
3. How do we measure forces?
4. What is the relation between impulse and momentum?
5. State the principle of the conservation of momentum.
6. Derive the equation for kinetic energy of a body.
7. What happens to the velocities of equal masses that collide, providing that the masses are perfectly elastic spheres?
8. Derive the expression for pressure in terms of molecular kinetic energy.
9. What characteristic of molecules of a gas does temperature measure?
10. Do all the molecules of a gas have the same speed and kinetic energy?
11. Is it possible for a gas to have molecules all traveling at the same speed?

CHAPTER 6

Energy Equation of Gases

6·1 Energy conversion. We had our first hint on how to change heat energy into mechanical shaft energy in Chap. 4. Let us review. Figure 6·1 shows the constant-pressure heating of 1 lb of gas trapped behind a leakproof frictionless piston in a cylinder. The gas exerts a constant force F on the piston while it moves from 1 to 2 and absorbs a heat input Q. Here we have energy entering as heat at one end of the cylinder and energy leaving as mechanical shaft work at the other, a heat engine.

The work done by the gas on the piston is

$$W = FL \tag{6·1}$$

but
$$F = PA \tag{6·2}$$

where P = gas pressure, psfa
A = cross-sectional area of the piston, sq ft
W = work, ft-lb

Substituting for F we get

$$W = PAL \tag{6·3}$$

but Fig. 6·1 shows that

$$AL = V_2 - V_1 \tag{6·4}$$
$$W = P(V_2 - V_1) \tag{6·5}$$

Plotting this change in pressure and volume for the 1 lb of gas, we get the horizontal line shown in the graph of Fig. 6·1. From the equation for W we see that the area under the pressure line also measures the value of W.

We find the P-V graph a useful tool to analyze what takes place in converting all or part of heat energy Q into useful shaft work W. Examining the P-V graph, we find that the gas was first in state 1 and then expanded at constant pressure to state 2. For *ideal* conditions the area W

under the process curve always measures the amount of mechanical work W done by the gas on the moving piston, or vice versa.

What happened to the gas temperature in changing from state 1 to state 2? Since the mass of gas stays constant, there are always the same number of molecules trapped in the cylinder. Allowing the volume to increase means that there would be fewer hits per second on the piston, causing a drop in pressure. But since the pressure stayed constant with increasing volume, it means that the added heat energy Q sped up the molecules so they kept making the same number of hits per second, per

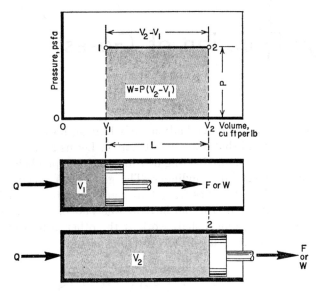

Fig. 6·1 In a constant-pressure process, gas takes in heat Q while it does mechanical work on a piston during the expansion.

unit area even though they had greater average distances to travel. This means that the gas temperature at state 2 is higher than at 1.

According to the First Law of Thermodynamics we should be able to account for all energy entering or leaving the gas during the process. From the simple analysis we made, we see that heat Q went partly to raising the internal energy of the gas from E_1 to E_2, and the remainder went to produce mechanical work W.

6·2 Reversibility. If we can control all heat flows into and out of the gas at will, if there is no friction between piston and cylinder, and if the gas can contract or expand smoothly without setting up any vortices or pressure waves inside itself, we shall achieve an ideal condition for making energy conversions. Most ideal thermodynamic calculations assume that such a condition exists.

When this is true, we can reverse the force F in Fig. 6·1 and have it compress the gas from V_2 to V_1 and eject heat Q. In this case W would be the work done *on* the gas *by* the piston. The temperature would also drop from the higher T_2 to the lower T_1. The ability to do this exactly is called *reversibility*. The first process, then, is a reversible constant-pressure expansion, and the second a reversible constant-pressure compression. These are only theoretically possible—ideals; actually we cannot attain them.

6·3 General gas equation. When we have a gas confined as in Fig. 6·1, do not forget that the general gas equation always applies:

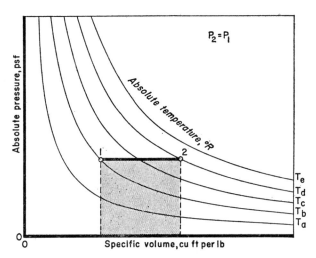

Fig. 6·2 Constant-pressure expansion 1–2 or compression 2–1 always involves a change in the temperature of the gas.

$PV = RT$. If we know two of the properties P, V, or T at any state, we can always figure the third, assuming we know what kind of gas we have. For all *reversible processes* the area under the process line on a P-V graph represents the mechanical work done (in foot-pounds) on or by the gas.

Before we go further, let us refresh our memory about Fig. 3·4 showing the plot of the general gas equation for air at constant temperatures over a range of 100 to 1500°R. To change the state of a gas we have to inject or remove energy as either work or heat. Figure 6·2 shows the same general relation (without numbers) of P, V, and T of a perfect gas, as a general counterpart of Fig. 3·4. The constant-pressure process 1–2 plotted in Fig. 6·2 shows that a change of temperature must be involved. The shaded area underneath represents the work that must be done to compress the gas from 2 to 1 or the work output realized when expanding the gas from 1 to 2.

6·4 Joule's experiment. To show that the internal energy of a gas is essentially independent of pressure and specific volume, Joule did his famous experiment. The equipment was relatively simple. He took two vessels and connected them through a pipe carrying a closed shutoff valve. One vessel was evacuated to remove practically all air. The other vessel was charged with air to any convenient pressure. He then placed both vessels in a water bath and waited long enough to ensure that vessels, entrapped air, and the water bath were all at the same temperature, Fig. 6·3.

He then opened the valve, letting air from the pressurized vessel flow freely into the empty one until the pressure was equal in each. Closely watching the thermometer he found no change in temperature. Even though the air doubled its original volume and halved its pressure, its temperature stayed constant. While some of the gas molecules moved into

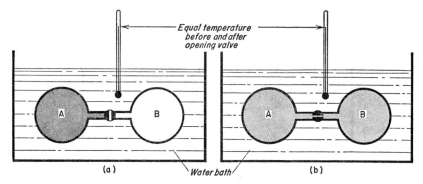

Fig. 6·3 Joule's simple experiment of allowing a gas to double its volume showed that its temperature was not affected by the change.

additional space, they did no mechanical work; neither did they receive or give up heat energy. So the molecules retained their original kinetic energy as shown by the constant temperature.

This rather coarse experiment developed the principle of independence of the temperature of a gas from its pressure and density. This idea is practical enough in most engineering machines.

Later experiments of different form showed this to be not strictly true. Actually, there is a slight mutual force of attraction between the fast-moving molecules of a gas. When the gas expands to increase its volume greatly, the molecules move farther apart on the average.

This means that they do mechanical work against their mutual forces of attraction in filling the larger volume. This slows them down. They simply redistribute the energy they have, exchanging a small part of their kinetic energy for potential mechanical energy, Chap. 1. This causes a small drop in temperature. Remember, temperature measures only kinetic energy, not potential energy, of molecules.

This effect becomes important only at high pressures. It is detected by the porous-plug experiment of Joule-Thomson, Fig. 6·4. Here high-pressure gas flows very slowly through a porous ceramic plug with many small holes to a lower pressure. Precise temperature measurements show

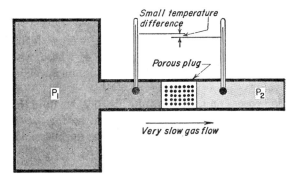

Fig. 6·4 Gas at high pressure P_1 seeping through a porous plug to low pressure P_2 shows slight drop in temperature at P_2.

small changes at the higher pressures. These effects are taken into account in designing h-p engines.

6·5 System. Let us get back to the gas in our cylinder and piston and put down what we have learned in more formal style. The gas in an ideal cylinder and piston can be called a closed system. Energy changes in this system ideally take place in only three ways: (1) heat in or out Q, (2) work in or out W, (3) gain or loss of internal energy E. Then according to the first law

$$Q = (E_2 - E_1) + W \tag{6·6}$$

This is the *simple energy equation*, also called the *nonflow energy equation*. Q and W are not properties of the gas but forms of energy in transit. On the other hand E is a property of the gas directly related to T. We must be careful in interpreting the signs in this equation. We shall arbitrarily assume that energy Q entering the gas will be plus and that work done by the gas W (energy leaving) will also be plus. When Q and W reverse their direction, their signs become minus.

When W is figured from P and V, the result is in foot-pounds. We convert this to Btu by dividing W by $J = 778$ ft-lb per Btu. Normally Q and E are given in Btu.

Note that the energy equation does its computation by a bookkeeping or accounting procedure. It makes no pretense of *explaining* just *how* energy converts from one form to another. If we accept the first law, then the energy changes must balance according to the basic equation.

6·6 Constant-pressure process.

We already discussed this process in detail, Figs. 6·1 and 6·2. But let us analyze it according to the simple energy equation

$$Q = (E_2 - E_1) + W \tag{6·7}$$

$$Q = c_v(T_2 - T_1) + P_1\frac{V_2 - V_1}{J} \tag{6·8}$$

The Joule experiment showed that internal energy depended only on temperature. We anticipated this in Chap. 4, where we learned that change in E can be figured by the constant-volume specific heat c_v, as in Eq. (6·8) above.

Plotting this process on a P-V graph, Fig. 6·2, we note that a considerable change in temperature must take place, so for expansion the gas stores internal energy, for compression it gives up internal energy.

Let us rewrite Eq. (6·8) remembering that $P_1 = P_2$:

$$Q = E_2 - E_1 + \frac{P_2 V_2 - P_1 V_1}{J}$$

$$Q = \left(E_2 + \frac{P_2 V_2}{J}\right) - \left(E_1 + \frac{P_1 V_1}{J}\right) \tag{6·9}$$

The quantity $(E + PV/J)$ appears so often in thermodynamics that it is given a special symbol H and the name *enthalpy*. This combination quantity appears in both the gas tables and steam tables—it is mostly a symbol of convenience rather than physical significance. Because of this we can now rewrite (6·9) as

$$Q = H_2 - H_1 \tag{6·10}$$

By definition, let us not forget that for the constant-pressure process

$$Q = c_p(T_2 - T_1) \tag{6·11}$$

Then we come to an important relation by combining (6·10) and (6·11):

$$H_2 - H_1 = c_p(T_2 - T_1) \tag{6·12}$$

Do not forget that these equations are all exact for reversible processes. Equation (6·12) holds for all perfect gases, since H is a combination property of a gas. Before we leave these equations note that

$$\Delta E = c_v \Delta T$$
and
$$\Delta H = c_p \Delta T$$

where Δ (delta) refers to change or difference in a property during a process.

For this process one form of Charles' law applies, since $P_1 = P_2$; then

$$\frac{V_1}{V_2} = \frac{T_1}{T_2}$$

6·7 Constant-volume process. In Fig. 6·5 we take our cylinder full of a perfect gas and lock the piston into place. This holds the gas at constant volume and does not allow energy to enter or leave the gas as mechanical work W, so $W = 0$. On the P-V graph the process line 1-2 appears vertical.

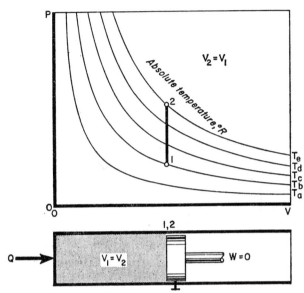

Fig. 6·5 A constant-volume gas process can only accept or reject energy as heat; it cannot do any form of mechanical work.

Since a true line has no width, there can be no area under it. This zero area ties in with zero W. In this process, energy can enter or leave the gas only as heat Q. The simple energy equation for $V_1 = V_2$ then boils down to

$$Q = E_2 - E_1 = c_v(T_2 - T_1) \tag{6·13}$$

Heating the gas from state 1 to state 2 by the injected Q raises the gas pressure. Cooling the gas from 2 to 1 by extracting heat Q lowers the pressure.

We find that this process follows one form of Charles' law (Art. 3·1):

$$\frac{P_1}{P_2} = \frac{T_1}{T_2} \tag{6·14}$$

In words, the pressure is proportional to the temperature when the volume stays constant. The graph shows this relation directly.

6·8 Constant-temperature process. Next let us learn what happens when we control our energy flows in and out of a perfect gas so its temperature stays constant. In Fig. 6·6 we let our gas expand from volume V_1 to V_2 so its temperature stays constant, $T_2 = T_1$. This means that the internal energy of the gas stays constant, or $E_2 = E_1$ or $E_2 - E_1 = 0$. Then the simple energy equation boils down to

$$Q = W \tag{6·15}$$

We can become excited about this, because here we can change all of Q into W without affecting the energy content of the gas—a profitable arrangement. Unfortunately, our practical reciprocating engines need more than one kind of process to produce mechanical work; more will be said about this later.

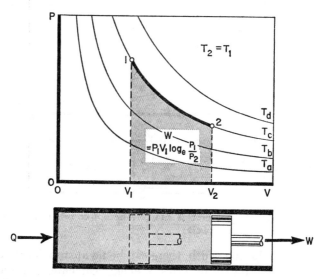

Fig. 6·6 A constant-temperature gas process acts as a transformer of energy; heat transferred exactly equals work done by the gas.

The reverse also holds true for this process; we can change mechanical work into heat without affecting the gas internal energy during compression. Pressure keeps changing during the process, rising during compression and falling during expansion. This makes it more difficult to calculate the work W. The area under the constant-T process curve, however, still measures the mechanical work W during compression or expansion.

Since T is constant, the gas follows Boyle's law during this process:

$$P_1 V_1 = P_2 V_2 \tag{6·16}$$

With this relation we can figure a formula for W by using calculus (see Appendix A) that takes the form

$$W = \frac{P_1 V_1}{J} \log_e \frac{P_1}{P_2} \tag{6·17}$$

Here $\log_e$ means logarithm (to the base number 2.7183) of the pressure ratio P_1/P_2. We can simplify any figuring with this formula by using the adjacent scales in Fig. 6·7. First reduce the pressure ratio to $P_1/P_2 = N$. Then by spotting N on the scales you can find $\log_e N$ directly above. We shall find out when to use $N^{1.3}$ and $N^{1.4}$ scales later.

How does the constant-temperature process work from the molecular standpoint? During gas expansion, vibrating molecules in the heating surface accelerate the gas molecules toward the piston. At the receding piston face the molecules decelerate as they rebound with lowered speed because they have given up some of their kinetic energy to the piston by

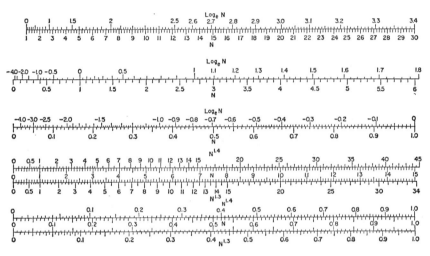

Fig. 6·7 Adjacent scales, upper three, give the natural logarithm for numbers from 0 to 30. The lower two pairs of scales give values of numbers to the 1.3 and 1.4 powers for the range from 0 to 15. For greater accuracy use tables of natural logarithms.

doing work on it. The molecules then return more slowly toward the heating surface (head end of cylinder).

These are the underlying motions or transfers superimposed on the chaotic and erratic colliding of the molecules with one another. The higher net speed of molecules from the heat-transfer surface to the piston counterbalances the lower net speed from the piston to the heat-transfer surface, so that the rms speed of the molecules, the temperature, stays constant.

During compression processes the molecules are accelerated by the advancing piston toward the heat-transfer surface (now a cooling surface), where they give up part of their kinetic energy to the slower vibrating molecules of the cylinder wall. This makes them rebound with lower speed toward the piston face. The two opposing motions balance out to keep the gas temperature constant.

6·9 Adiabatic process.

So far we have dealt with processes that in turn have kept ΔV, ΔP, ΔE, and W at zero. Now let us see what happens when we keep $Q = 0$. This reduces the simple energy equation to

$$0 = E_2 - E_1 + W$$
or
$$W = E_1 - E_2 \qquad (6 \cdot 18)$$

This means that we can produce mechanical work, Fig. 6·8, by letting the gas in a cylinder expand and draw on the internal energy of the gas. In reverse, by doing mechanical work of compression on the gas we can store

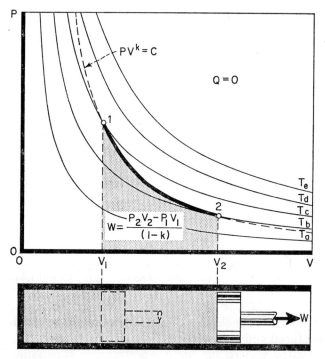

Fig. 6·8 An adiabatic gas process loses internal energy by doing work or gains it by absorbing work energy; heat-transferred energy = zero.

this energy as a rise in internal energy. These are adiabatic processes, with no gain or loss of energy as heat.

We can figure W directly by the simple energy equation, since

$$W = E_1 - E_2 = c_v(T_1 - T_2) \qquad (6 \cdot 19)$$

Figure 6·8 shows that the adiabatic-process curve crosses the constant-temperature curves, as it must with internal energy varying during the process.

ENERGY EQUATION OF GASES

Work W also equals the area under the process curve as long as it is reversible. A reversible adiabatic P-V curve for a perfect gas always has the relation

$$PV^k = C \tag{6.20}$$

where $k = c_p/c_v$ and C is a constant depending on the gas (see Appendix B). For *air* at temperatures near 900 F, $k = 1.4$; for temperatures around 3000 F, $k = 1.3$; see Art. 4·9. We can find another formula for figuring W as the area under the curve. Again this must be done by calculus to get

$$W = \frac{P_2 V_2 - P_1 V_1}{J(1-k)} \tag{6.21}$$

The general gas equation *always* holds regardless of the process, so

$$P_2 V_2 = RT_2 \quad \text{and} \quad P_1 V_1 = RT_1 \tag{6.22}$$

Substituting these in the work equation (6·21) we get

$$W = \frac{RT_2 - RT_1}{J(1-k)} = \frac{R}{J}\frac{T_2 - T_1}{1-k} \tag{6.23}$$

For an adiabatic reversible process, since $P_1 V_1^k = P_2 V_2^k$, we can find relations between P and T and between V and T as follows:

$$P_1 V_1^k = P_2 V_2^k \tag{6.24}$$

$$\frac{RT_1}{V_1} V_1^k = \frac{RT_2}{V_2} V_2^k$$

$$T_1 V_1^{k-1} = T_2 V_2^{k-1}$$

$$\frac{T_1}{T_2} = \left(\frac{V_1}{V_2}\right)^{1-k} = \left(\frac{V_2}{V_1}\right)^{k-1} \tag{6.25}$$

This gives us the relation between T and V for *reversible adiabatic perfect-gas* processes. Now let us find the relation between P and T by substituting in Eq. (6·24) from the general gas equation:

$$P_1 \left(\frac{RT_1}{P_1}\right)^k = P_2 \left(\frac{RT_2}{P_2}\right)^k$$

$$P_1^{1-k} T_1^k = P_2^{1-k} T_2^k$$

$$\frac{T_1}{T_2} = \left(\frac{P_1}{P_2}\right)^{(k-1)/k} \tag{6.26}$$

These fractional exponents look forbidding, but they quickly disappear when we make numerical substitutions. Figure 6·7 will help to find the value of some of the exponential quantities without using logarithm tables.

6·10 Polytropic process. We have looked at specific processes where a state, property, or energy transfer has been kept constant or at zero.

In the most general case for a process, all factors vary at once in some regular fashion; we call this a polytropic process.

Test of regularity usually consists of examining the process on a P-V graph and finding whether a simple formula will relate all the states. Many practical processes in i-c engines and compressors behave this way, so that the formula $PV^n = C$ describes the process curve.

This equation closely parallels the one for the adiabatic process, the exponent $k = c_p/c_v$ being replaced by n, which has no simple relation to the gas specific heats. This being so, we have the interesting situation where all the equations for the adiabatic process also apply to the polytropic by simply substituting n for k and keeping in mind that Q is not zero.

We can figure Q by referring to the simple energy equation:

$$Q = E_2 - E_1 + W \tag{6.27}$$

Substituting the equivalent expressions we get

$$Q = c_v(T_2 - T_1) + \frac{R}{J}\frac{T_2 - T_1}{1 - n}$$

$$= \left[c_v + \frac{R}{J(1 - n)}\right](T_2 - T_1)$$

$$= \frac{c_v - nc_v + R/J}{1 - n}(T_2 - T_1)$$

Remembering that $c_v + R/J = c_p$ and $k = c_p/c_v$

$$Q = \frac{c_p - nc_v}{1 - n}(T_2 - T_1)$$

$$= c_v \frac{k - n}{1 - n}(T_2 - T_1) \tag{6.28}$$

We could replace the quantity $c_v(k - n)/(1 - n)$ by c_n and call it the polytropic specific heat; then

$$Q = c_n(T_2 - T_1)$$

6·11 Process comparison. The curves for all types of processes have definite relations to one another. Figure 6·9a shows how the processes are related for expansion or heating or both when the gas is originally in state 1. Figure 6·9b gives similar information for a gas when compressed or cooled or both.

Remember that the area under these reversible-process curves measures the mechanical work involved, either produced in expansion or absorbed in compression. The labels on each curve show the value of n when the process is considered as a special case of the general reversible polytropic.

Theoretically a polytropic process can take any direction on the P-V plane, but Fig. 6·9 shows those of practical importance.

Examples: Let us study a few examples to get the feel of the dimensions involved. Assume we have 0.1 lb of air at 1500 F (1960°R). Then from Fig. 4·9 we find that $c_v = 0.21$ Btu per lb and $c_p = 0.28$ Btu per lb, so $k = c_p/c_v = 0.28/0.21 = 1.33$.

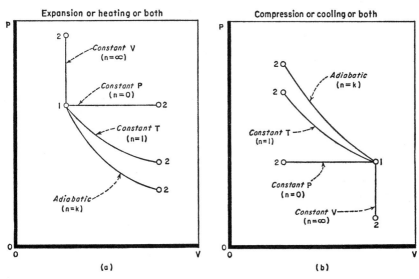

Fig. 6·9 (a) Most important gas processes that yield work or absorb heat or both can be considered as special cases of the polytropic process, with *n* taking special values. (b) The same can be done with processes of compression or cooling.

Also from Chap. 4 we learned that $R/J = c_p - c_v = 0.28-0.21 = 0.07$ Btu per lb air and $R = 778 \times 0.07 = 54.5$. This gives us all the constants we need, so let us study each process in turn:

Constant volume. We have the 0.1 lb of 1960°R air trapped in a cylinder with a fixed volume of 1 cu ft. We heat the air to 2960°R. (1) What is the initial pressure P_1? (2) What is the final pressure P_2? (3) What heat has been transferred into the gas? (4) What mechanical work has been done?

$$V = 1 \text{ cu ft}/0.1 \text{ lb}$$
$$= 10 \text{ cu ft per lb}$$
$$PV = RT$$
$$P_1 = \frac{RT_1}{V}$$
$$= 54.5 \times \frac{1960}{10}$$
$$= 10{,}680 \text{ psfa}$$

or
$$= \frac{10{,}680}{144}$$
$$= 74.2 \text{ psia} \qquad (1)$$

$$P_2 = P_1 \frac{T_2}{T_1}$$
$$= 74.2 \times \frac{2960}{1960}$$
$$= 112 \text{ psia} \tag{2}$$
$$Q = c_v(T_2 - T_1)$$
$$= 0.21(2960 - 1960)$$
$$= 210 \text{ Btu per lb}$$
$$= 210 \times 0.1 \text{ lb} = 21 \text{ Btu} \tag{3}$$
$$W = 0 \quad \text{when } V_2 = V_1 \tag{4}$$

Constant Pressure. We have 0.1 lb of 1960°R air initially in a volume of 1 cu ft; we expand the volume in a cylinder to 2 cu ft at constant pressure. (1) What is the initial gas pressure P_1? (2) What is the final gas pressure P_2? (3) What is the final gas temperature T_2? (4) What heat Q has been transferred into the gas? (5) What work W has been done by the gas on the piston? (6) What has been the rise in internal energy ΔE of the gas?

As found for the constant-volume process,

$$P_1 = 74.2 \text{ psia} \tag{1}$$

By definition
$$P_2 = P_1$$
$$= 74.2 \text{ psia} \tag{2}$$
$$V_2 = 2 \text{ cu ft}/0.1 \text{ lb}$$
$$= 20 \text{ cu ft per lb}$$
$$T_2 = \frac{P_2 V_2}{R}$$
$$= 74.2 \times 144 \times \frac{20}{54.5}$$
$$= 3920°R \tag{3}$$
$$Q = c_p(T_2 - T_1)$$
$$= 0.28(3920 - 1960)$$
$$= 549 \text{ Btu per lb}$$

or
$$Q = 549 \times 0.1$$
$$= 54.9 \text{ Btu} \tag{4}$$
$$W = \frac{P_1(V_2 - V_1)}{J}$$
$$= 74.2 \times \frac{144(20 - 10)}{778}$$
$$= 137.5 \text{ Btu per lb}$$

or
$$W = 137.5 \times 0.1 \text{ lb}$$
$$= 13.8 \text{ Btu} \tag{5}$$
$$E_2 - E_1 = Q - W$$
$$= 54.9 - 13.8$$
$$= 41.1 \text{ Btu} \tag{6}$$

ENERGY EQUATION OF GASES

We can check this by the specific-heat equation, remembering that we have 0.1 lb:

$$E_2 - E_1 = c_v(T_2 - T_1)0.1$$
$$= 0.21(3{,}920 - 1{,}960)0.1$$
$$= 41.2 \text{ Btu} \quad (6)$$

a close enough agreement, seeing this is slide-rule computation and we are working with only three significant figures.

Constant Temperature. We have 0.1 lb of 1960°R air initially in a volume of 1 cu ft, we expand this volume in a cylinder to 2 cu ft at constant temperature. (1) What is the initial gas pressure P_1? (2) What is the final gas temperature T_2? (3) What is the final gas pressure P_2? (4) What heat Q has been transferred to the gas? (5) What work W has been done by the gas on the piston? (6) What is the change in internal energy ΔE of the gas?

As found for the constant-volume process,

$$P_1 = 74.2 \text{ psia} \quad (1)$$

By definition

$$T_2 = T_1 = 1960°R \quad (2)$$

$$P_2 = \frac{RT_2}{V_2}$$

$$= 54.5 \times \frac{1960}{20}$$

$$= 5{,}340 \text{ psfa}$$

or

$$P_2 = \frac{5{,}340}{144}$$

$$= 37.1 \text{ psia} \quad (3)$$

$$E_2 - E_1 = c_v(T_2 - T_1) = 0 \quad (6)$$

since the temperature stays constant.

$$Q = W = \frac{P_1 V_1}{J} \log_e \frac{P_1}{P_2}$$

$$= 74.2 \times 144 \times (10/778) \log_e (74.2/37.1)$$
$$= 137.4 \times \log_e 2$$
$$= 137.4 \times 0.69$$
$$= 94.8 \text{ Btu per lb}$$

or

$$Q = W$$
$$= 94.8 \times 0.1$$
$$= 9.48 \text{ Btu}$$

Adiabatic Process. We again have 0.1 lb of 1960°R air initially in a volume of 1 cu ft; we expand this volume in a cylinder to 2 cu ft adiabatically (without loss or gain of energy as heat). (1) What is the initial gas pressure P_1? (2) What is the final gas pressure P_2? (3) What work W

has been done by the gas? (4) What heat Q has been transferred in or out of the gas? (5) What has been the change in internal energy ΔE of the gas?

As found for the constant-volume process,

$$P_1 = 74.2 \text{ psia} \tag{1}$$

$$P_2 V_2^k = P_1 V_1^k$$

$$\begin{aligned}
P_2 &= P_1 \left(\frac{V_1}{V_2}\right)^k \\
&= 74.2(10/20)^{1.33} \\
&= 74.2 \times 0.5^{1.33} \\
&= 74.2 \times 0.398 \\
&= 29.5 \text{ psia}
\end{aligned} \tag{2}$$

(*Note:* 0.398 was found by using the log tables. You can estimate this by using Fig. 6·7; $0.5^{1.3} = 0.407$ and $0.5^{1.4} = 0.379$. Interpolating for 1.33 you will get 0.3986.)

$$\begin{aligned}
W &= \frac{P_2 V_2 - P_1 V_1}{J(1-k)} \\
&= \frac{144(29.5 \times 20 - 74.2 \times 10)0.1 \text{ (lb)}}{778(1-1.33)} \\
&= \frac{14.4(590 - 742)}{-256.7} \\
&= 8.5 \text{ Btu}
\end{aligned} \tag{3}$$

In an adiabatic process

$$Q = 0 \tag{4}$$

$$W = E_1 - E_2 \quad \text{since } Q = 0$$

so $$E_1 - E_2 = 8.5 \text{ Btu} \tag{5}$$

This can be cross-checked by finding the temperature T_2 and then using it in the specific-heat equation:

$$\begin{aligned}
T_2 &= \frac{P_2 V_2}{R} \\
&= 144 \times 29.5 \times \frac{20}{54.5} \\
&= 1559°R \\
E_1 - E_2 &= c_v(T_1 - T_2)0.1 \\
&= 0.21(1960 - 1559)0.1 \\
&= 8.4 \text{ Btu}
\end{aligned}$$

This is good agreement with W for computation by slide rule.

6·12 Comparison. Figure 6·9a gives us a graphical comparison of four different processes starting from the same state at 1. These show that W is zero for constant V and grows progressively for the adiabatic, constant

T and constant P, when we estimate the comparative areas under the curves.

The examples we just worked out all started from the same state and expanded through the same volume change. This is how their W's compare:

Adiabatic	8.5 Btu
Constant T	9.5 Btu
Constant P	13.8 Btu

Their Q's compare like this:

Adiabatic	0.0 Btu
Constant T	9.5 Btu
Constant P	54.9 Btu

REVIEW TOPICS

1. Demonstrate the conversion of heat to work by means of the constant-pressure gas process.
2. Using the same process demonstrate the conversion of work to heat.
3. What is meant by reversibility?
4. What is the significance of the area under a process line on a P-V graph?
5. Describe Joule's experiment. What does it signify?
6. Write and explain the basic energy equation.
7. What is the enthalpy of a gas?
8. How is the internal energy of a gas related to its temperature?
9. How is the enthalpy of a gas related to its temperature?
10. Write the energy equation for a constant-pressure process.
11. Write the energy equation for a constant-volume process.
12. What is the P-T relation of a gas undergoing a constant-volume process?
13. Write the energy equation for a constant-temperature process.
14. What is the P-V relation for the constant-temperature process?
15. What is meant by an adiabatic process?
16. Write the energy equation for an adiabatic process.
17. What are the property relations of a gas for an adiabatic process?
18. Write the work formula for a reversible adiabatic process.
19. Write the P-V relation for a reversible polytropic process.
20. Write the work equation for a reversible polytropic process.
21. Write the energy equation for a reversible polytropic process.

PROBLEMS

1. We have 0.05 lb of 1200-F air trapped in a bottle with fixed volume of 0.75 cu ft. Then we heat the air to 1600 F. (a) What is the initial pressure?

(b) What is the final pressure? (c) What heat has been injected into the gas? (d) What mechanical work has been done?

2. Assume the same properties as in Prob. 1, but assume that the gas is hydrogen and calculate the same factors.

3. A cylinder with a movable piston holds 0.15 lb of hydrogen initially in a volume of 0.75 cu ft at a temperature of 1300 F. Then the volume is expanded to 2.5 cu ft while gas pressure stays constant. (a) What is the initial gas pressure? (b) What is the final gas pressure? (c) Find the final gas temperature. (d) What heat has been transferred into the gas? (e) What mechanical work has been done on the piston by the gas? (f) What has been the change in internal energy of the gas?

4. The same as Prob. 3, but the gas is nitrogen.

5. A cylinder with a movable piston holds 0.08 lb of air at 1000 F initially in a volume of 0.9 cu ft which then expands to 2.1 cu ft while the gas stays at constant temperature. (a) What is the initial gas pressure? (b) What is the final gas temperature? (c) What is the final gas pressure? (d) What heat has been transferred into or out of the gas? (e) What work has been done by the gas? (f) What is the change in internal energy of the gas?

6. A cylinder with a movable piston holds 1.2 lb of hydrogen at 1300 F initially in a volume of 4.0 cu ft which then shrinks to 0.8 cu ft while the gas temperature remains constant. (a) What is the initial gas pressure? (b) What is the final gas temperature? (c) What is the final gas pressure? (d) What heat has been transferred to or from the gas? (e) What work has been done on or by the gas? (f) What change in internal energy of the gas has taken place?

7. A cylinder and piston with an initial volume of 3 cu ft hold 0.04 lb of hydrogen at 50 F. A reversible adiabatic compression then reduces the volume to 1 cu ft. (a) Find the initial gas pressure. (b) What is the final gas pressure? (c) What work has been done by or on the gas? (d) What heat has been transferred into or out of the gas? (e) What is the change in internal energy of the hydrogen?

8. A cylinder and piston with an initial volume of 2.5 cu ft hold 0.15 lb of air at 80 F. A reversible polytropic compression reduces the volume to 0.9 cu ft with a process of $PV^{2.3} = C$. (a) What is the initial air pressure? (b) What is the final air pressure? (c) What work has been done on or by the gas? (d) What heat has entered or left the air? (e) What is the change in internal energy of the air?

CHAPTER 7

The Carnot Engine

We have studied a variety of processes that a perfect gas may undergo when held in a cylinder behind a frictionless leakproof piston. We are almost ready to see how these processes can be put together so they can steadily transform heat to work or work to heat. But first we shall have to expand our ideas about reversibility.

7·1 Thermal reversibility. In Art. 6·2 we talked about reversibility, which we should now pinpoint as *mechanical reversibility*. This refers to the ability to reverse any process so all the gas properties (P, T, V, E) through which the gas passed during the original process would be *exactly* retraced during reversal. In addition, any energy as heat Q or work W expended or absorbed during the original process would be *exactly* repaid during the reversal. This can never be realized practically. Later we shall talk about irreversible processes.

Another important idea is *thermal reversibility*. This also is only an ideal condition, one that cannot be attained. By experience we learn that we can transfer heat from one body to another only when their temperatures

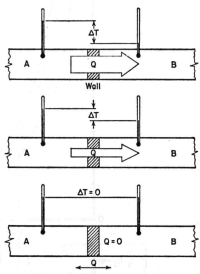

In reversible processes Q flows either way (theoretically) when $T_A = T_B$, that is, $\Delta T = 0$.

Fig. 7·1 Actual heat transfer depends on the temperature difference making Q flow.

differ, Q flowing from the hotter to the colder. Some writers propose that this be called the Zero Law of Thermodynamics—meaning that this should be propounded before the first law and that it is a principle of equal importance.

Figure 7·1 shows two gases A and B separated by a wall. Heat Q flows from A to B because of their temperature difference ΔT. As ΔT decreases, Q shrinks, till it disappears just as ΔT equals zero. For a process to be thermally reversible Q may be transferred in either direction when $\Delta T = 0$. At first glance this looks like a silly idea, but it has importance in setting up an ideal benchmark against which to measure practical performances of heat engines. When we apply this notion in the following process analyses, close your mind to the fact that it cannot be done and look upon it as an ultimate limit—the best we could do in a Utopia.

7·2 Carnot cycle. In our studies of gas processes in a cylinder and piston we learned that they can convert heat to work. A single process is

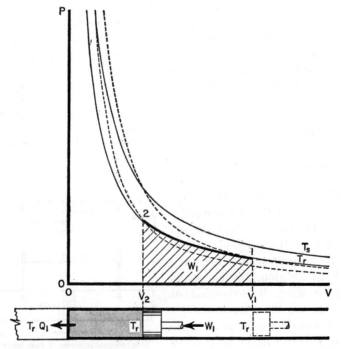

Fig. 7·2 The Carnot cycle works between a pair of constant-T curves and a pair of adiabatic curves; the receiver absorbs rejected Q_1.

not of much practical use—we need an arrangement that will cause this conversion as a repeating procedure, something that will alternate the piston motion back and forth so it could turn a wheel through a mechanical linkage. This means that the gas should alternately expand and contract.

We do this by making the gas go through a regular series of processes. Such a series is called a cycle, because the processes continually repeat in making the engine work. Sadi Carnot, in 1824, was the first to propose a cycle that proved to be a fundamental analysis of great value. The odd fact about his proposal—it was made when heat was thought to be a tangible substance. This mistake, fortunately, did not affect the validity

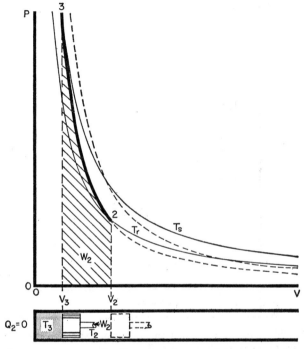

Fig. 7·3 In the second process the Carnot engine completes compression stroke adiabatically, so W_2 raises the gas temperature to T_s.

of the analysis in showing how the cycle changes heat to mechanical work and vice versa.

Carnot needs three main things to make his cycle work: (1) a perfect gas confined in a cylinder and piston, (2) a source of energy at temperature T_s, (3) a receiver of energy at temperature T_r. In Fig. 7·2 the gas in the cylinder is initially at T_r and V_1. The graph has two constant-temperature curves for the gas at T_r and T_s. Crossing these are two adiabatic lines (dotted), processes through which the gas can expand or contract without transferring any heat Q.

Now let us analyze the Carnot cycle, step by step:

First Process, Fig. 7·2. We bring the cylinder head (and so the gas) in contact with the receiver at temperature T_r. Work input W_1 compresses the gas from V_1 to V_2 and rejects heat Q_1 to the receiver at constant T_r as a mechanically and thermally reversible process. The area under the process 1–2 on the P-V graph measures the amount of work done on the gas, W_1. Since the temperature stays constant, $Q_1 = W_1$ according to the simple energy equation.

Second Process, Fig. 7·3. At state 2 we remove the receiver and continue compressing the gas in a reversible adiabatic process. It takes work

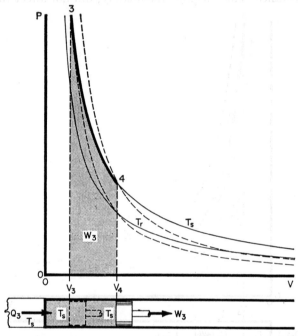

Fig. 7·4 In the third process the Carnot engine begins expansion stroke by absorbing the heat input Q_3 at a constant temperature T_s.

W_2 to compress the gas to state 3, that is, volume V_3. During the process 2–3 the gas temperature rises from T_r to T_s and $Q_2 = 0$. All the work goes to raising the internal energy of the gas:

$$W_2 = E_3 - E_2 = c_v(T_3 - T_2) = c_v(T_s - T_r)$$

Third Process, Fig. 7·4. We now bring the cylinder in contact with the source at temperature T_s and let the gas expand reversibly at constant T_s to V_4. During process 3–4 the gas absorbs heat Q_3 and does work W_3. Since the temperature and internal energy of the gas do not change, $Q_3 = W_3$.

Fourth Process, Fig. 7·5. This process brings us back to our starting point and closes the cycle. To reach this we remove the source from the cylinder and let the gas expand adiabatically and reversibly to state 1. Again for an adiabatic process we have $Q_4 = 0$ but

$$W_4 = E_1 - E_4 = c_v(T_s - T_r)$$

Summary. For all processes the area under the curves in Figs. 7·2 to 7·5 measure the amount of work done on or by the gas. Let us review what has happened during the cycle: (1) The piston made two full equal strokes between fixed end points, V_1 and V_3. (2) During the compression stroke,

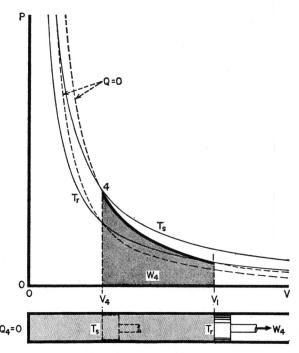

Fig. 7·5 In the fourth process the Carnot engine completes expansion stroke by expanding the gas adiabatically to receiver temperature T_r.

work W_1 and W_2 was done on the gas. (3) During the expansion stroke, work W_3 and W_4 was done by the gas. (4) During part of the compression stroke, heat $Q_1 = Q_r$ was rejected to the receiver at constant temperature T_r. (5) During part of the expansion stroke, heat $Q_3 = Q_s$ was added to the gas by the source at constant temperature T_s.

By plotting all four processes on the P-V graph, Fig. 7·6, we can estimate what has happened energywise. Remembering that areas under process lines measure work done on or by the gas, we see that the compression work done *on* the gas (hatched area) is less than the expansion work done *by* the gas (gray area). The net area 1–2–3–4 measures the net work done by the cycle:

$$\text{Net } W = W = W_3 + W_4 - W_1 - W_2$$

when we regard all quantities as positive. We could have written this equation as the sum of all works and let the negative signs of the first and second processes take care of the subtraction.

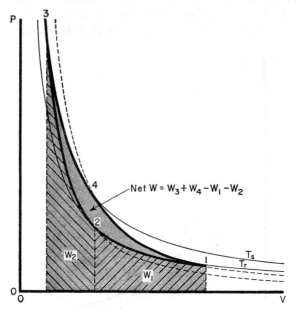

Fig. 7·6 Net area enclosed by process curves measures net mechanical work produced by Carnot engine cycle from Q_3 input.

Heat has been added to the gas by the source and rejected by the gas to the receiver, so the net heat energy input to the gas is

$$\text{Net } Q = Q = Q_s - Q_r$$

Since the gas has gone through four processes of the cycle and ended up in exactly the same state (P_1, V_1, T_1), it has the same internal energy it started with. So all the net transient forms of energy must be equal or

$$Q = W$$

This means that, although the gas has drawn a total heat energy of Q_s from the source and rejected a smaller energy Q_r to the receiver, it has "retained" a net Q and converted all of it to useful work W, often called the *shaft work*.

Figure 7·6 shows one feature of reciprocating engines not generally appreciated: The large flows of work energy into and out of the gas yields a relatively small net work output. Note how small the net area is compared with W_1, W_2, W_3, and W_4.

7·3 Adiabatic processes. To get back on the main track of the discussion, we find that the adiabatic processes, the second and fourth, involve works that are both equal *numerically;* that is,

$$W_2 = W_4 = c_v(T_s - T_r)$$

Since W_2 is work input and W_4 is work output, they cancel each other as far as contributing to net W of the cycle. On the other hand the first and third processes show that

$$W_1 = Q_r \quad \text{and} \quad W_3 = Q_s$$

since they are both constant-T processes. Then we can write

$$W = W_3 - W_1 = Q_s - Q_r = Q \tag{7·1}$$

7·4 Cycle efficiency. The Carnot cycle being an ideal reversible cycle is one of the most efficient, but the thermal efficiency may be far from 100 per cent. Thermal efficiency, like any efficiency, is the ratio of energy output to input. For the Carnot cycle, the output is the useful net W, but the energy input is the amount drawn from the source Q_s, so the thermal efficiency is

$$e_t = \frac{W}{Q_s} = \frac{Q}{Q_s} = \frac{Q_s - Q_r}{Q_s} = \frac{W}{W + Q_r} \tag{7·2}$$

Since W_3 equals Q_s, we can get some idea of the efficiency by comparing W in Fig. 7·6 with W_3 in Fig. 7·4. For the relative temperatures chosen, we can see that W is only a small part of W_3, so the thermal efficiency is well below 100 per cent.

Let us find out what controls the thermal efficiency of an ideal cycle. In Fig. 7·7 we choose constant-T lines T_r, T_1, T_2, and T_3 and plot a pair of adiabatic processes, AB for compression and CD for expansion. We choose a constant-T line T_r for rejecting heat to the receiver. The area underneath, W_r, measures the heat rejected from the cycle, which stays constant in the following discussion.

When we use a source with temperature T_1, we get net work W_a and heat rejection of W_r. Total heat input to the cycle $Q_s = Q_r + W_a$; this also equals the area under the T_1 curve.

Next, suppose we use a higher temperature for our source T_2 but the same receiver temperature T_r. We then get a larger net $W = W_a + W_b$ but the same $Q_r = W_r$. Then $Q_s = Q_r + W_a + W_b$. Since $e_t = W/(W + Q_r)$ and W increases, the efficiency must increase.

If we raise the source temperature further to T_3, we get the additional net work W_c, so that $W = W_a + W_b + W_c$. Since Q_r does not change, we have further raised the thermal efficiency of the cycle.

This demonstrates one important principle: As the source temperature increases and the receiver temperature stays constant, the thermal efficiency increases. In fact, we aim to show that the efficiency of a heat engine depends primarily on the temperature of the source from which it

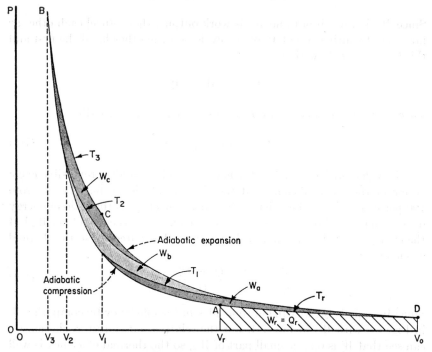

Fig. 7·7   Carnot engine output per cycle can be increased by raising the source temperature from T_1 to T_2 to T_3; this requires using succeedingly longer piston strokes.

receives heat energy and the temperature of the receiver to which it rejects unusable energy.

A second important principle we learn from the Carnot cycle: With average available receiver temperatures around 60 F (river, lake, ocean, atmosphere) it is impossible for a heat engine to convert completely all heat energy received into net work energy output. Some heat energy must always be rejected (as unavailable for conversion to work) to the receiver.

7·5   Temperature influence. Let us study three cycles all based on a common constant-T heat-rejection process as suggested by Fig. 7·7. We shall use state numbering 1–2–3–4 as in Fig. 7·6. Table 7·1 shows the

THE CARNOT ENGINE

Table 7·1 Performance of three Carnot cycles with common receiver temperature of 500°R, all using 1 lb of air as the gas

Cycle	State	P	V	T	$W_r = Q_r$	$W_s = Q_s$	$W = Q$	e_t	Q/T
	1	15.0	12.34	500	23.73	.....	.....	......	0.04747
	2	30.0	6.17	500					
A	3	56.75	3.91	600	.....	28.48	4.75	0.1666	0.04747
	4	28.40	7.82	600					
B	3	97.4	2.66	700	.....	33.23	9.50	0.2860	0.04747
	4	48.7	5.32	700					
C	3	155.4	1.91	800	.....	37.98	14.25	0.3751	0.04747
	4	77.4	3.83	800					

results. The reversible adiabatic compression 2–3 and expansion 4–1 cancel each other as far as work energy is concerned, and we can ignore them in our analysis.

Remember that all properties are related by the basic gas equation $PV = RT$. States 3 and 2 and states 4 and 1 are related by $P_1V_1^k = P_2V_2^k$, the adiabatic gas equation. We assume that our gas is air with $k = 1.4$, ratio of specific heats. This equation can be put into the form

$$T_1V_1^{k-1} = T_2V_2^{k-1}$$

by substituting for P from the basic gas equation. The works for the two constant-T processes in each cycle are figured from the corresponding equation: $W = (P_1V_1/J) \log_e (P_1/P_2)$.

Table computations were started by assuming a state 1 of 15.0 psia and 500°R and compressing to a state 2 of 30.0 psia and 500°R. All the other values were figured from these as a base with a slide rule. You can check them with the above formulas to get exercise in understanding the principles and equations involved.

All three cycles A, B, and C use the same receiver temperature and compression process at 500°R. This always rejects 23.73 Btu per lb of air per cycle for all cycles. Cycle A with a source temperature of 600°R injects 28.48 Btu per lb of air per cycle and produces a net output of 4.75 Btu for a thermal efficiency of 16.66%. Cycle B with a 700°R source injects 33.23 Btu for a net output of 9.50 Btu at a thermal efficiency of 28.60%. Cycle C with 800°R source injects 37.98 Btu for a net output of 14.25 Btu at an efficiency of 37.51%.

Figure 7·8 summarizes this data. It shows that the net work is directly proportional to the source temperature when receiver temperature stays constant or directly proportional to the temperature drop through which the cycle operates; that is,

$$W = Q = C(T_s - T_r)$$

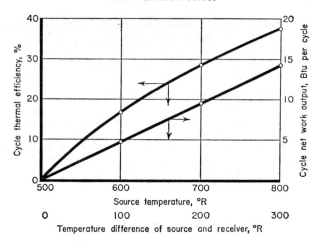

Fig. 7·8 For constant receiver temperature both the cycle thermal efficiency and work output rise with source temperature.

where C is a proportionality constant. The thermal efficiency of the cycle also rises with temperature differential, showing the advantage of using a high temperature where practicable.

7·6 Receiver temperature. Let us see what effect a variable receiver temperature has on cycle performance. We shall assume that we have a constant source temperature of 800°R and use the series of cycles listed in Table 7·1. We shall use successive receiver temperatures of 800, 700, 600, and 500°R. We can then use the Q's already calculated in Table 7·1 to figure Table 7·2.

Table 7·2 Cycle performance with source temperature of 800°R

Receiver temp,°R	Q_r	$W = Q$	Cycle e_t
800	37.98	0	0
700	33.23	4.75	0.125
600	28.48	9.50	0.250
500	23.73	14.25	0.375

Figure 7·9 summarizes these data and shows that both thermal efficiency and net work W are inversely proportional to the receiver temperature, but again they are proportional to the temperature differential of the cycle. Using as low a receiver temperature as possible is important to cycle capacity and efficiency. This again demonstrates that cycle performance depends wholly on temperature.

Q_s and W_s also depend on temperature, so we can write $Q_s = CT_s$. Then it follows that

$$e_t = \frac{C(T_s - T_r)}{CT_s} = \frac{T_s - T_r}{T_s}$$

This is the Carnot efficiency and represents the best that can be done by any cycle running between the same temperature levels. Remember that these temperatures are absolute and not degrees Fahrenheit.

We have developed some important thermodynamic basics in this chapter that should be remembered whenever dealing with thermodynamic problems. In Table 7·1 we figured a ratio Q/T that proved to be

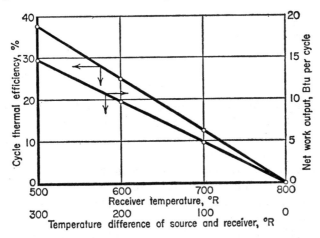

Fig. 7·9 For a constant source temperature both the cycle thermal efficiency and work output decrease with rise in receiver temperature.

constant at 0.04747 for all temperature levels. We shall use this in the next chapter to develop basic ideas about *entropy*.

REVIEW TOPICS

1. What is meant by mechanical reversibility?
2. What is meant by thermal reversibility?
3. What statement can be called the Zero Law of Thermodynamics?
4. What is an engine cycle composed of?
5. Describe the Carnot cycle.
6. How many compression strokes are in the Carnot cycle?
7. How many compression processes are in the Carnot cycle?
8. How many expansion strokes are in the Carnot cycle?
9. How many expansion processes are in the Carnot cycle?
10. Is the net work output of the Carnot cycle small compared with the total mechanical work done on and by the gas?

11. Does the net work output equal the net heat input?
12. What feature about the adiabatic processes in the Carnot cycle is the same?
13. Write two equations for figuring net W and Q.
14. Write the thermal efficiency for the Carnot cycle.
15. What factors control the thermal efficiency of the Carnot cycle?
16. Can all the heat withdrawn from the source of a Carnot engine be completely converted to mechanical work?
17. Does the thermal efficiency of the Carnot cycle depend on the length of the piston strokes?
18. Does the work output of a Carnot cycle depend on the length of the piston strokes?
19. Does the thermal efficiency of the Carnot cycle depend on the pressure of the working fluid?

PROBLEMS

1. A Carnot cycle uses a perfect gas and works between a 2000-F source and an 80-F receiver. What will be its thermal efficiency?
2. If the reversible adiabatic compression process of a Carnot cycle requires 150 Btu per lb of mechanical work input, what work does the reversible adiabatic expansion process produce?
3. A Carnot cycle (Fig. 7·6) has 0.1 lb of air at 15 psia in the cylinder at state 1. With the air and receiver at 80 F the air is compressed to 45 psia at constant temperature at state 2. From state 2 to 3 the air is compressed reversibly and adiabatically to 600 F at state 3. From state 3 the air expands at constant temperature, absorbing heat from the 600-F source, to state 4. Calculate: (a) pressure and temperature of the air at each of the four states marking the end points of the Carnot-cycle processes, (b) the heat and work transfers for each of the processes, (c) the net work and net heat of the cycle, (d) the thermal efficiency of the cycle two ways.
4. Check the figures listed in Tables 7·1 and 7·2.

CHAPTER 8

Entropy—Index of Heat Flow

In our study of the Carnot cycle we learned a very important fact: Thermal efficiency of a heat engine depends only on the absolute temperatures of the source and receiver. The higher the source temperature or the lower the receiver temperature, the greater the thermal efficiency of the cycle. Thermal efficiency of a heat engine is completely independent of the pressure of the gas (working fluid). We will look into this in more detail in the next chapter.

8·1 Second law. A fact of even more fundamental importance that we learned was that even in an ideal cycle, such as the Carnot, it was impossible to convert *all* the energy received *as heat* from the source into mechanical work. Invariably, we have to reject some of that energy as heat to a receiver at a lower temperature than the source. The part of the heat that was converted into mechanical work is *available energy;* the remainder of the heat that had to be rejected to the receiver is *unavailable energy.*

The importance of this fact is recognized by stating it as the *Second Law of Thermodynamics: All energy received as heat by a heat-engine cycle cannot be converted into mechanical work.*

There is one condition under which all the heat energy from a source could theoretically be converted to mechanical work. Remembering that, for the Carnot cycle, $e_t = (T_s - T_r)/T_s$, we can see that, if T_r is zero, then the thermal efficiency becomes 100%. But we have not any natural region or receiver at this temperature handy, so for practical purposes we can accept the second law as a fact.

It has often been proposed that we should refrigerate the receiver to lower its temperature and raise the engine thermal efficiency. But we shall find that the additional work obtained must all be used to lower the receiver temperature below the atmospheric temperature, so the net gain becomes zero. This holds for ideal reversible processes. Practically, there would be an overall loss because actual processes are not reversible. In the next chapter we will study the reversed Carnot cycle as applied to refrigeration.

8·2 Carnot-cycle review. This cycle is so important to an understanding of thermodynamics that it will pay us to review it again from a slightly different angle. Figure 8·1 shows three constant-T curves, T_a, T_b, and T_c crossed by two reversible adiabatic process curves. Do not forget that the mechanical work areas under any reversible adiabatic process curves are always the same between two given temperature levels, regardless of the pressure or specific volume of the gas. You should thoroughly understand this by now because for this process $W = \Delta E = c_v(T_2 - T_1)$; in other words, the work depends only on the temperature difference for the process, since $Q = 0$.

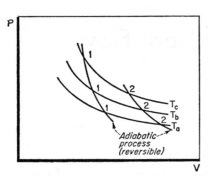

Fig. 8·1 Constant-T processes bounded by a pair of adiabatics can be related through equations.

Now let us turn to the constant-T process. Three are shown in Fig. 8·1. Remembering that the universal gas equation $P_1V_1 = RT_1$ always holds, we can see that

$$Q = \frac{W}{J} = \frac{P_1V_1}{J}\log_e \frac{P_1}{P_2} = \frac{RT_1}{J}\log_e \frac{P_1}{P_2} = \frac{RT_1}{J}\log_e \frac{V_2}{V_1} \quad (8\cdot1)$$

where the process starts and stops at any two states 1 and 2 [see Eq. (6·17)].

Now in a Carnot cycle, pairs of states 1 and 2 can be related at any number of temperature levels by pairs of reversible adiabatic curves as in Fig. 8·1. The states along a reversible adiabatic are related by

$$PV^k = \frac{RT}{V}V^k = RTV^{k-1} \quad (8\cdot2)$$

Then we can write the following relations:

$$T_a V_{1a}{}^{k-1} = T_b V_{1b}{}^{k-1}$$
$$T_a V_{2a}{}^{k-1} = T_b V_{2b}{}^{k-1}$$

Transposing these two equations we find that

$$\frac{T_a}{T_b} = \left(\frac{V_{1b}}{V_{1a}}\right)^{k-1} = \left(\frac{V_{2b}}{V_{2a}}\right)^{k-1}$$

This means that
$$\frac{V_{1b}}{V_{1a}} = \frac{V_{2b}}{V_{2a}}$$

Transposing this equation we find that
$$\frac{V_{2a}}{V_{1a}} = \frac{V_{2b}}{V_{1b}} = \frac{V_{2c}}{V_{1c}} = V_r \qquad (8\cdot3)$$

This shows that the volume ratios V_r are equal for all constant-T processes bound by a common pair of reversible adiabatic processes as in Fig. 8·1. Applying the heat equation (8·1) we get

$$Q_a = \frac{1}{J} RT_a \log_e \frac{V_{2a}}{V_{1a}} = \frac{1}{J} RT_a \log_e V_r$$

$$Q_b = \frac{1}{J} RT_b \log_e \frac{V_{2b}}{V_{1b}} = \frac{1}{J} RT_b \log_e V_r$$

$$Q_c = \frac{1}{J} RT_c \log_e \frac{V_{2c}}{V_{1c}} = \frac{1}{J} RT_c \log_e V_r$$

Since $\frac{1}{J} R \log_e V_r$ is a constant C, we can write the general relation
$$Q = CT$$

Choosing any cycle at random, say between temperatures T_a and T_b, the thermal efficiency would be

$$e_t = \frac{W}{Q_b} = \frac{Q_b - Q_a}{Q_b} = \frac{CT_b - CT_a}{CT_b} = \frac{T_b - T_a}{T_b} \qquad (8\cdot4)$$

This again brings us to the conclusion that the thermal efficiency of an ideal heat-engine cycle depends only on the temperatures of the source and the receiver.

8·3 Entropy. Now we are ready to find another property of gases and vapors that proves helpful in analyzing engine cycles. To understand it, let us start with a heat engine using helium as the working fluid (we could use any gas). Figure 8·2 shows the constant-T curves for 500, 600, and 700°R. Three reversible adiabatic curves 1–4–7, 2–5–8, and 3–6–9 cross the constant-T curves at nine state points which are listed in the table of Fig. 8·2.

Table 8·1 lists the work and heat transfers at constant temperature between the state points in Fig. 8·2. When the heat transfers are divided by the temperatures at which they take place, we see that the ratios are equal between pairs of reversible adiabatic processes. We let

$$\frac{Q}{T} = S \qquad (8\cdot5)$$

We also just learned that the volume ratios of the constant-T processes between common pairs of reversible adiabatics are equal. For example in Fig. 8·2,

$$\frac{V_2}{V_1} = \frac{V_5}{V_4} = \frac{V_8}{V_7} = V_r$$

$$\frac{15}{10} = \frac{11.37}{7.59} = \frac{9}{6} = 1.5$$

between adiabatics 1–4–7 and 2–5–8.

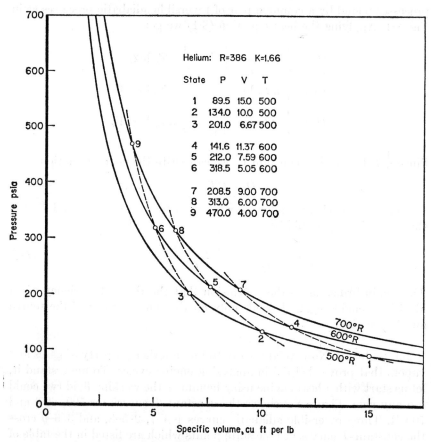

Fig. 8·2 With three constant-T curves and three adiabatics we can make up any number of Carnot cycles. Their thermal efficiency depends only on the temperature limits.

We deliberately chose the volume ratio for the constant-T processes between the adiabatics 2–5–8 and 3–6–9 to be the same as between the adiabatics 2–5–8 and 1–4–7. As Table 8·1 shows, this makes the heat-temperature ratio S all equal for the constant-T processes. All the Q's and W's were figured by Eq. (8·1).

When we combine the constant-T processes, such as 3–2 and 2–1, into one process 3–1, note in Table 8·1 that $S_{32} + S_{21} = S_{31}$. We have apparently hit on some kind of relationship.

Table 8·1 Heat transfers and works for constant-T processes of Fig. 8·2

$Q_{21} = W_{21} = 100.6$ Btu	$Q_{21}/T_a = 100.6/500 = 0.2012 = S_{21}$
$Q_{54} = W_{54} = 120.7$ Btu	$Q_{54}/T_b = 120.7/600 = 0.2012 = S_{54}$
$Q_{87} = W_{87} = 140.8$ Btu	$Q_{87}/T_c = 140.8/700 = 0.2012 = S_{87}$
$Q_{31} = W_{31} = 201.2$ Btu	$Q_{31}/T_a = 201.2/500 = 0.4024 = S_{31}$
$Q_{64} = W_{64} = 241.4$ Btu	$Q_{64}/T_b = 241.4/600 = 0.4024 = S_{64}$
$Q_{97} = W_{97} = 281.6$ Btu	$Q_{97}/T_c = 281.6/700 = 0.4024 = S_{97}$
$Q_{32} = W_{32} = 100.6$ Btu	$Q_{32}/T_a = 100.6/500 = 0.2012 = S_{32}$
$Q_{65} = W_{65} = 120.7$ Btu	$Q_{65}/T_b = 120.7/600 = 0.2012 = S_{65}$
$Q_{98} = W_{98} = 140.8$ Btu	$Q_{98}/T_c = 140.8/700 = 0.2012 = S_{98}$

Let us transpose our ratio $Q/T = S$ into $Q = TS$. We know T to be a property of a gas; now let us assume that S is a property also. If we plot the two properties as coordinates on a graph as in Fig. 8·3a, we see that for the constant-T process 3–2 at 500°R we draw a horizontal line at 500 with a length S_{32}. Then by our equation $Q = TS$ the area under the line equals the heat transferred during the constant-T process Q_{32}. In Fig. 8·3b we do the same thing for constant-T process 3–1 and have a horizontal line S_{31} long with the area under it equaling the heat transferred Q_{31}.

Figure 8·3c shows that the difference S_{21} must be the length from 0.2012 to 0.4024 on the S scale, since we showed above that these quantities are additive. These lengths are really *differences* in quantities; for instance, $S_{31} = S_1 - S_3$ and $S_{32} = S_2 - S_3$. The quantity S_3 corresponds to the state 3 in Fig. 8·2, S_2 to state 2, and S_1 to state 1. We can conclude that S is a property.

We can project this reasoning to all nine state points in Fig. 8·2, and we would then come out with a graph as in Fig. 8·3d. Let us identify the property S as *entropy*. The state points 3, 6, and 9 have an entropy of zero. But since we picked these states arbitrarily in Fig. 8·2, we can see that the zero entropy value is just arbitrary and not absolute. In general we deal with *changes* in entropy and not the absolute values of entropy.

Theoretically, it would seem that a gas has zero entropy at zero pressure, temperature, and specific volume. Then as we add heat Q, all the properties including the entropy increase. Wherever we have information on the behavior of a gas from zero temperature to higher levels, we shall be able to calculate the absolute entropy of a gas. There still is some information missing on all gases, however, though scientists have approached absolute zero within fractions of a degree.

We can make direct measures of properties of a gas, like pressure, temperature, density, specific volume. But entropy cannot be measured directly; as we showed, it is calculated from changes in other properties during processes. It is especially useful in calculating heat transfers during processes.

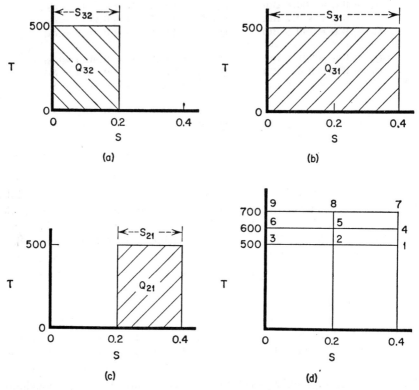

Fig. 8·3 The area on T-S graphs measures only the heat transferred during the heating or cooling process.

In deriving the property entropy, we used the constant-T process. But how do we figure entropy (more accurately entropy changes) for the other processes? They must all be based on heat transferred at constant temperature.

8·4 Adiabatic entropy change. During a reversible adiabatic process there is no heat transferred, so $Q = 0$. Then since $S = Q/T$, the entropy must be zero. In Fig. 8·3d we can identify the adiabatic processes 3–6–9, 2–5–8, and 1–4–7 as corresponding to those shown on the P-V graph in Fig. 8·2. The complete adiabatic process on the T-S graph is a vertical line extending to zero temperature. This ties in with zero heat transfer during an adiabatic process; a vertical line has no area under it, so $Q = 0$.

8·5 Constant-P entropy change. Next, let us consider the constant-P process, Fig. 8·4a. On the P-V graph it appears as a horizontal line, and we know the temperature changes during the process. But figuring entropy depends on knowing the amount of heat transferred at constant

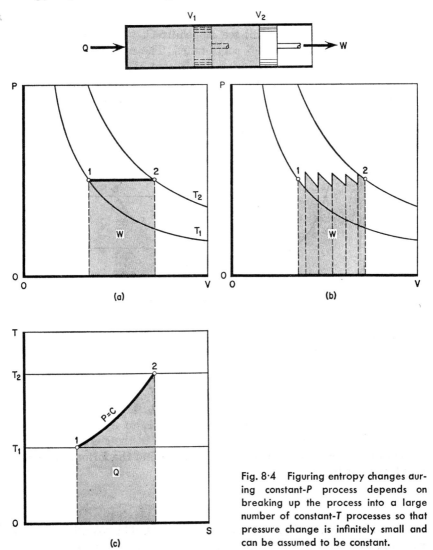

Fig. 8·4 Figuring entropy changes during constant-P process depends on breaking up the process into a large number of constant-T processes so that pressure change is infinitely small and can be assumed to be constant.

temperature—what do we do when the temperature varies? Figure 8·4b shows the reasoning involved. We simply break up the constant-P process into a very large number of constant-T processes so we can consider the temperature constant during a very large number of small volume changes. If we take a large enough number of elements, the saw-tooth path of Fig. 8·4b would become effectively a straight line as in Fig. 8·4a. By this line of reasoning we find that the equation for $S = Q/T$ of the constant-P process can be figured as

$$S_2 - S_1 = c_p \log_e \frac{T_2}{T_1} \tag{8·6}$$

For the derivation of Eq. (8·6) see Appendix C. Figure 8·4c shows the curve of this equation on T-S coordinates. The area under the curve measures the heat transferred during the process. We can see now that we have a pair of companion graphs; P-V, Fig. 8·4a, gives us a graphical measure of work W and T-S gives us a measure of Q.

8·6 Constant-V entropy change. So far we have paired up the P-V and T-S graphs for (1) the constant-T process, (2) the adiabatic process,

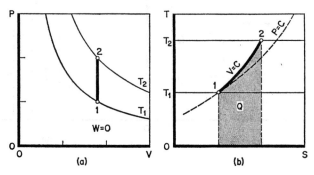

Fig. 8·5 In the constant-V process no work W can be done as shown on (a), but the heat transferred Q appears on a T-S graph as in (b).

(3) the constant-P process. Next, let us see how they compare for the constant-V process. Figure 8·5 shows the P-V graph we met before; $W = 0$ and the only energy change can be as Q. But since T is variable while Q flows, we need a more complex formula to figure $S_2 - S_1$. By assuming that we add very small amounts of Q to the gas at very small steps of increasing levels of constant T, we can find the formula we need. This turns out to be

$$S_2 - S_1 = c_v \log_e \frac{T_2}{T_1} \tag{8·7}$$

For the derivation of Eq. (8-7) see Appendix D. This parallels the one we found for the constant-P process, differing only in using c_v instead of c_p. Figure 8·5b shows the T-S graph for a constant-V process. The dotted curve shows how it compares with a constant-P process passing through state 1. Since no work W is done in a constant-V process, the amount of heat Q that can be absorbed between two given temperatures T_1 and T_2 is less than in a constant-P process. This makes the constant-V curve slope more steeply than the constant-P process. This reduces the amount of area (equivalent to Q) under the curve.

8·7 Polytropic entropy change. Finally, we should know the entropy changes for a polytropic process in which all factors change simultaneously. This proves to be a parallel to the previous two equations:

$$S_2 - S_1 = c_n \log_e \frac{T_2}{T_1} \qquad (8·8)$$

where c_n is the polytropic specific heat and equals $c_v(k - n)/(1 - n)$. This usually gives us a process curve on the T-S graph similar to the constant-V or -P curves at some different slope. For the derivation of Eq. (8·8) see Appendix E.

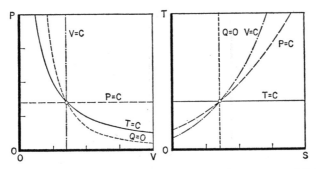

Fig. 8·6 Process curves passing through one state point of a gas have different shapes on companion P-V and T-S graphs.

Figure 8·6 sums up the relative process-curve trends on P-V and T-S coordinates that pass through one state point. The polytropic process can have a curve going in any direction through the state point. We shall be using these curves quite often, so it is wise to memorize them for future use. See Appendix F for a summary of perfect-gas formulas.

8·8 Using T-S graph. As we learned, the P-V graph gave us a picture of net work available in the Carnot cycle. Later on we shall see that this applies to many different types of cycles as well. The P-V graph, however, gives it to us only from the mechanical work standpoint. Any heat transfers Q must be inferred from the simple energy equation

$$Q = E_2 - E_1 + \frac{W}{J}$$

On the other hand, the T-S diagram gives us a picture of net energy available in the Carnot cycle also but gives it from the heat-transferred standpoint. For reversible processes the net areas on both types of graphs are equal for the Carnot cycle, one in terms of heat and the other in terms of mechanical work.

Figure 8·7 shows how the Carnot cycle appears on T-S graphs. In Fig. 8·2 we can assume that a Carnot cycle works between the state points 1–2–8–7. Then if all four processes are reversible, the net area enclosed measures the mechanical shaft work output. Remember, areas under the process curves (not outlined in Fig. 8·2) measure the amount of work done on or by the gas during the process.

Figure 8·7 shows the corresponding processes on T-S coordinates. The area Q_r under the constant-T process 1–2 measures the amount of heat rejected to the receiver during the first part of the compression process.

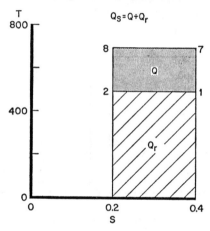

Fig. 8·7 The Carnot cycle on a T-S graph has a simple rectangular shape; work done depends on net Q transferred.

The adiabatic compression process 2–8 involves no heat transfer, so it shows no lower area by being a vertical line on this graph.

The total area under the constant-T process 8–7 equal to $Q + Q_r$ is the heat transferred from the source to the gas, Q_s. The expansion process ends with an adiabatic process 7–1. Again this involves no heat transfer, and it has no area underneath.

The two areas under the receiver and source temperatures are coincident. For the Carnot cycle we know that the net heat absorbed by the gas is equal to $Q_s - Q_r = Q$. But we already showed that $W = Q$ for the Carnot cycle, so the two net areas on the P-V and T-S graphs are equal in quantity but different in units; one is in foot-pounds and the other in Btu.

REVIEW TOPICS

1. What is available energy?
2. What is unavailable energy?

ENTROPY—INDEX OF HEAT FLOW

3. Write the Second Law of Thermodynamics.

4. What practical limitation prevents a Carnot engine from operating at 100 per cent thermal efficiency?

5. Show that the volume ratios V_r are equal for all constant-T processes between one pair of reversible adiabatic processes.

6. Show that the ratio of Q/T is constant for all constant-temperature processes between any pair of reversible adiabatic processes.

7. Define entropy. Is it a property of a working fluid?

8. Shall we generally deal in the absolute entropy of a fluid?

9. Can entropy be measured by any instrument directly?

10. Write the equation for heat transferred during a reversible adiabatic process. Is this the same as a constant-entropy process?

11. Write the equation for entropy change during a constant-pressure process.

12. Write the equation for entropy change during a constant-volume process.

13. Write the equation for entropy change during a polytropic process.

14. What do areas mean on T-S graphs?

PROBLEMS

1. Find the change in entropy of the gas in Prob. 1 of Chap. 6.
2. Find the change in entropy of the gas in Prob. 2 of Chap. 6.
3. Find the change in entropy of the gas in Prob. 3 of Chap. 6.
4. Find the change in entropy of the gas in Prob. 4 of Chap. 6.
5. Find the change in entropy of the gas in Prob. 5 of Chap. 6.
6. Find the change in entropy of the gas in Prob. 6 of Chap. 6.
7. Find the change in entropy of the gas in Prob. 7 of Chap. 6.
8. Find the change in entropy of the gas in Prob. 8 of Chap. 6.

CHAPTER 9

The Reversed Carnot Cycle

First we shall study the Carnot cycle some more to discover further uses and limitations it may have. We already know that the thermal efficiency of the cycle depends only on the temperature limits between which it works. Let us see what effects pressure, volume, and piston stroke have on the performance of the engine.

9·1 Design factors. In Fig. 9·1 we again use helium as the working fluid. Table 9·1 shows the states at the end points of the processes and gives work inputs, outputs, and heats transferred. In Fig. 9.1a and b we have the P-V and T-S diagrams for an engine working between 500 and 700°R temperatures. The shaded area shows the net W and net Q (which are equal) developed by the cycle, 40.2 Btu per lb of helium.

The thermal efficiency of the cycle equals 28.6%. Designers strive to get this factor as high as possible, but other conditions must be considered to get the best overall engine. Engine piston stroke must be manageable; in our ideal engine this appears as the volumetric displacement of the piston $V_d = V_1 - V_3$. This is the volume swept out by the piston in the cylinder. The relation of V_d to V_1 controls maximum pressure in the cycle at V_3.

In practical design considerations, designers measure this effect by finding the *mean effective pressure* P_m. When P_m acts for one full stroke on the piston, it produces the same mechanical work as the net work of the entire cycle:

$$P_m = \frac{W}{V_d} = \frac{W}{V_1 - V_3} \qquad (9\cdot1)$$

Figure 9·1a shows P_m for the cycle and the corresponding W that equals the cycle W above. We can regard P_m acting on the piston during the

THE REVERSED CARNOT CYCLE 113

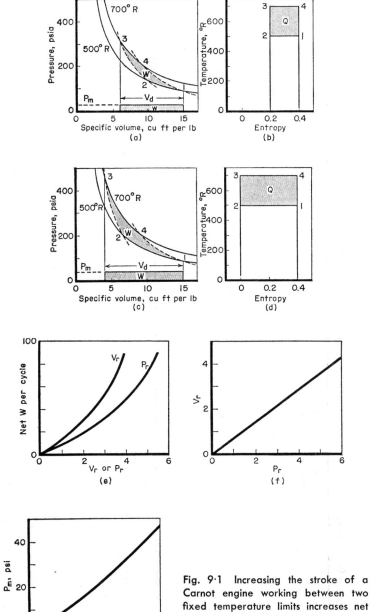

Fig. 9·1 Increasing the stroke of a Carnot engine working between two fixed temperature limits increases net work produced per cycle but does not affect thermal efficiency. Mean effective pressure P_m depends on W and piston displacement.

Table 9-1 Working fluid: helium, $c_p = 1.25$ Btu per lb, $c_v = 0.754$ Btu per lb, $k = 1.66$, $R = 386$

State	P, psia	V, ft³/lb	T, °R	Process, Btu per lb $Q = W$	Process, Btu per lb W	ΔS		State	P, psia	V, ft³/lb	T, °R	Process, Btu per lb $Q = W$	Process, Btu per lb W	ΔS
1	89.5	15.00	500					1	89.5	15.00	500			
				−100.6		−0.201						−201.2		−0.402
2	134.0	10.00	500					2	201.0	6.67	500			
					−150.8	0							−150.8	0
3	313.0	6.00	700					3	470.0	4.00	700			
				140.8		0.201						281.6		0.402
4	208.5	9.00	700					4	208.5	9.00	700			
					150.8	0							150.8	0

Adiabatic-process $W = c_v(T_s - T_r) = 0.754(700 - 500) = 150.8$ Btu per lb helium

Net $W = $ net $Q = 140.8 - 100.6 = 40.2$ Btu per lb per cycle

$P_m = \dfrac{W}{V_1 - V_3} = \dfrac{40.2 \times 778}{15.0 - 6.0} = 3{,}475$ psf $= 24.13$ psi

Compression ratio $V_r = \dfrac{V_1}{V_3} = \dfrac{15.0}{6.0} = 2.5$

Pressure ratio $P_r = \dfrac{P_3}{P_1} = \dfrac{313}{89.5} = 3.5$

$e_t = \dfrac{W}{Q_s} = \dfrac{40.2}{140.8} = 0.286$ or 28.6%

$e_t = \dfrac{T_s - T_r}{T_s} = \dfrac{700 - 500}{700} = 0.286$

Net $W = $ net $Q = 281.6 - 201.2 = 80.4$ Btu per lb per cycle

$P_m = \dfrac{W}{V_1 - V_3} = \dfrac{80.4 \times 778}{15.0 - 4.0} = 5{,}690$ psf $= 39.49$ psi

Compression ratio, $V_r = \dfrac{V_1}{V_3} = \dfrac{15.0}{4.0} = 3.75$

Pressure ratio $P_r = \dfrac{P_3}{P_1} = \dfrac{470}{89.5} = 5.25$

$e_t = \dfrac{W}{Q_s} = \dfrac{80.4}{281.6} = 0.286$ or 28.6%

$e_t = \dfrac{T_s - T_r}{T_s} = \dfrac{700 - 500}{700} = 0.286$

working or expansion stroke to develop the net work W of the cycle. During the return or "compression" stroke of the piston the pressure is zero, so no work is done on the gas during that stroke.

We have two ratios important in describing an engine: (1) overall compression ratio and (2) overall pressure ratio. Overall compression ratio V_r for the engine is the ratio of gas volumes in the cylinder at the extremes of the piston stroke; that is, $V_r = V_1/V_3$. Note that this differs from the volume ratio for a single process.

The overall pressure ratio P_r is the ratio of gas pressures at the extremes of the piston stroke; that is, $P_r = P_3/P_1$. This factor gives us some idea of how heavy the engine cylinder and other parts must be made to contain the working gas. The higher the pressure, the heavier and more expensive the engine.

9·2 Raising W. In Fig. 9·1c and d we use the same engine but raise the compression ratio from 2.5 to 3.75. This boosts the pressure ratio from 3.5 to 5.25 and $W = Q$ from 40.2 to 80.4 Btu per lb of gas per cycle. W doubles per cycle, but thermal efficiency remains unchanged at 28.6% because the source and receiver temperatures have not been altered.

Mean effective pressure P_m has risen from 24.17 to 39.55 psi; note that this pressure is neither gage nor absolute. Doubling W and Q has not doubled P_m because V_d increases with increasing V_r. Figure 9·1e, f, and g shows how these various engine factors are related. So much for the Carnot engine as a shaft-work producer; now let us look at the reversed Carnot cycle.

9·3 Refrigerator. The Carnot cycle proves quite versatile in doing various jobs. Figure 9·2a outlines the problem when we wish to cool a space below surrounding temperatures. As soon as we establish a temperature difference between two areas, heat transfers from the hotter to the colder. Here heat Q leaks through the insulation from the 550°R atmosphere into the 400°R cold room. To maintain the lower temperature we must remove the total leakage heat Q_c as soon as it enters. We can do this by using a Carnot cycle as in Fig. 9·2b but *reverse* the order of the processes.

At state 1 gas in the cylinder is in contact with the atmosphere; compressing the gas to state 2 rejects heat to the atmosphere at a constant temperature of 550°R. The total area under the process 1–2 in Fig. 9·2c measures the heat rejected to the atmosphere: $Q_r = Q + Q_c$.

Compressed gas is then allowed to expand adiabatically from state 2 to state 3. Since it does work on the piston, the gas temperature drops to 400°R at state 3. The gas is then placed in contact with the cold room air at 400°R and allowed to expand at constant-T to state 4. The gas does work W_c on the piston and absorbs heat Q_c from the cold room during this

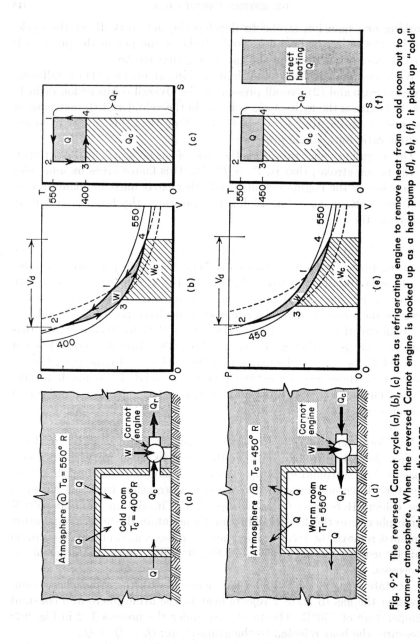

Fig. 9·2 The reversed Carnot cycle (a), (b), (c) acts as refrigerating engine to remove heat from a cold room out to a warmer atmosphere. When the reversed Carnot engine is hooked up as a heat pump (d), (e), (f), it picks up "cold" energy from the air to warm the room.

process. From state 4 the piston then adiabatically compresses the gas, raising its temperature to 550°R at state 1, the starting point of the cycle.

The total work done by the piston in compressing the gas exceeds the work done by the gas on the piston during expansion, so the cycle loop measures the net mechanical shaft work *input* to the refrigerating engine, (usually called compressor). The overall energy-flow equation for the reversed Carnot cycle is

$$Q_r = Q_c + W \tag{9.2}$$

This shows that, by expending shaft work (valuable *available* energy), we can raise low-temperature energy and deliver it to a high-temperature region or receiver.

Consider this from a different viewpoint: Energy naturally degrades itself from high-temperature regions to low-temperature areas by spontaneous heat transfer. We can never completely stop this energy transfer, but we can limit it by interposing suitable insulating materials between the regions. To reverse the direction of transfer we use high-grade available energy as in the reversed Carnot cycle. This is another way of stating the Second Law of Thermodynamics.

9·4 Performance. As in the direct Carnot cycle we are interested in developing some criterion of performance for the refrigerating application. We can use a parallel to thermal efficiency but find we must use a different name. Most efficiencies are defined as the ratio of output to input.

For the reversed Carnot cycle the useful function or output is the refrigeration or heat removed from the low-temperature region Q_c. Useful input is the shaft work W. When we divide Q_c by W, we have a quotient usually larger than 1.0. Instead of calling this ratio an efficiency, we name it the *coefficient of performance* (COP). In words, it is the number of units of energy removed from the cold room for each energy unit of shaft work input.

The T-S diagram, Fig. 9·2c, shows that the COP has sole dependence on temperature. From the simple areas we can write

$$\begin{aligned}
\text{COP} &= \frac{Q_c}{W} = \frac{Q_c}{Q_r - Q_c} \\
&= \frac{T_c(S_4 - S_3)}{T_a(S_1 - S_2) - T_c(S_4 - S_3)} \\
&= \frac{T_c}{T_a - T_c}
\end{aligned} \tag{9.3}$$

since $S_4 - S_3 = S_1 - S_2$, according to Fig. 9·2c.

By letting T_c equal various percentages of T_a and assuming the latter as 1.00, we see how the COP varies. Figure 9·3 shows the relationship.

Here we see that, when T_c is only slightly lower than T_a, the COP is high but, as the difference grows, the COP drops rapidly. When T_c is less than half of T_a, the COP is less than 1.0. In other words, the colder we wish to keep an area, the more work we must put into the refrigerating cycle per unit energy removed. A slight degree of cooling needs only a small W input, giving a high COP.

Figure 9·2b indicates the V_d for the refrigerating cycle. Since the COP depends only on temperature limits, we find that raising the V_d only raises the amount of heat removed per cycle with a proportionate increase in W. This effect parallels the one we discussed for the direct cycle.

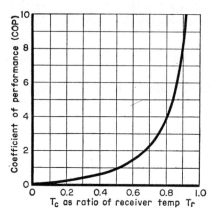

Fig. 9·3 The coefficient of performance for the reversed Carnot engine climbs fast with rising cold room temperature.

9·5 Heat pump. Under certain conditions the reversed Carnot cycle can be used as an efficient heating unit. Figure 9·2d shows the controlling conditions. Here a warm room at 550°R leaks heat Q to a colder atmosphere at 450°R. To maintain the warmer temperature we must have a heat input to the warm room of Q_r equal to the total heat leakage.

We supply Q_r by using a reversed Carnot cycle as a *heat pump*. This works as indicated in Fig. 9·2e. At state 1 we put the gas in contact with the room air and compress the gas at constant T of 550°R to state 2; this discharges heat Q_r into the room, Fig. 9·2f.

The gas then expands adiabatically to state 3, where its temperature has dropped to 450°R. At state 3 the gas is in contact with the cold atmosphere at 450°R and the gas expands at constant T to state 4; the gas does work W_c on the piston and absorbs heat Q_c. From state 4 the gas is adiabatically compressed until it reaches state 1 at 550°R, the starting point of the cycle.

Figure 9·2f shows that work $W = Q$ picks up low-temperature energy Q_c to deliver a total of higher temperature energy Q_r. What is the advantage of this heat pump? Suppose Q is in the form of electric energy; if we used it directly in this form for resistance heating, we would have to use an amount Q shown in the right-hand area. But with the heat pump we need only the smaller Q to deliver the same total heat energy to the room. This is a more efficient use of high-cost available energy which salvages low-temperature energy from the cold atmosphere.

The heat pump increases its attractiveness when the temperature difference $T_r - T_c$ decreases. This follows from the improvement in the COP and lower leakage Q. Of course, when cheap fuel is available, the heat pump may not offer any cost advantage. This is true in many areas today.

9·6 Refrigerated receiver. Now let us show why there is no advantage to refrigerating the receiver of a Carnot engine producing shaft work.

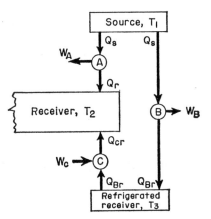

$$W = Q_e = Q \frac{T_s - T_r}{T_s} \quad (1)$$

$$W_A = Q_s \frac{T_1 - T_2}{T_1} \quad (2)$$

$$W_B = Q_s \frac{T_1 - T_3}{T_1} \quad (3)$$

$$Q_{Br} = Q_s - W_B \quad (4)$$

$$Q_{Br} = Q_s - Q_s \frac{T_1 - T_3}{T_1} \quad (5)$$

$$Q_{Br} = Q_s \frac{T_3}{T_1} \quad (6)$$

$$\text{COP} = \frac{T_3}{T_2 - T_3} = \frac{Q_{Br}}{W_C} \quad (7)$$

$$Q_{Br} = W_C \frac{T_3}{T_2 - T_3} \quad (8)$$

$$Q_s \frac{T_2}{T_1} = W_C \frac{T_3}{T_2 - T_3} \quad (9)$$

$$W_C = Q_s \frac{T_3}{T_1} \frac{T_2 - T_3}{T_3} \quad (10)$$

$$W_C = Q_s \frac{T_2 - T_3}{T_1} \quad (11)$$

$$W_B - W_A = Q_s \frac{T_1 - T_3}{T_1} - Q_s \frac{T_1 - T_2}{T_1} \quad (12)$$

$$W_B - W_A = Q_s \left(\frac{T_1 - T_3}{T_1} - \frac{T_1 - T_2}{T_1} \right) \quad (13)$$

$$W_B - W_A = Q_s \frac{T_2 - T_3}{T_1} \quad (14)$$

$$W_B - W_A = W_C \quad (15)$$

Fig. 9·4 Cooling a receiver for a Carnot engine does not raise the overall thermal efficiency.

We can do this by considering the energy-flow equations accompanying Fig. 9·4. Here we have an energy source at the high temperature T_1. We have a natural receiver (atmosphere) at lower temperature T_2 and a refrigerated receiver at lowest temperature of T_3. Engine A draws energy

Q_s from the source, produces shaft work W_A, and rejects unavailable energy Q_r to the receiver at T_2.

Engine B draws the same amount of energy Q_s from the source, produces a larger amount of shaft work W_B, and rejects smaller unavailable energy Q_{Br} to the refrigerated receiver at T_3.

Refrigerating engine (compressor) C picks up energy Q_{Br} from the refrigerated receiver at T_3 and with the aid of work input W_C rejects energy Q_{Cr} to the natural receiver, at T_2.

Now let us find how the three works W_A, W_B, and W_C compare. Equation (1) states the relation between engine work output and heat energy input from the source and the thermal efficiency.

Equation (2) applies Eq. (1) to engine A; Eq. (3) to engine B. Equation (4) is the energy-flow balance about engine B. Substituting for W_B in Eq. (4) from (3) we get Eq. (5). Factoring out Q_s, we get Eq. (6). Equation (7) states the COP for engine C. Solving for Q_{Br} we get Eq. (8). Substituting from (6) we get Eq. (9). Solving for W_C, we get Eqs. (10) and (11).

Subtracting Eq. (2) from (3) we get Eq. (12). Clearing, we get Eqs. (13) and (14). Substituting from (11) we get Eq. (15). This tells us that the additional amount of work produced by engine B over A exactly equals the amount of work input required by the refrigerating engine C. This is true for ideal engines working with reversible processes. Actual engines with their numerous irreversibilities would require a W_C much greater than $W_B - W_A$.

9·7 Irreversibility. So far we have been studying the reversible Carnot cycle. Now let us see what effects practical irreversibilities—temperature drop, turbulence, and friction—have on cycle performance. The most important effect is thermal irreversibility, that is, temperature drop. All experience teaches us that we cannot transfer heat without a temperature difference—despite this, we assume reversible heat transfer at zero temperature differential in setting up the Carnot cycle. This gives us an ideal bench mark of unattainable perfection against which to measure practical cycles.

Figure 9·5a indicates the direction of temperature drops ΔT needed in a "practical" Carnot cycle to transfer energy during heat-admission and -rejection processes. The gas temperature must be less than the source temperature and higher than the receiver temperature. The amount of difference depends on the speed with which the heat must be transferred—the greater the flow, the greater the differential. Flow is also controlled by the amount of area of the heat-transfer surface.

Figure 9·5b shows the temperature curves for source T_s, for gas when receiving heat Q_s from the source T_{gs}, for gas when rejecting heat Q_r to the receiver T_{gr}, and the receiver temperature T_r. Ideally the Carnot cycle could produce the net work W shown by the gray area in cycle 1–2–3–4. But because of the temperature drops, the gas actually works on the cycle outlined by the heavy curves and produces the smaller net W defined by the crosshatching. This P-V diagram shows the incentive for keeping temperature differentials as small as possible.

Figure 9·5c shows the same story on T-S coordinates. The gray area Q shows the ideal amount of heat convertible to W for the given T_s and T_r. But the crosshatched area shows the actual amount attainable. The

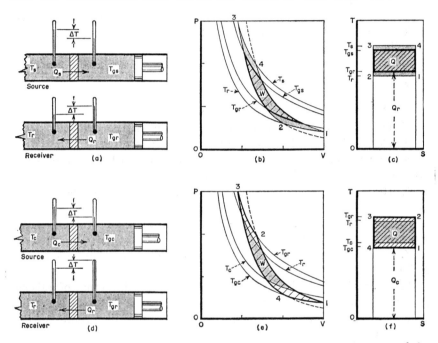

Fig. 9·5 Temperature drops needed to transfer heat reduce the work output of the Carnot engine (a), (b), (c); raise the work input needed to run the reversed engine for refrigeration as in (d), (e), (f). Reducing ΔT's improves W output or input.

corresponding unavailable energy is then Q_r beneath the lower heavy line. The ideal Q_r would be the area under process 1–2, an area smaller than the actual Q_r.

Now, let us see what happens in the reversed Carnot cycle as a refrigerating engine. Figure 9·5d shows the temperature differentials ΔT needed at source and receiver. Here the gas temperature needs to be higher than the receiver on the high-temperature end of the cycle, lower than the source on the low-temperature end—just the reverse of the engine cycle as far as the general temperature level is concerned.

Figure 9·5e shows the effect on the P-V cycle. The gray area shows the ideal net work input W needed to run the cycle, but the larger crosshatched area gives the actual work needed to run the cycle between the same source and receiver temperature limits.

Figure 9·5*f* shows the relations on *T-S* coordinates. Q is the heat equivalent of the net work input W needed to run the refrigerating cycle. The gray area shows the ideal amount needed; the crosshatched area shows the actual larger amount needed because of the temperature differentials during the heat transfers. The actual cycle withdraws a smaller amount Q_c from the refrigerated area per cycle to deliver to the receiver at temperature T_r than the ideal cycle.

So we see that in both the direct engine and reversed refrigeration cycles, the necessary temperature differentials cause a decrease in cycle overall performance.

REVIEW TOPICS

1. Define the mean effective pressure of an engine cycle.

2. Define the overall compression ratio of an engine cycle.

3. Define the overall pressure ratio of an engine cycle.

4. How does the net work output per cycle vary with the compression ratio of an engine? With the pressure ratio?

5. How does the compression ratio of an engine vary with the pressure ratio?

6. How does the mean effective pressure of an engine vary with the pressure ratio?

7. Describe the reversed Carnot cycle for removing heat from a cold region.

8. State the Second Law of Thermodynamics in terms of raising energy to a higher temperature level.

9. Define the COP of a refrigerating cycle. Write the formula for the reversed Carnot cycle.

10. How does the COP of a refrigerating cycle vary with the cold-room and receiver temperatures?

11. How do compression and pressure ratios affect the performance of the reversed Carnot cycle?

12. Describe the use of the reversed Carnot cycle as a heat pump.

13. What advantage does a heat pump have over direct heating?

14. Show why refrigerating the receiver of an engine does not offer any advantage in overall increased thermal efficiency.

15. Demonstrate the effect of thermal irreversibility on the performance of a Carnot engine and refrigerator.

CHAPTER 10

Process Irreversibilities

We discussed thermal irreversibilities in Chap. 9. These are caused by the need for a temperature difference to make heat flow. They invariably cause a loss in engine- or refrigeration-cycle performance. While we cannot eliminate temperature difference, we do have a strong incentive to make it as small as possible. In practical engine design we must consider the size of the engine, amount of heat-transfer surface, speed of the shaft, cost of materials, cost of fuels, in other words—design economics.

10·1 Internal versus external. The temperature drop we were talking about in reference to the visionary Carnot engine is external to the engine proper, so we call it an *external* thermal irreversibility. The engine also has an *internal* thermal irreversibility—heat can flow into the body of the working gas only when the cylinder head (where heat enters) is hotter than the gas. In addition, the gas nearer the piston face must be cooler than the gas next to the cylinder head during heat addition Q_s. This means that there must be a temperature drop ΔT_s through the mass of gas, Fig. 10·1, so it can never actually reach a uniform temperature.

During heat rejection Q_r, the gas next to the piston face must be hotter than the gas next to the cylinder head or the heat will not flow from the gas to the receiver. Making a detailed analysis of this condition becomes very difficult. We see from this that the gas temperature can never be uniform. To get around this practical difficulty some writers assume that a Carnot cycle works at a very slow speed, so the temperature differences can be made so small as to be neglected. Obviously such a slow-speed engine would not be practical.

10·2 Pressure variations. Our ideal-engine analysis assumes that the working gas at any one instant exerts the same pressure in all directions

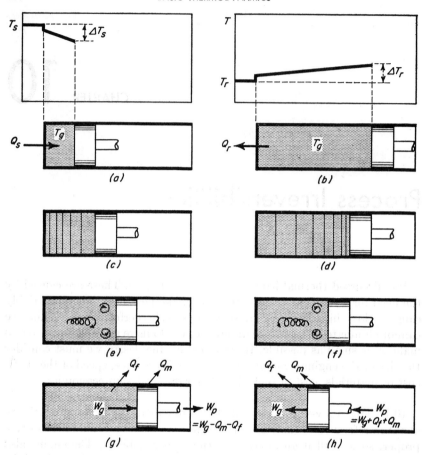

Fig. 10·1 Irreversibilities take several forms that prevent actual processes from reaching the ideal conditions. Irreversible effects: *Internal thermal*—temperature drop needed to transfer energy as heat causes loss of available energy by never letting gas temperature rise as high as the source temperature (a) or as low as the receiver temperature (b). In "practical" cycles gas temperature is never uniform throughout its mass; temperature must drop in direction of heat transferred. *Internal mechanical*—because actual gas cannot expand uniformly during quick changes, it sets up pressure waves that reduce the pressure on the piston below ideal during expansion and raises piston pressure above ideal during compression, (c) and (d). *Internal mechanical*—unequal pressure distribution and temperature variations in gas set up vortices and eddy currents that prevent some of the molecular energy doing useful work on the moving piston during expansion; increase work that must be done by piston on gas during compression, (e) and (f). *External mechanical*—friction between the gas and cylinder wall and between the piston and cylinder wall dissipates energy to heat, reducing net piston work during gas expansion and raising the amount needed during gas compression, (g) and (h).

throughout its mass. This is reasonably true when the gas stays at constant volume and temperature, but when either property changes, local gas pressures vary during the change. In Fig. 10·1c when the piston starts moving to the right at a "practical" speed, the gas at the piston face momentarily becomes more rarefied and molecules in the body of the gas need time to move into the additional space made available by the piston. This sets up pressure waves in the gas that reduce the overall working force exerted on the piston during the working stroke. The severity of the pressure waves depends on cylinder dimensions, gas temperature, and piston speed variation.

In Fig. 10·1d we see the condition for the compression stroke. When the piston starts moving to the left, the gas at the piston face momentarily squeezes into a smaller specific volume than the rest of the mass. This again sets up pressure waves in the gas that tend to raise the overall working force that must be exerted by the piston. We call this form of nonideal-gas behavior an *internal mechanical irreversibility*.

10·3 Gas chaos. Uneven pressure and temperature variations in a gas undergoing changes cause the molecules to be pushed in directions other than the random ones they would take in an undisturbed gas. We see the effect as eddies and vortices in the gas, Fig. 10·1e and f. This means that some of the molecules acquire momentary kinetic energies during expansion and compression that are not transmitted uniformly to the rest of the gas and so cannot act on the piston. In a sense the gas develops local "hot spots" that retain energy in the gas instead of passing it on to either piston or receiver.

When the gas momentarily settles to a condition of equilibrium at the ends of a piston stroke, the kinetic energies of the vortices and eddy currents dissipate by spreading out to the rest of the gas and warming it to a higher temperature. In this way the gas retains more energy than it would under ideal conditions and reduces the thermal efficiency of the cycle.

We shall see later how we evaluate these losses in efficiency. The creation of unwanted gas motions in a cylinder is an *internal mechanical irreversibility*. Later we shall see that these types of irreversibility are sometimes "built into" an engine to gain certain practical advantages.

10·4 Friction. In ideal engines we assume that there is no such thing as friction. Looking at just the cylinder and piston we actually have two areas of friction: between (1) gas and cylinder wall and (2) piston and cylinder wall. A practical reciprocating engine would also have frictional effects at the crankpin and main bearings and other points of moving contact in its mechanical linkages. Figure 10·1g shows the effect of energy transfer during the working stroke of gas expansion. The gas does total work W_g during the expansion. Most of it is expended upon the piston, but a small part goes to moving the expanding gas along the interior surface of the cylinder. This rubbing or frictional effect warms the surface of the cylinder above its normal temperature. The cylinder radiates this energy to the surroundings as heat.

Rubbing of the oscillating piston against the cylinder interior wall also warms the cylinder wall, the frictional heat being radiated to the surroundings. We need only apply the first law to find the effect of these energy dissipations. The work of the gas goes to supply the frictional heat dissipated and the net work developed by the piston on the piston rod; in equation form

$$W_p = W_g - Q_f - Q_m \qquad (10\cdot1)$$

where W_p = net work developed by piston
W_g = gross work developed by gas
Q_f = heat dissipated by friction between gas and cylinder interior wall
Q_m = heat dissipated by friction between piston and cylinder wall

So we see that mechanical work developed by the piston is less than the total work done by the gas. This gives a strong incentive to reduce frictional losses to a minimum.

Figure 10·1h shows the energy balance during the compression stroke. Here the piston does work in compressing the gas. But part of this energy goes to supplying frictional losses and the remainder to actually compressing the gas; in equation form

$$W_p = W_g + Q_f + Q_m \qquad (10\cdot2)$$

Thus the piston must do more work during compression than in an ideal engine. This type of loss is called an *external mechanical irreversibility.*

When we see the number and variety of irreversibilities, it is easy to understand why practical engines cannot too closely approach the ideal in their thermal performance. It takes careful engineering to minimize these losses, as well as a thorough understanding of practical gas behavior and how it contrasts to the ideal concept.

Regardless of whether a piston compresses a gas or an expanding gas pushes a piston, the irreversibilities always cause a net loss in energy exchange.

10·5 Entropy of energy. One of our major interests in thermodynamic processes is the estimating of energy loss. This can be measured directly as the conversion of *available* energy to *unavailable* energy by various irreversibilities. The most serious irreversibility usually is temperature loss during heat transfer.

To do this we can use the property entropy. We have already learned that entropy is an abstract mathematical property of a fluid, Chap. 8. When a fluid absorbs energy as heat, its entropy increases; when it gives up energy as heat, its entropy decreases. We can extend the idea of entropy by thinking of it as a property of energy. If we have a quantity of energy Q at a constant temperature T, then we can say that energy entropy change is

$$\Delta S = \frac{Q}{T}$$

The energy Q could then be shown graphically on T-S coordinates as we did for gases.

In Fig. 10·2 we have a Carnot engine taking energy from a furnace at 3000°R. Energy $Q = 300$ Btu, and if it is absorbed reversibly by the engine so $T_g = 3000°R$, then the entropy change is

$$\Delta S = {}^{300}\!/_{3000} = 0.10$$

We can plot this on the graph in Fig. 10·2 as a bar. The horizontal axis of the graph is not graduated in entropy units, but the width of the bars is measured in entropy units; the area of the bar measures 300 Btu. If the lowest natural available receiver temperature is 500°R, then energy at this temperature with a ΔS or 0.10 unit will equal

$$T \Delta S = 500 \times 0.10 = 50 \text{ Btu}$$

We can superimpose this as the crosshatched section on the first bar. The meaning comes easily: Of the total 300 Btu at 3000°R, 50 Btu is unavailable energy and 250 Btu is available for conversion to mechanical shaft work. The upper part of the bar is a T-S cycle diagram of the Carnot engine working between temperature limits of 3000 and 500°R. The thermal efficiency of the cycle is

$$e = \frac{T_s - T_r}{T_s} = \frac{Q - Q_r}{Q}$$
$$= {}^{250}\!/_{300} = 0.833 \text{ or } 83.3\%$$

Next let us assume that, because of external thermal irreversibility, T_g must be at 2000°R to accept the 300 Btu of heat from the 3000°R source. We find that the energy entropy change in the engine gas is now ${}^{300}\!/_{2000} = 0.15$ unit. It has gained 0.05 unit of entropy in dropping 1000°R of temperature. The unavailable energy now has grown to $500 \times 0.15 = 75$ Btu, and the available energy has shrunk to

$$300 - 75 = 225 \text{ Btu}$$

The corresponding thermal efficiency has decreased to ${}^{225}\!/_{300} = 0.75$ or 75%.

In the third case, we let the gas temperature drop to 1000°R. The entropy change rises to 0.30 and unavailable energy to 150 Btu, making the thermal efficiency drop to 50%. In the last case we let the 300 Btu transfer directly to the receiver at 500°R, raising the entropy change to 0.60 unit. All the energy, of course, becomes unavailable.

This gives us a graphic example of the penalty of running heat engines at low temperatures. For example, furnace temperatures in our modern

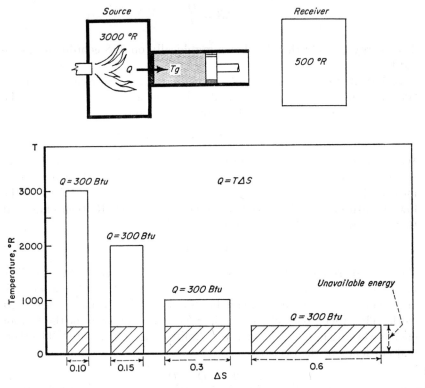

Fig. 10·2 We can assume that entropy change is also a property of energy instead of only a property of a thermodynamic fluid. As we drop the temperature of a given quantity of energy, its entropy increases as well as the unavailable portion.

steam generators run about 3000°R, but the energy picked up by the steam is at 1050 F (1510°R). The entropy of the energy more than doubles, and the actual thermal efficiency is less than half the theoretical potential efficiency, figured from furnace temperature. At present, we are forced into this position because of lack of materials that can withstand temperatures above 1050 F continuously. We do somewhat better with our i-c engines, which use hot gases to work directly on the pistons, but cooling them with water jackets forces the degrading of some high-temperature energy.

10·6 Irreversible processes. Now that we know the pitfalls of making engines run efficiently, how can we account for irreversibilities in calculating the various processes? Constant-pressure, -volume, and -temperature processes by definition say what happens to these properties during a process. When we plot these processes on P-V and T-S diagrams, however, the areas under the curves no longer measure the amount of work done or the amount of heat transferred. The actual energy transfers will be either more or less than the theoretical amounts.

These deviations from theoretical are found by comparing practical tests with theoretical predictions. To emphasize the nontheoretical character of irreversible processes, they are usually plotted with dotted lines on P-V and T-S diagrams. This automatically warns against placing any significance in the areas underneath.

10·7 Adiabatic processes. You may have noticed that we did not mention an irreversible constant-entropy process above. Such a process is only accidental and has not much practical significance. You will recall that a reversible adiabatic process is a constant-entropy process which involves no transfer of heat. The solid curves 1–2 in the graphs of Fig. 10·3 show the ideal constant-entropy processes for expansion and compression of a gas.

An irreversible adiabatic process also involves no transfer of heat, but it does not take place at constant entropy. For an expansion, Fig. 10·3a and b, the process is often defined by the pressures P_1 and P_2 between which it works. The initial state at P_1, T_1, V_1 is the same for reversible and irreversible processes. But the two processes follow different paths on the graph and so have different end states, though they have the same pressure P_2.

Because some of the energy stays in the gas during the irreversible process (see Fig. 10·1e and f), the gas has more internal energy at state 2′ than the reversible one at state 2. So the temperature at 2′ is higher than at 2.

The higher temperature gives 2′ a higher specific volume since $V = RT/P$. Its entropy is higher also; let us see why. Both processes end at the same pressure, but the irreversible one ends at a higher temperature. Figure 10·3b shows that a pressure curve on T-S coordinates rises with increasing entropy and temperature. It follows that a gas with a higher temperature at a given pressure must also have a higher entropy. As we develop our study of thermodynamics, we shall find that an irreversible adiabatic process always involves an increase in entropy.

Referring back to Fig. 10·2, we can see that any process or cycling involving an increase in entropy will degrade available energy to unavailable energy. If there is a limited amount of high-temperature energy, high irreversibility means that we shall get only a small amount of it as available energy. To stave off the fateful day of high-grade energy exhaustion, we should make every effort to decrease irreversibilities where economically sound.

Figure 10·3c and d compares the reversible and irreversible adiabatic compression processes. Here again we see that energy staying in the gas

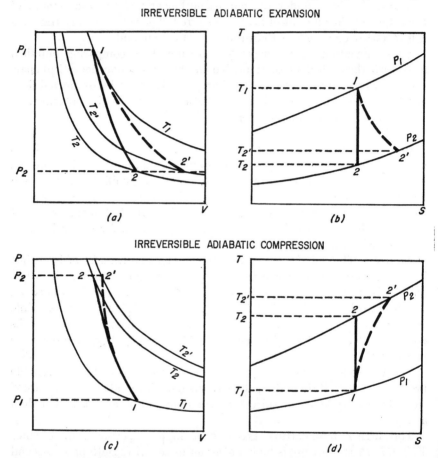

Fig. 10·3 During irreversible adiabatic processes no heat enters or leaves the gas, but the entropy of the gas always increases whether it is expanded or compressed.

because of irreversible effects causes the end temperature to be higher; $T_{2'}$ is higher than T_2 even though $P_{2'} = P_2$. This again increases both the final specific volume and entropy of the gas.

10·8 Irreversible cycle. Now that we understand irreversible processes, let us see how they affect the "practical" Carnot cycle. Figure 10·4 shows a Carnot cycle made up of a reversible constant-T heat-absorbing expansion process 1–2, an irreversible adiabatic expansion 2–3, a constant-T reversible heat-rejecting compression process 3–4, and an irreversible adiabatic compression 4–1.

Even though some of these processes are irreversible, the four state points are always the process terminal points. This means that the gas neither accumulates nor loses any net energy in regularly passing through the cycle of processes. Heat enters and leaves the cycle only during the constant-T processes. The difference in heat absorbed and rejected must equal the mechanical work output of the gas.

The easiest analysis can be made by studying the T-S diagram in Fig. 10·4b. The area under process 1–2 measures heat absorbed Q_s, since the process is reversible. The areas under the irreversible adiabatics 2–3 and

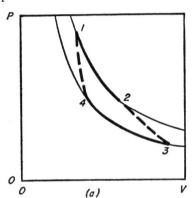 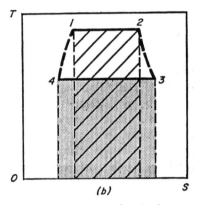

Fig. 10·4 In a Carnot cycle with irreversible processes the net areas of cycle diagrams do not measure the net work done or heat transferred.

4–1 have no significance. But the area under the reversible process 3–4 measures the heat rejected to the receiver, Q_r. Then, by the first law, the net work of the cycle is

$$W = Q = Q_s - Q_r$$

This means that the net work output must equal the difference of the crosshatched area under 1–2 and the gray area under 3–4. This difference does *not* equal the net area enclosed by 1–2–3–4. In fact, the difference is smaller than the enclosed area. It is possible for the net work to be zero if the irreversibilities become large enough.

By the same reasoning it should be clear that the net area on the P-V plane, Fig. 10·4a, does not measure the net work output of the cycle either. Net areas measure net work only when all the processes are reversible. Remember this in all future cycle analyses that we shall make.

REVIEW TOPICS

1. What is external thermal irreversibility?
2. What is internal thermal irreversibility?

3. How can we approach thermal reversibility in an engine?
4. Give an example of mechanical internal irreversibility in the gas in an engine cylinder.
5. Give some examples of external mechanical irreversibilities in an engine.
6. Describe the entropy of energy concept.
7. How are irreversible processes drawn on P-V and T-S charts? Why?
8. What is the difference between reversible and irreversible adiabatic processes?
9. Does the First Law of Thermodynamics apply to irreversible cycles?

PROBLEMS

1. One-tenth pound of a gas held in a cylinder by a piston does 30 Btu per lb of gas work on the piston and produces friction on the cylinder wall of 1 Btu per lb of gas. Friction between piston and cylinder absorbs 3 Btu per lb of gas. (a) How much work is delivered by the piston? (b) How much energy is given up by the gas?

2. A piston does work on $\frac{1}{5}$ lb of gas in a cylinder at the rate of 40 Btu per lb of gas. Friction between gas and cylinder dissipates 2 Btu per lb of gas and between piston and cylinder dissipates 4 Btu per lb of gas. (a) How much work was done by the piston? (b) How much energy was absorbed by the gas?

3. Fire in a boiler furnace burns at 2500 F. If the steam absorbs energy from the fire at 900 F and expands reversibly in a "Carnot engine" to an exhaust temperature of 80 F, what is the loss in available energy of the total energy originally released in the furnace?

4. Combustion in the cylinder of a Carnot engine burns at 2500 F, and the gases expand reversibly to a receiver temperature of 80 F. Is there any loss in available energy?

5. Draw the T-S diagram of a Carnot engine that receives and rejects energy as reversible constant-temperature processes but has irreversible adiabatic expansions and compressions and has a net work output of zero.

CHAPTER 11

Basic Engine Cycles

We have paid much attention to the theoretical Carnot cycle and learned how it can convert heat to mechanical work, how it can cool areas below the surrounding natural temperatures, and how it can use low-level energy to heat an area. The Carnot engine is relatively simple and can operate at the highest efficiency—but no one has tried to build one.

There are several reasons: While simple in basic needs, it is mechanically difficult to bring the hot source and cold receiver to the cylinder head in turn; each cycle produces only a small amount of net mechanical work. The net area on the P-V diagram in Fig. 11·1 shows the small net work realized for an expansion ratio of 6.3; this would make the engine very heavy on a pound per horsepower basis.

11·1 Stirling cycle. Realizing this weight disadvantage Robert Stirling, in 1827, set about designing a more practical engine. We shall not study the design of the actual engine (now obsolete) but only the series of processes it attempted to use. As you might anticipate, the engine fell far short of the ideal reversible cycle it used for a model, but it worked, ran at good efficiency, and developed a fair output per cycle.

With the poor material available in the early days the heat-transfer surfaces at the furnace heat source corroded and ultimately failed. But the engine had only about half the weight per horsepower that a theoretical Carnot engine would need. Stirling engines served practical uses for a time.

Figure 11·1 compares the Carnot, Stirling, and Ericsson cycles; we shall discuss the latter later. The P-V and T-S diagrams show that, for the same source and receiver temperatures and total piston displacements,

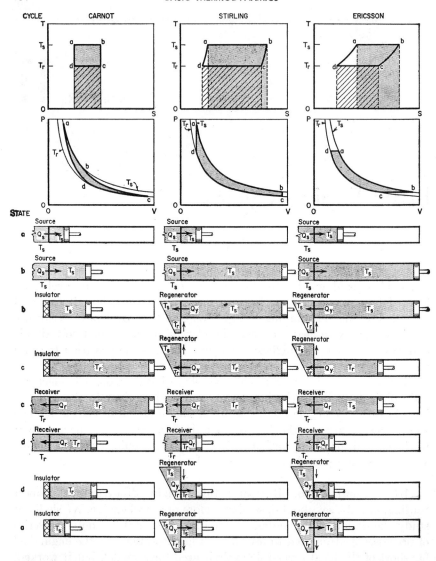

Fig. 11·1 These three cycles work between the same temperature levels and with the same piston displacement. Though they all perform with the same thermal efficiency (for reversible processes), they produce different net work outputs per cycle.

the Stirling cycle produced more than double the output of the Carnot cycle.

We already know the Carnot cycle, but Fig. 11·1 lists all the steps needed to make it work: application of source during heat input, application of insulator during constant-entropy expansion, application of receiver during heat rejection, application of insulator during constant-entropy compression.

In the Stirling cycle we also absorb heat energy from the source during a constant-T expansion ab, but for the full piston stroke. But now we have a cylinder full of hot gas at state b. We must get rid of some of this internal energy before we can return the piston to the head end. If we applied the receiver at this time, we would lose a lot of the energy by irreversible heat transfer through the temperature difference $T_s - T_r$.

To avoid this, Stirling got the brilliant idea of removing some of this energy *reversibly* by using a regenerator. This unit absorbs heat Q_y and stores it for future use. Figure 11·2 outlines the practical form of a

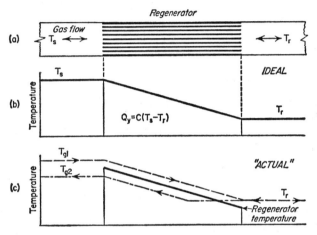

Fig. 11·2 Hot gas entering the regenerator at the left cools down as it gives up and stores internal energy in the unit. Cool gas reentering from the right warms up as it reabsorbs the stored internal energy.

regenerator. It consists of refractory brickwork, metal mesh, or small tubes through which the hot gas flows. As the hot gas enters from the left at T_s, it gives up heat energy to the regenerator material and drops in temperature as it advances. By the time the gas leaves at the right end, it has cooled to a lower temperature T_r.

When the gas flow is reversed, it re-enters the regenerator at the right, reabsorbs the energy as heat, and rises in temperature as it progresses through the unit. For the ideal reversible condition it would leave the regenerator with the same temperature it originally entered, T_s. The center graph, Fig. 11·2b, shows the reversible temperature changes in the regenerator for either direction of gas flow. As a matter of interest, Fig. 11·2c shows roughly how actual gas and regenerator temperatures would behave.

We indicate the regenerator in Fig. 11·1 by the wedge-shaped figure. It is moved over the conducting face of the cylinder to absorb heat Q_y

reversibly while it lowers the gas temperature from T_s to T_r during the process bc. The gas volume stays constant during this process, and the pressure drops.

We then bring the cylinder head in contact with the receiver and reject the unavailable energy Q_r for the full piston stroke during the constant-T process cd. Next, at d we place the low-temperature end of the regenerator in contact with the cylinder head and move it down, so stored heat Q_y is reversibly reinjected into the gas during the constant-V process da. This raises the gas temperature from T_r to T_s and also the gas pressure, and the cycle is ready to repeat.

The enclosed gray area $abcd$ on the P-V graph measures the amount of mechanical work done by the cycle. It is more than double the amount of work done by the Carnot cycle, at left.

In the T-S graph for the Stirling cycle the gray area under ab measures the heat input Q_s. The area under the constant-V process bc measures the amount of heat Q_y stored in the regenerator. The hatched area under cd measures the unavailable energy rejected Q_r. The area under the constant-V process da measures the heat Q_y reinjected into the gas by the regenerator and equals that under bc.

The total heat injected into the gas is the area under dab and the total heat rejected by the gas is the area under bcd, so the net area $abcd$ measures the net heat input to the cycle. By the simple energy equation we know that this heat equals the net work output of the cycle.

The thermal efficiency of the cycle is equal to $(Q_s - Q_r)/Q_s$. Since both heats are transferred reversibly at constant T, the efficiency then also equals $(T_s - T_r)/T_s$, the same as the Carnot cycle efficiency.

11·2 Ericsson cycle. About the middle of the nineteenth century, John Ericsson designed a hot-air engine to use a regenerator also. Cycle-wise it differed from the Stirling engine in making the regenerative heat transfers during constant-P processes instead of during constant-V processes.

Figure 11·1 shows the T-S and P-V charts for the reversible Ericsson cycle and how they compare with the Carnot and Stirling cycles. Thermal efficiencies of all three are equal for identical source and receiver temperatures. The work output per cycle for the Ericsson is about 60 per cent greater than the Carnot, but only 70 per cent of the Stirling. The Ericsson cycle has the practical advantage of having a lower peak pressure than the other two; in Fig. 11·1 it is about one-third less. This helps in that it requires a lighter engine construction. Ericsson built thousands of his hot-air engines, most of them of small capacity and not using the regenerative feature, but today they are only historical curiosities.

11·3 Regenerative heating. Let us make sure we understand the basic duty and advantage of the regenerator in the Stirling and Ericsson cycles. It primarily retains a certain amount of internal energy within the engine cycle by heat transfer during certain processes. This is done to lower and raise gas temperatures reversibly. In this way we avoid degrading high-grade available energy into low-grade unavailable energy by allowing heat to be transferred irreversibly through a temperature drop.

As we well know, heat can be transferred actually only by a temperature difference, so the practical advantage lies in *reducing* the amount of temperature loss in transferred heat energy. For instance, in Fig. 11·2c, instead of letting all the energy Q_y in the gas at the end of the working stroke be degraded from $T_{g1} = T_s$ to T_r, we can recover part of it at a lower temperature T_{g2} during the reverse flow. All actual regenerative heating processes only partly realize the theoretical advantages.

The regenerative principle is used in modern steam-electric and gas-turbine plants. These plants, however, do not have a storage-type regenerator. Instead, heat exchangers keep a certain amount of internal energy circulating within the plant energy cycle. The basic aim is the same: to reduce the amount of irreversible heat transfer.

11·4 Joule or Brayton cycle. In the latter part of the nineteenth century, two men independently proposed engine designs to work on the cycle shown in Fig. 11·3. James Joule of England and George Brayton of Boston, Mass., intended this cycle for reciprocating engines, but it actually is the basis for the modern simple-cycle gas-turbine plant.

Again, since the original engine designs are obsolete, let us study this cycle, assuming that a fixed amount of gas (working fluid) is trapped in a cylinder behind a piston as in the Carnot engine of Fig. 11·1. In Fig. 11·3 we have the cycle for three different pressure ratios (P_a/P_d) but all work between the same source and receiver temperatures.

In all of them we have the gas at its highest pressure in state a. We then apply the source at temperature T_s and heat the gas at constant pressure until the gas temperature rises to T_s at state b. Notice that this process is externally thermally irreversible, since the heat added Q_s enters the gas through a temperature drop. We may, however, regard the constant-pressure heating process as internally reversible. We use the temperature drop for practical advantage, realizing that the thermal efficiency suffers accordingly.

At state b we remove the source and put on the insulating cap over the cylinder head. We then let the gas expand at constant S to the lower pressure of the cycle at state c. This converts some of the internal energy of the gas to work output at the piston.

Next, we remove the insulating cap and apply the receiver at temperature T_r. We find the gas temperature higher than the receiver temperature, so the energy Q_r rejected by the gas to the receiver drops irreversibly through a large temperature drop. Again we do this for practical advantage, despite the adverse effect on thermal efficiency. We transfer Q_r during the constant-pressure compression process cd, until the gas temperature drops to T_r.

At state d we remove the receiver and again apply the insulating cap. Then compressing the gas at constant S we raise the gas temperature to some intermediate level at state a. From here we repeat the cycle.

Figure 11·3 shows the P-V and T-S diagrams for this cycle. In the T-S charts the sloping constant-P lines for heat absorption and rejection in

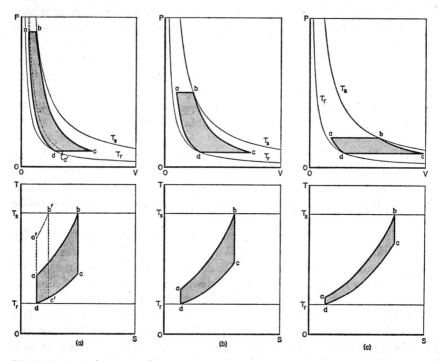

Fig. 11·3 Joule (or Brayton) cycle performance depends on both temperature and pressure ratio. Because it does not transfer heat at constant temperature, the thermal efficiency of the cycle is lower than the Carnot for the same temperature limits.

the cycle show how far it departs from the Carnot cycle with its constant-T heat transfer processes, Fig. 11·1. The enclosed area $abcd$ on the P-V chart measures the net work output of the cycle; on the T-S chart it shows the net heat converted to work.

For fixed temperature limits we see that the work developed per cycle depends on the pressure ratio. State d is the same for all cycles shown in Fig. 11·3; the other three states vary with the pressure ratio. Figure 11·3c has a pressure ratio of 2, Fig. 11·3b a pressure ratio of 5, and Fig. 11·3a a pressure ratio of 9. As the pressure ratio rises through this range, the work output rises. But if we continue raising the pressure ratio as suggested by the dotted cycle $a'b'c'd$ in Fig. 11·3a, we see that the work shrinks. Figure 11·4 shows that the work output becomes a maximum at a pressure ratio of about 11.

When we study the heat-transfer processes in Fig. 11·3 we see that, as the pressure ratio rises, the *average* temperature of the gas rises during heat absorption and drops during heat rejection. This means that the thermal efficiency of the Joule cycle rises with pressure-ratio increase. In fact, at infinite pressure ratio as state a approaches and equals state b,

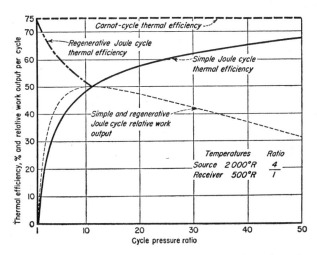

Fig. 11·4 The Joule-cycle work output varies with pressure ratio and reaches a maximum at about 11. The regenerative cycle has higher efficiency in the lower pressure-ratio range than the simple cycle.

we have the equivalent of a Carnot cycle which runs at the maximum possible efficiency for the given temperature limits. But simultaneously, the net area $abcd$ approaches zero (see Fig. 11·3a) and the net work output disappears. Figure 11·4 shows the variation of thermal efficiency with pressure ratio.

11·5 Regenerative Joule cycle. For the lower pressure ratios, where the temperature of state c is higher than at state a, we have the opportunity to apply a regenerator and improve the efficiency of the Joule cycle. Figure 11·5a shows that at state c we would apply the regenerator to the cylinder head and reversibly absorb the heat Q_y until the gas temperature dropped to $T_x = T_a$. Then we would remove the regenerator and apply the receiver and reject Q_r.

After the constant-S compression to a we again apply the regenerator in Fig. 11·5b and the gas reabsorbs heat Q_y while expanding to state z where

$T_z = T_c$. We finally apply the source and inject the heat Q_s during the rest of the constant-P expansion. The net enclosed area $abcd$ again measures the net heat converted to work and equals $Q_s - Q_r$. Heat Q_y is a constant amount kept within the engine cycle with the aid of the regenerator.

Here again, note that the regeneration raises the average temperature of the gas during heat input and lowers its average temperature during heat rejection. This makes the temperatures approach the ideal Carnot limits for maximum efficiency. In effect, the regeneration prevents the degradation of some available energy by rejection to the receiver. Any process which prevents such rejection automatically increases cycle efficiency.

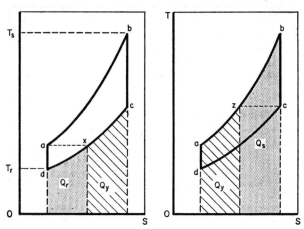

Fig. 11·5 The regenerative Joule cycle lessens the loss from irreversible heat transfer to boost the overall thermal efficiency.

Figure 11·4 shows how the regenerative thermal efficiency of the Joule cycle varies with pressure ratio. At zero pressure ratio it equals the Carnot efficiency. It then drops to equal the simple Joule cycle efficiency at the maximum work output per cycle. Beyond this pressure ratio the regenerator cannot be applied because T_a becomes higher than T_c, Fig. 11·3.

REVIEW TOPICS

1. What is the principal shortcoming of a Carnot engine?
2. What is the function of a regenerator?
3. Describe the Stirling cycle using both P-V and T-S charts.
4. What is the thermal efficiency of the ideal Stirling cycle?
5. Describe the Ericsson cycle using P-V and T-S charts.
6. What is the thermal efficiency of the Ericsson cycle?
7. How do the work outputs per cycle compare for the Carnot, Stirling, and Ericsson engines?

8. Describe the Joule (Brayton) cycle using P-V and T-S diagrams.

9. Show how the work output per cycle of the Joule cycle varies with pressure ratio for fixed temperature limits.

10. How is the Joule cycle made to approach the Carnot cycle in thermal efficiency? What limitation appears?

11. Describe the regenerative Joule cycle with the aid of a T-S graph.

12. Can the regenerative Joule cycle be used at all pressure ratios?

CHAPTER 12

Compressed-air Cycles

Before looking at the problems of practical heat engines let us study the operation of compressed-air systems. These involve relatively simple types of reciprocating machinery and will help us in understanding heat engines later on.

12·1 Compressed air. This has many practical uses: (1) driving air engines for drilling, grinding, sawing, etc.; (2) driving reciprocating pistons in mechanical hammers; (3) atomizing liquids as in paint spraying; (4) conveying solid and powdered materials in pipelines; (5) transmitting control-system pressures to remote locations; (6) pumping water by air lift; (7) cleaning surfaces by air blast.

A compressed-air system transmits energy; in some situations it has more flexibility and safety than other energy-transmitting systems that might be used. An air compressor produces pressurized air that carries energy to the point of use. In some industrial operations compressors pressurize gases other than air. Generally the conditions to be met and the results parallel those for air compressors. In this chapter we shall cover only reciprocating compressors—other types will be discussed later.

12·2 Air compressors. Figure 12·1 shows the main parts of a simplified single-cylinder single-acting air compressor with zero clearance. A flywheel receives energy in the form of mechanical shaft work through the rotating shaft on which it is mounted. Through the connecting rod it moves the piston inside the cylinder back and forth through intake and compression-discharge strokes.

At each end of the strokes the piston comes to a dead stop briefly, though the flywheel rotates at a continuous rate. Figure 12·1f shows that piston speed accelerates rapidly until it reaches maximum velocity at midstroke, when it starts decelerating for the last half of the stroke. These rapid accelerations and decelerations make it necessary to use a flywheel. Rapid change in rate of energy transmission to the piston puts a varying load on the driving engine or motor supplying the shaft energy input. The flywheel stores considerable energy as rotating kinetic energy. When this energy is drawn on and replenished as needed during one revolution, the instantaneous fluctuations in shaft speed become smaller.

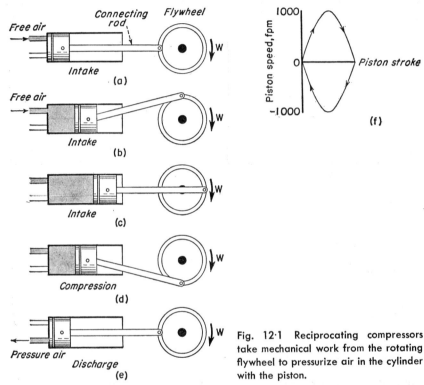

Fig. 12·1 Reciprocating compressors take mechanical work from the rotating flywheel to pressurize air in the cylinder with the piston.

In Fig. 12·1a the piston stands at the head end of its stroke with the intake valve open to atmosphere ready to admit "free" air. In b the piston moves at its highest speed toward the crank end while free air enters through the inlet valve to fill the cylinder.

In c the piston stands at the crank end of its stroke. The inlet valve has just closed, and the cylinder is filled with free air (at atmospheric pressure). In d the piston moves at its highest speed toward the head end, compressing the entrapped air into a diminishing volume and raising its pressure. When the air pressure rises to the rated discharge pressure, the outlet valve automatically opens and the pressurized air begins to discharge to the compressed-air system. In e the piston stands again at the head end of the cylinder and has just finished discharging the last of the pressurized air. The outlet valve closes while the inlet valve opens, and we are ready for the next cycle as in *a*.

12·3 Air processes. From Fig. 12·1 we see that the flywheel through the connecting rod moves the piston back and forth. In Fig. 12·2 we study what the piston does on the air being compressed. Again we use *P-V* coordinates to analyze the actions of the gas in the cylinder, but this time the chart does not picture a complete thermodynamic cycle. In the latter the processes of the cycle always refer to 1 lb of the working

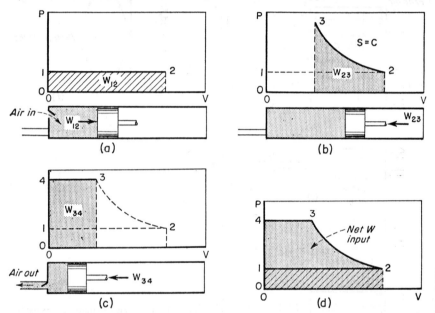

Fig. 12·2 Pressure variations plotted on a *P-V* graph of air in a cylinder measure the amount of work done on air by the piston.

fluid. In Fig. 12·2 we shall find that only one true process occurs in the cylinder during which the mass of air in the cylinder stays constant.

In Fig. 12·2a the piston moves from V_1 to V_2. Atmospheric air enters the cylinder through the inlet valve to fill the displacement volume swept out by the piston. During this induction stroke we assume that the entering free air presses on the piston with atmospheric pressure $P_1 = P_2$. But this represents a force acting through a distance, so the air does mechanical work on the piston:

$$W_{12} = P_2(V_2 - V_1) = P_2 V_2 \tag{12·1}$$

since $V_1 = 0$ with a zero-clearance compressor as we have here.

Note particularly that we did *not* say that the entering air during the induction stroke moved the piston. The connecting rod supplied the force to move the piston; the air merely added a small contributing force. In so far as the air did work, it gave up energy to the piston. In this study we are interested only in what happens to the free air.

At V_2 the inlet air valve closes. Then, during the first part of the return stroke we have a true process with a constant mass of air being compressed at constant entropy from V_2 to V_3 as in Fig. 11·2b. The area under the process curve then measures the mechanical work done on the air by the piston in pressurizing it to P_3, or

$$W_{23} = \frac{P_3V_3 - P_2V_2}{1 - k} \qquad (12\cdot2)$$

See Art. 6·9 for the basic equation of work done by a reversible adiabatic (constant-S) process. When the process expands a gas from a higher to a lower temperature, Eq. (12·2) gives us a positive number for the work done W. But when the process compresses the gas from a lower to a higher temperature as we do here, the equation gives us a negative number for the work W. Do not let this confuse you; this simply means that the work is done *on* the gas, not by the gas. Always keep this in mind when combining equations for several processes as we shall do in a little while.

Notice that in writing the equation, whether for compression or expansion, the quantities for the initial state are subtracted from the quantities for the final state of the process. When the equation is written this way, the positive and negative answers immediately show whether work has been done by the gas or on the gas.

In an actual compressor we usually know P_2, P_3, and V_2, but not V_3. To use this equation easily we have to eliminate V_3. We do this by remembering that for an isentropic compression (see Art. 6·9)

$$\frac{T_3}{T_2} = \left(\frac{P_3}{P_2}\right)^{(k-1)/k} \qquad (12\cdot3)$$

By multiplying Eq. (12·2) by P_2V_2/P_2V_2 we get

$$W_{23} = \frac{P_2V_2}{1-k}\left(\frac{P_3V_3}{P_2V_2} - 1\right) \qquad (12\cdot4)$$

Then using Boyle's law, $PV = RT$, we change the equation to

$$W_{23} = \frac{P_2V_2}{1-k}\left(\frac{RT_3}{RT_2} - 1\right) \qquad (12\cdot5)$$

Canceling R's and substituting from Eq. (12·3) we have

$$W_{23} = \frac{P_2V_2}{1-k}\left[\left(\frac{P_3}{P_2}\right)^{(k-1)/k} - 1\right] \qquad (12\cdot6)$$

This looks complicated, but as soon as we make numerical substitutions, the equation is easily solved.

In Fig. 12·2c when the pressure reaches P_3, the discharge valve automatically opens, and as the piston continues moving to the left, it forces the pressurized air out of the cylinder into the pressurized system at P_3.

During the discharge stroke we again assume that the air pressure stays constant, this time at $P_3 = P_4$. The discharge air impresses a resisting force on the piston and makes it do work W_{34} on the air. The piston transfers this energy to the air by mechanical work, which equals

$$W_{34} = P_3(V_3 - V_4) = P_3 V_3 \tag{12·7}$$

since $V_4 = 0$. At V_4 the discharge valve closes and the inlet valve opens as in a to repeat the cycle for the next revolution of the flywheel.

In Fig. 12·2d we put all the cylinder events on a common chart. The gray areas represent work done on the air by the piston, V_2 to V_4. The hatched area under 1–2 represents work done by the air on the piston. So the difference in areas, 1–2–3–4, measures the net work done by the piston in compressing the air from P_2 to P_3. In equation form this works out to

$$W = W_{23} - W_{34} + W_{12}$$
$$= \frac{P_3 V_3 - P_2 V_2}{(1-k)} - P_3 V_3 + P_2 V_2 \tag{12·8}$$

remembering that minus means work done on the gas, plus is work done by the gas. Combining terms, we get

$$W = \frac{k(P_3 V_3 - P_2 V_2)}{(1-k)} \tag{12·9}$$

This equation parallels Eq. (12·2), differing only by the constant k in the numerator. Since (12·2) was transformed into Eq. (12·6), we can immediately write a parallel equation for (12·9) in the form

$$W = \frac{kP_2 V_2}{1-k}\left[\left(\frac{P_3}{P_2}\right)^{(k-1)/k} - 1\right] \tag{12·10}$$

This evaluates the net work done on the air by an ideal zero-clearance compressor as in Fig. 12·2d, when the compression is done at constant entropy, that is, as a reversible adiabatic. To use this equation for a reversible polytropic compression simply substitute n for k. Mechanical work done on the air figured by all these equations will be in foot-pounds; to convert to Btu simply divide by $J = 778$.

12·4 Intake and discharge. Let us take a closer look at the intake stroke 1–2 in Fig. 12·2. We make the simplifying assumption that the incoming free air acts with constant pressure on the moving piston. If we accept the molecular kinetic theory of gases, this cannot be true.

COMPRESSED-AIR CYCLES

Air pressure on the piston varies with the piston speed. In Fig. 12·3a with zero piston speed v_p the pressure or force acting on the piston is a maximum. This pressure is produced by gas molecules approaching the piston with a speed v_m and rebounding with an equal but opposite speed.

But suppose the piston was moving with the same speed as the molecules, so $v_p = v_m$; then the molecules could not catch up with the piston, and there would not be any pressure on the piston because of no impact.

From Art. 5·8 we learned that the pressure or force was

$$F = \frac{2mv}{t}$$

where v is the speed of the molecule *relative to* the wall or the piston.

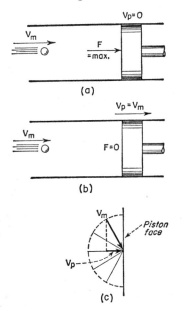

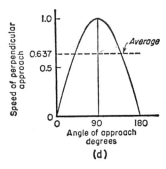

Fig. 12·3 Molecular study shows that air pressure on a moving piston differs from pressure exerted on a stationary piston.

Let us estimate the extent of the error caused by assuming constant pressure during intake. At 60 F, air molecules have an average speed of about 90,000 fpm. But, as shown in Fig. 12·3c, molecules can approach any one point on the piston face through an angle varying from 0 to 180°. However, as we learned in Art. 5·8 it is only the component of the motion perpendicular to the face that produces force or pressure. So for v_m, only v_p produces pressure. Any one point is struck by billions of molecules coming from all directions in 1 sec.

We then can assume a distribution curve of pressure-producing velocities at one point as shown in Fig. 12·3d. This is a sine curve; the average of a sine curve is 0.637 of the maximum value. So, for an air molecule speed of 90,000 fpm, the average pressure-producing velocity will be about 90,000 × 0.637 = 57,000 fpm. Recognizing m and t as constants, the relative force produced by 60-F air molecules is equal to 2 × 57,000, or 114,000 units on a stationary wall or piston.

The average piston speed of a high-speed air compressor is about 700 fpm. For such a unit the relative speed of approach of a molecule to the piston would be 57,000 − 700 = 56,300 fpm. The corresponding relative force on the piston is then 2 × 56,300 = 112,600 units. For this case then, the actual pressure can be only 112,600/114,000 = 0.988 of the atmospheric pressure, or the error is about 1.2%. The slower the piston speed, the smaller the error.

On the other hand, the energy of the air molecules depends on the square of the velocities. Before impact, the molecules have an internal energy of about $57{,}000^2 = 32.5 \times 10^8$ units.

After impact, the absolute velocity of the molecules drops to

$$57{,}000 - 2 \times 700 = 55{,}600 \text{ fpm}$$

Their internal energy after impact is then $55{,}600^2 = 30.9 \times 10^8$ units. The internal energy of the entering air then drops on the order of 30.9/32.5 = 0.951 or 95.1% of the atmospheric air. The error in assuming constant internal energy of the air is of the order of 4.9%.

In the reverse stroke of discharge in process 3–4 of Fig. 12·3, the pressure cannot stay constant either, as we assumed. Here the piston approaches the advancing molecules and speeds them up more than if the piston was stationary. This raises both the pressure and the internal energy of the discharging air. When we study actions in an actual compressor, we shall find that the air pressure during intake is lower and during discharge higher than we assume. The molecular study we just made shows why. For many design problems the constant-pressure assumption is good enough for our needs because it simplifies calculations.

12·5 Constant-T compressor. In most compressed-air systems no effort is made to conserve the internal energy of the air. By the time it reaches its point of use the air has cooled down to atmospheric temperature. This condition makes it practical to use a constant-T instead of a constant-S compression process. Figure 12·4a shows how the two compressors compare. The area of the constant-T compressor is smaller than that of the constant-S compressor. We can get the same amount of pressurized air for less work input, a desirable gain in economy.

Net work input for the constant-T compressor is

$$W = P_3 V_3 + P_2 V_2 \log_e \frac{P_3}{P_2} - P_2 V_2 \qquad (12\cdot 11)$$

The first and last terms are the discharge and intake works, the middle term is the work of constant-T compression (see Art. 6·8).

Since $P_3V_3 = P_2V_2$ when $T_3 = T_2$, the net work of a constant-T compressor is

$$W = P_2V_2 \log_e \frac{P_3}{P_2} \tag{12.12}$$

This is in foot-pounds for an ideal compressor with zero clearance.

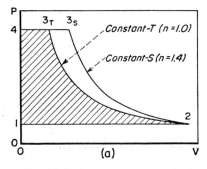 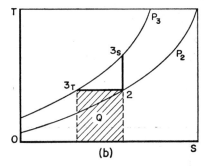

Fig. 12·4 A compressor using the constant-T process needs least total work input.

Figure 12·4b shows the comparable processes on T-S coordinates. It is impossible to show the intake and discharge strokes on these coordinates. The constant-S process 2–3_s is vertical and so involves zero heat transfer. On the other hand the constant-T process 2–3_T involves the transfer of heat Q from the air during compression. And since the internal energy of the air does not change, the heat transferred $Q = W$, the net work input to the compressor.

Actual compressors usually run with a polytropic process intermediate to the two shown in Fig. 12·4b. The process line would end on P_3 somewhere between 3_s and 3_T. For a reversible process, the area under the curve measures the heat transferred during the compression, which differs from total work.

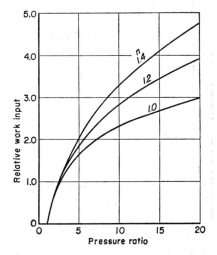

Fig. 12·5 Compressor work input depends on both the compression process and the pressure ratio but not on inlet air pressure.

12·6 Work inputs. Equations (12·10) and (12·12) give us the works for the two extremes of heat transfer during compression. We can figure intermediate W's by substituting n for k in (12·10). For a compressor with fixed P_2 and V_2 working on air, W depends solely on the pressure ratio

P_3/P_2. Assuming that P_2V_2 equals 1, we can study how W varies with pressure ratio as in Fig. 12·5 for different values of n. When $n = k = 1.4$, we have constant-S compression. When $n = 1.0$, we have constant-T compression; for intermediate values we have a polytropic compression. Figure 12·5 shows that work input rises at a decreasing rate as the pressure ratio increases.

To get work input in foot-pounds from Fig. 12·5 we need only to multiply the relative work input by the product P_2V_2, remembering that P must be in pounds per square foot and V in cubic feet per pound. Notice that P_2 can be any pressure; it is not limited to atmospheric.

12·7 Compressor clearance. The zero-clearance compressor of Fig. 12·1 is only a simplified ideal. Practically it would be difficult to build a

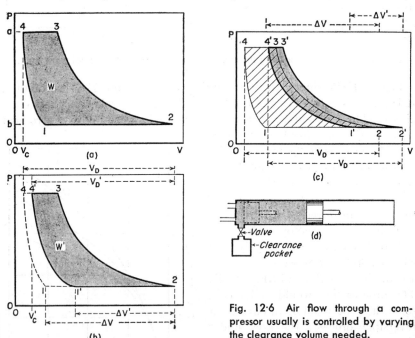

Fig. 12·6 Air flow through a compressor usually is controlled by varying the clearance volume needed.

compressor without clearance and not have trouble from mechanical collision of cylinder head and piston. Other practical needs, such as room for valves and ports, make it essential to provide clearance volume at end of the compression-discharge stroke.

Figure 12·6a shows the actions in an ideal compressor with clearance. At 2 the cylinder is filled with air at atmospheric pressure. From 2 to 3 the entrapped air is compressed at any desired process. At 3 the discharge valve automatically opens, and from 3 to 4 the pressurized air leaves the cylinder to enter the compressed-air system. At 4 the discharge valve closes and the pressurized clearance air expands as the piston moves to the right toward 1.

When the clearance air pressure drops to P_1 of the free air, the inlet valve automatically opens. Then from 1 to 2 free air flows into the cylinder as the piston recedes to its crank-end position, and the cycle repeats.

The area under 2–3–4 measures the work done on the air by the piston. The area under 4–1–2 measures the work done on the piston by the air. The difference of these areas 1–2–3–4, shown in gray, measures the net work W done on the air by the compressor. How do we calculate this work?

By studying Fig. 12·6a we see that W is the difference of two zero-clearance cycles, a–b–2–3 and a–b–1–4. Using the polytropic form of Eq. (12·10),

$$W = \frac{nP_2V_2}{1-n}\left[\left(\frac{P_3}{P_2}\right)^{(n-1)/n} - 1\right] - \frac{nP_1V_1}{1-n}\left[\left(\frac{P_4}{P_1}\right)^{(n-1)/n} - 1\right] \quad (12\cdot13)$$

Since $P_2 = P_1$ and $P_4 = P_3$, we can clear the equation to the form

$$W = \frac{nP_2(V_2 - V_1)}{1-n}\left[\left(\frac{P_3}{P_2}\right)^{(n-1)/n} - 1\right] \quad (12\cdot14)$$

The only difference between this equation and (12·10) is that the volume used in the numerator is a difference of two volumes. But in either case this volume represents the volume of air drawn into the cylinder for compression during one cycle.

12·8 Varying air output. An air compressor produces a given mass of compressed air per cycle. To vary the rate of air output we could vary the speed of the compressor to vary the number of cycles per minute.

Another method would be to vary the volumetric displacement of the piston from V_D to V_D' as shown in Fig. 12·6b. This varies the amount of air drawn in per cycle from ΔV to $\Delta V'$ and the amount of work input from W to W'. It also changes the clearance volume from V_c to V_c'.

Neither of the two methods proves practical for air-flow control. The method generally used varies the amount of clearance volume as indicated in Fig. 12·6d. One or more clearance pockets connect to the head end of the cylinder through valves. The piston always sweeps out a fixed volumetric displacement V_D in the cylinder, Fig. 12·6c.

With the clearance pocket shut off from the cylinder, the cycle 1–2–3–4 measures the work input and shows the ΔV drawn in. With the clearance pocket connected through the open valve, the clearance volume increases from V_4 to V_4'. But V_D remains the same in the cylinder. On the diagram it shifts to the right. Note that the chart processes no longer line up with the cylinder cross section below. Now with the piston at the crank end, the total volume has increased from V_2 to V_2'.

The gas and the piston then pass through the cycle $1'$–$2'$–$3'$–$4'$. This shrinks the volume of air intake from ΔV to $\Delta V'$, and the work input drops from the hatched area to the gray area. Figure 12·6c shows what happens when the clearance volume increases by four times. If we have several clearance pockets connected to the cylinder, we could vary the clearance by small steps and have finer control over rate of air flow.

12·9 Multistage compressors. We have been studying single-stage compression up to now, but gases can be compressed in steps or stages in

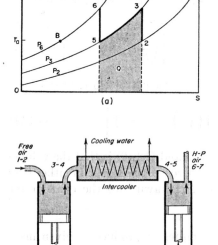

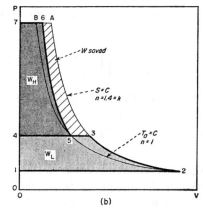

Fig. 12·7 Two-stage compression with intercooling cuts the shaft work input and avoids too high a temperature of discharged air.

a series of cylinders. Figure 12·7c shows the arrangement of a two-stage compressor with intercooler; 12·7b shows the P-V relations with zero clearance.

Free air enters the l-p cylinder during the admission stroke 1–2. Air compresses to an intermediate pressure at constant-S from 2 to 3. During 3–4 the air discharges to the intercooler. The air gives up heat Q to the intercooler water and cools back to atmospheric temperature. This reduces the specific volume of the air. When the air enters the h-p cylinder on the admission stroke, it occupies less volume than it had during discharge from the l-p cylinder. From 5 to 6 the compression is completed at constant S and the air discharges to the h-p system during 6–7.

In Fig. 12·7b the 2–A line is at constant-S for a single-stage compressor working directly between atmospheric and h-p system pressure levels. Process 2–B is at constant T, needing minimum work input, the ideal we try to approach by this two-stage intercooled compression. The hatched area 3–5–6–A measures the work input saved by intercooling. The area 1–2–3–4 measures the work input W_L to the l-p cylinder, and 4–5–6–7 the work input W_H to the h-p cylinder.

How do we choose the intercooler pressure P_3? As we make P_3 approach P_2, the work-saved area 3–5–6–A becomes a narrow crescent, paralleling 2–A, and finally disappears. State 5 must always lie on the constant-T process 2–B. As we make P_3 approach P_6, the work-saved area becomes a shorter and shorter oblique rectangle until it disappears.

The proper level for P_3 can be fixed by taking the formula for total work input $W = W_L + W_H$ and finding where it becomes a minimum. The complete equation is

$$W = \frac{nP_2V_2}{1-n}\left[\left(\frac{P_3}{P_2}\right)^{(n-1)/n} - 1\right] + \frac{nP_3V_5}{1-n}\left[\left(\frac{P_6}{P_3}\right)^{(n-1)/n} - 1\right] \quad (12\cdot15)$$

Note that

$$P_2V_2 = P_3V_5 = RT_a$$

since they all lie on the constant-T_a line which is usually atmospheric temperature. By usual differential analysis (see Appendix G) of the work equation we find that for minimum W the compressor pressures must be in the relation

$$P_3 = \sqrt{P_2P_6}$$

For this condition the work done in each cylinder is equal:

$$W_L = W_H$$

This is shown in Fig. 12·7b, where W *saved* is at the maximum.

Figure 12·7a shows the state variation for the two-stage intercooled compressor on T-S coordinates. Remember that on the T-S chart we cannot show the admission and discharge actions in the cylinder; all we can see is the variation in state for a fixed mass of air.

From 2–3 we have the constant-S compression of 1 lb air from P_2 to P_3. During 3–5 the air cools at constant pressure, giving up heat Q in the intercooler. Then during 5–6 the air compresses at constant S to final pressure P_6. Note: $T_5 = T_2$, the atmospheric temperature. (For minimum work input that we show here $T_6 = T_3$.) By passing the air leaving the l-p cylinder through the intercooler, we get the air in the same state 5 that it would have had if it had been compressed at constant T. This means that Q in Fig. 12·7a equals W_L in Fig. 12·7b.

If we had an aftercooler following the h-p cylinder to cool the h-p air back to atmospheric temperature at constant pressure as at B in Fig. 12·7a, the area under 6–B would be equal to W_H, work in the h-p cylinder.

12·10 Three-stage compressor. As the final air-pressure level rises, it becomes more economical to use extra stages of compression. Also, for given total pressure rise we can decrease the work input by increasing the number of stages and intercooling between them.

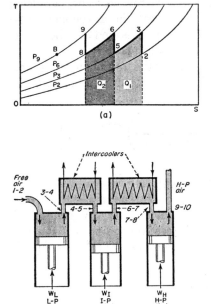
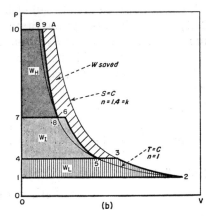

Fig. 12·8 Three-stage compression with two intercooling stages reduces the work input more than for two-stage; used on higher pressure air.

Figure 12·8 shows the circuit and P-V and T-S diagrams for three-stage compression. Here we have the problem of finding the intermediate pressure levels that give minimum total work input. By an analysis similar to that for Fig. 12·7 these pressures must be

$$P_3 = \sqrt[3]{P_2{}^2 P_9} \qquad P_6 = \sqrt[3]{P_2 P_9{}^2}$$

For this condition,

$$W_H = W_I = W_L$$

For the conditions stated in Fig. 12·8, which are the optimum, heat Q given up in the l-p intercooler equals W_L, work input to the l-p cylinder and heat Q_2 equals W_I, work input to the i-p (intermediate-pressure) cylinder. If we had an aftercooler following the h-p cylinder to cool the h-p air to B in Fig. 12·8a, the area under 9–B would equal the work of the h-p cylinder W_H.

Aftercoolers are used in some compressed-air systems to condense out the water vapor carried in with the free air. If this vapor condenses out in the h-p air lines as the air cools, it can cause water hammer and give considerable trouble in the air-driven tools.

12·11 Air engines. Compressed air acts as an energy-transmitting medium. Compressors "generate" the energy in the air; engines use the energy in the air to reproduce mechanical work. In a literal sense we can call air engines reversed compressors. They have the same essential working elements, a piston reciprocating in a cylinder. They differ in type and control of admission and exhaust valve. Compressors need a "flap" valve that works entirely from differences in gas pressure on its two sides. For normal operation they need no linkages with the piston. On engines,

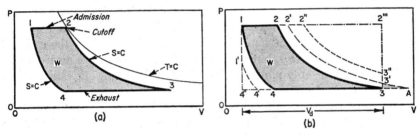

Fig. 12·9 (a) An ideal air engine expands pressurized air to atmospheric pressure to produce work W per cycle. (b) Actual air engines use variable cutoff to vary the net air work output.

however, the valves must be operated through linkages which are usually synchronized mechanically with piston position.

Figure 12·9a shows the ideal P-V variations in an air engine cylinder. At 1 the cylinder holds pressurized air in its clearance volume. The admission valve opens to let in air from the h-p line at constant pressure. At 2 (cutoff) the admission valve closes and the air expands at constant S to state 3, atmospheric pressure. At 3 the exhaust valve opens and the piston forces the air from the cylinder at constant atmospheric pressure to the atmosphere. At 4 the exhaust valve closes and the remaining air is compressed adiabatically (constant S) back to 1 to repeat the cycle. For internally reversible processes and flows, the net area 1-2-3-4 measures the net work W done by the air on the piston. The analysis follows just the reverse of that done for the compressor with clearance (Art. 12·7).

The cycle in Fig. 12·9a shows the ideal having maximum efficiency, because we expand and compress the air at constant S. We can, however, get more work from a particular cylinder by varying the timing of valve openings and closings as shown in Fig. 12·9b.

We could delay cutoff to 2' to take in a greater mass of air before expansion, then expand at constant S to 3' at the end of the piston stroke. Since the air still has a higher pressure than the atmosphere, some of the air blows out of the cylinder when the exhaust valve opens until the pressure drops to atmospheric at 3. The piston then forces the rest of the air out of the cylinder at constant pressure to 4, where the valve closes and compresses the clearance air. By this operation we get a net work measured by the area 1–2'–3'–3–4, an increase measured by area 2–2'–3'–3.

We lose some work by incomplete expansion to atmospheric pressure. The amount is small as shown by the small toe, 3'–A–3. But this loss grows the longer we delay cutoff. At 2" we would expand to 3" with a great gain in work per cycle but increasing loss in efficiency because of more pressure air blown off to the atmosphere at the beginning of the exhaust.

We can also increase work per cycle by delaying closing the exhaust valve as in Fig. 12·9b. If we closed at 4', the clearance air would compress to 1'. Opening the admission valve then forces h-p air into the cylinder to raise the pressure to 1, but no work would be done on the momentarily stationary piston—any motion of h-p air that does no work causes a loss in efficiency of energy use.

The greatest work per cylinder (at the least efficiency) would be realized by the full-admission cycle 1–2'''–3–4''. Here we simply let the h-p air act on the piston for its full stroke V_D. Then open the exhaust valve to let the full-pressure air blow to atmosphere until the cylinder pressure drops to atmospheric. Then we return the piston to force out the remaining air at atmospheric pressure for the full stroke. Opening the admission valve at 4'' forces h-p air into the cylinder with the piston stationary. This cycle is used often on small tools or on tools needing maximum piston pressure for the full stroke.

12·12 Compressed-air system. Now that we know the major components of a compressed-air system, let us see how they work together. Figure 12·10a shows the hookup for a single-stage compressor, an aftercooler and receiver shown as a single unit, and a single-stage air engine. Free air enters the compressor, flows through the aftercooler-receiver then through the engine, and discharges back to atmosphere.

For easy analysis let us consider this a closed air circuit as shown by the dotted line. For steady flow the *mass* of air exhausting from the engine always equals that entering the compressor, even though their temperatures differ. We can consider the atmosphere as a heat exchanger in the equivalent closed circuit.

The receiver acts as a storage reservoir for the h-p air—it keeps a reserve of air for the engine, whose demand may vary widely. This allows the compressor to run at a steadier pace.

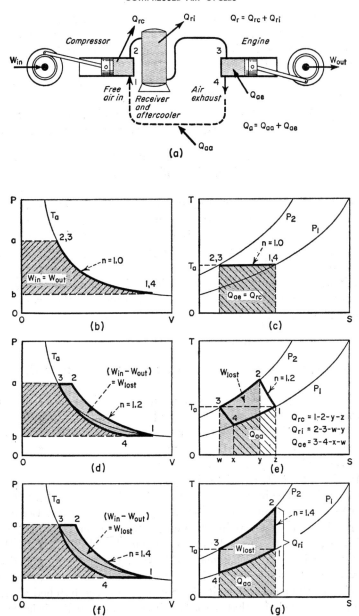

Fig. 12·10 A simple compressed-air system could work on constant-temperature processes b and c, on polytropic processes d and e, or on constant-entropy processes f and g. As n rises above 1.0, work lost keeps growing quite rapidly.

12·13 Energy flow into and out of the system. This takes two forms, work and heat. The prime energy source, Fig. 12·10a, for the system is W_in to the compressor shaft. The compressor cylinder and receiver-aftercooler usually reject heat to the atmosphere, Q_r. The engine produces the major product of the system, W_out. Usually the engine and atmosphere introduce heat into the "closed" system as Q_a.

According to the First Law of Thermodynamics, these four energy flows must always be in balance. For steady air flow through the system,

$$\text{Energy in} = \text{energy out}$$
$$W_\text{in} + Q_a = W_\text{out} + Q_r$$

12·14 Constant-T system. Figure 12·10b and c shows the events for an ideal constant-T compressor and engine with an atmospheric temperature T_a. W_in, the gray area, raises the air pressure from 1 to 2, while Q_{rc}, the shaded area, flows through the compressor-cylinder walls to the atmosphere. This system does not need an aftercooler. Air expands in the engine from 3 to 4 while it absorbs Q_{ae} from the atmosphere, the hatched area, through the engine cylinder walls and produces shaft work W_out, the hatched area. For reversible processes,

$$W_\text{in} = W_\text{out} = Q_{ae} = Q_{rc}$$

This assumes, of course, that there are no pressure drops in the system.

12·15 Polytropic system. This system, shown in Fig. 12·10d and e, gets nearer to the conditions of a practical system, with polytropic process having $n = 1.2$. The gray area W_in compresses air from 1 to 2, while the air rejects heat Q_{rc} to the atmosphere through the compressor cylinder walls. Q_{rc} equals area 1–2–y–z in Fig. 12·10e. In the aftercooler-receiver the air rejects heat Q_{ri} equal to the gray area 2–3–w–y.

In the engine, the air expands from 3 to 4 while absorbing heat Q_{ae} equal to area 3–4–x–w and produces work W_out, the hatched area. Because the pressurized air draws on its internal energy to produce the work, we find that its temperature drops below T_a, even though Q_{ae} enters the air during expansion. (In the constant-T cycle, Q_{ae} prevents the drop in temperature.) As the exhaust air returns through the atmosphere from the engine to the compressor, it absorbs heat Q_{aa}, the hatched area.

The P-V graph shows that work flow into the system W_in is larger than work output W_out. The net cycle area for the system 1–2–3–4 measures the work lost by the system—external temperature irreversibilities cause this. We have, however, considered all the processes as internally reversible.

This contrasts to the engine cycles we have studied. For these, net areas measured shaft work output, but for compressed-air systems, net area measures work lost. Remember, completely available energy, shaft work, runs our compressed-air systems; high-temperature heat (internal energy) runs our engine cycles.

To continue our analysis, from the circuit flow diagram we know that

$$W_{in} + Q_{ae} + Q_{aa} = W_{out} + Q_{rc} + Q_{ri}$$
$$W_{in} - W_{out} = Q_{rc} + Q_{ri} - Q_{ae} - Q_{aa}$$

From the P-V graph we learn that

$$W_{lost} = W_{in} - W_{out}$$

Substituting from the energy flow balance,

$$W_{lost} = Q_{rc} + Q_{ri} - Q_{ae} - Q_{aa}$$

The last equation means that the net area 1–2–3–4 on the T-S graph also measures the net work lost, even though this is in Btu rather than foot-pounds as on the P-V graph.

12·16 Constant-S system. This system, shown in Fig. 12·10f and g, uses adiabatic compression and expansion (zero heat transfer). All heat added to the cycle comes from heating the engine exhaust by Q_{aa}. Heat rejected from the cycle Q_{ri} leaves through the aftercooler.

Adiabatic expansion of the air in the engine causes a maximum temperature drop of the exhaust. Adiabatic compression causes a maximum temperature rise of the compressed air. These effects combine to cause the greatest work loss of any compressed-air system, when pressurized air must be cooled back to atmospheric temperature. The energy analysis parallels the one we just made for the polytropic system. This shows that net areas on both P-V and T-S graphs measure the work lost.

If the pressure parts of an adiabatic system can be thoroughly insulated to prevent loss of Q_{ri} and the aftercooler dispensed with, this would be an efficient energy-transmitting system, with no work lost. The compressor would work along 1–2 in Fig. 12·10f and g, and the engine along 2–1. There would be no areas on the T-S graph, and the two areas for the compressor and the engine would be equal on the P-V graph.

12·17 System analysis. We have examined in detail the behavior of components of a compressed-air system but have not taken the time to put any quantities in our performance formulas. We cannot understand formulas fully until we see how the quantities behave for given conditions. Let us do this now by working out an example for the system shown in Fig. 12·11.

We have here a two-stage air compressor, intercooler, aftercooler, and a single-stage engine. The air compressor takes in free air at 14.7 psia and 60 F. The coolers reduce the temperature of the compressed air back

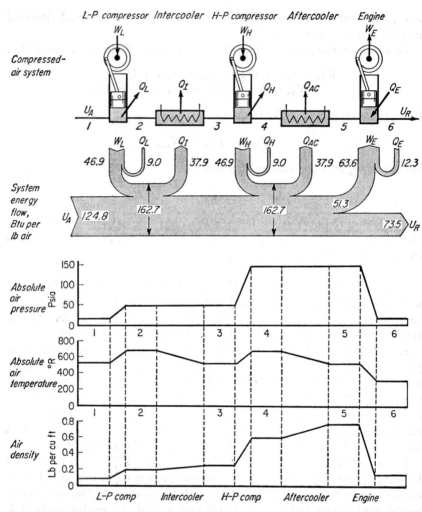

Fig. 12·11 A two-stage air compressor with inter- and aftercooling feeding a single-stage air engine taps the internal energy of free air to produce work output at the air-engine shaft.

to 60 F after each stage. The engine exhausts the spent air back to the atmosphere at 14.7 psia.

Both compressors and engine work on reversible polytropic processes with $n = 1.3$. To simplify calculation we assume that both compressors and engine have zero clearance.

In Table 12·1 we figure the state of the air entering and leaving each component of Fig. 12·11 and the energy entering and leaving each component and the total system.

Table 12·1 How to figure compressed-air-system performance (Fig. 12·11)

1. Polytropic compression and expansion: $n = 1.3$
2. Air specific heats: $c_p = 0.24$; $c_v = 0.1715$
3. Polytropic specific heat: $c_n = c_v \left(\dfrac{k-n}{1-n}\right) = 0.1715 \left(\dfrac{1.4-1.3}{1-1.3}\right)$
   $= -0.0572$ Btu/lb/F
4. $P_2 = P_3$ chosen for minimum compressor work, so
5. $W_L = W_H$
6. $P_1 = P_6 = 14.7$ psia; $14.7 \times 144 = 2{,}116$ psfa
7. $P_4 = P_5 = 147$ psia; $147 \times 144 = 21{,}160$ psfa
8. $P_2 = P_3 = \sqrt{P_1 P_4} = \sqrt{2{,}116 \times 21{,}160} = 6{,}695$ psfa
9. $T_1 = T_3 = T_5 = 60 + 460 = 520°R$
10. $V_1 = \dfrac{RT_1}{P_1} = \dfrac{53.3 \times 520}{2{,}116} = 13.10$ cu ft/lb air
11. $D_1 = \dfrac{1}{V_1} = \dfrac{1}{13.10} = 0.07635$ lb/cu ft air
12. $U_A = c_p T_1 = 0.24 \times 520 = 124.8$ Btu/lb air
13. $\dfrac{P_2}{P_1} = \dfrac{6{,}695}{2{,}116} = 3.165$
14. $\dfrac{P_4}{P_3} = \dfrac{21{,}160}{6{,}695} = 3.165$
15. $W_L = \dfrac{nP_1V_1}{1-n}\left[\left(\dfrac{P_2}{P_1}\right)^{(n-1)/n} - 1\right] = \dfrac{1.3 \times 2{,}116 \times 13.10}{1-1.3}(3.163^{(1.3-1)/1.3} - 1)$
    $= -36{,}540$ ft-lb/lb air $= -36{,}540/778 = -46.9$ Btu per lb air
16. $\dfrac{T_2}{T_1} = \left(\dfrac{P_2}{P_1}\right)^{(n-1)/n} = \left(\dfrac{6{,}695}{2{,}116}\right)^{(1.3-1)/1.3} = 1.3043$
17. $T_2 = 1.3043 T_1 = 1.304 \times 520 = 677.8°R$
18. $V_2 = \dfrac{RT_2}{P_2} = \dfrac{53.3 \times 677.8}{6{,}695} = 5.395$ cu ft/lb
19. $D_2 = \dfrac{1}{V_2} = \dfrac{1}{5.395} = 0.1854$ lb/cu ft
20. $S_2 - S_1 = c_n \log_e \dfrac{T_2}{T_1} = -0.0572 \times \log_e 1.3043 = -0.0152$
21. $Q_L = c_n(T_2 - T_1) = -0.0572(677.8 - 520) = -9.0$ Btu/lb
22. $Q_I = c_p(T_3 - T_2) = 0.24(520 - 677.8) = -37.9$ Btu/lb
23. $S_3 - S_2 = c_p \log_e \dfrac{T_3}{T_2} = 0.24 \log_e \dfrac{520}{677.8} = 0.24(-0.2655) = -0.0637$
24. $V_3 = \dfrac{RT_3}{P_3} = \dfrac{53.3 \times 520}{6{,}695} = 4.14$ cu ft/lb air
25. $D_3 = \dfrac{1}{V_3} = \dfrac{1}{4.14} = 0.2415$ lb/cu ft air
26. $W_H = W_L = -46.9$ Btu/lb air
27. $\dfrac{T_4}{T_3} = \left(\dfrac{P_4}{P_3}\right)^{(n-1)/n} = 3.165^{0.2308} = 1.3043$
28. $T_4 = 1.3043 T_3 = 1.3043 \times 520 = 677.8°R = T_2$
29. $S_4 - S_3 = c_n \log_e \dfrac{T_4}{T_3} = -0.0572 \log_e 1.3043 = -0.0152$
30. $Q_H = c_n(T_4 - T_3) = -0.0572(677.8 - 520) = -9.0$ Btu/lb air
31. $Q_{AC} = c_p(T_5 - T_4) = 0.24(520 - 677.8) = -37.87$ Btu/lb

Table 12.1 How to figure compressed-air-system performance (Fig. 12.11)
(Continued)

32. $V_4 = \dfrac{RT_4}{P_4} = \dfrac{53.3 \times 677.8}{21{,}160} = 1.707$ cu ft/lb air

33. $D_4 = \dfrac{1}{V_4} = \dfrac{1}{1.707} = 0.586$ lb/cu ft air

34. $V_5 = \dfrac{RT_5}{P_5} = \dfrac{53.3 \times 520}{21{,}160} = 1.310$ cu ft/lb air

35. $D_5 = \dfrac{1}{V_5} = \dfrac{1}{1.310} = 0.7635$ lb/cu ft air

36. $W_E = \dfrac{nP_5 V_5}{1-n}\left[\left(\dfrac{P_6}{P_5}\right)^{(n-1)/n} - 1\right] = \dfrac{1.3 \times 21{,}160 \times 1.31}{1 - 1.3}\left[\left(\dfrac{2{,}116}{21{,}160}\right)^{0.2308} - 1\right]$

    $= 49{,}500$ ft-lb/lb air $= \dfrac{49{,}500}{778} = 63.6$ Btu/lb air

37. $S_5 - S_4 = c_p \log_e \dfrac{T_5}{T_4} = 0.24 \log_e \dfrac{520}{677.8} = -0.0637$

38. $\dfrac{T_6}{T_5} = \left(\dfrac{P_6}{P_5}\right)^{(n-1)/n} = \left(\dfrac{2{,}116}{21{,}160}\right)^{0.2308} = 0.5878$

39. $T_6 = 0.5878 T_5 = 0.5878 \times 520 = 305.3°$R

40. $Q_E = c_p(T_6 - T_5) = -0.0572(305.3 - 520) = 12.28$ Btu/lb air

41. $U_R = c_p T_6 = 0.24 \times 305.3 = 73.3$ Btu/lb air

42. $V_6 = \dfrac{RT_6}{P_6} = \dfrac{53.3 \times 305.3}{2{,}116} = 7.695$ cu ft/lb air

43. $D_6 = \dfrac{1}{V_6} = \dfrac{1}{7.695} = 0.130$ lb/cu ft air

44. $S_6 - S_5 = c_n \log_e \dfrac{T_6}{T_5} = -0.0572 \log_e \dfrac{305.3}{520} = 0.03037$

12·18 Energy flow through a system. This flow, shown in Fig. 12·11, usually holds some surprises for the novice. We give an *energy-flow* diagram below the system diagram in Fig. 12·11. These energies are all calculated in the table and determine widths of the diagram.

The l-p compressor has two energy inflows, shaft work W_L and the energy carried by the entering air U_A. Because the compression is polytropic, the compressing air rejects heat Q_L, usually to the cylinder jacket-cooling water. The rest of the incoming energy leaves as increased energy of the compressed air.

We do not know the true energy carried by the incoming air U_A, but we can figure a pseudo-energy by assuming that the specific heat $c_p = 0.24$ Btu per lb, F of air holds constant all the way from absolute zero temperature. The flow diagram, Fig. 12·11, then shows that 124.8 Btu per lb of air enters the system. Of 46.9 Btu per lb input by W_L, 9.0 Btu immediately leaves as heat through the cylinder walls and 37.9 Btu enters the air. The air leaves the l-p cylinder with 162.7 Btu per lb.

The intercooler removes the heat put in by the l-p compressor as heat Q_I, and cools the air back to atmospheric temperature. The partly compressed air leaves the intercooler with no more energy than it originally had as free air—not a seemingly profitable arrangement energywise.

The three lower graphs of Fig. 12·11 show how the pressure, temperature, and density of the air vary as it flows through the system. These show that at state 3 all we have gained is increased pressure and density. Since the air temperature is the same as atmospheric, the air molecules at 3 move with the same average speed, hence have no increased energy. But since the density has been raised, this means that a given number of molecules have been packed into a smaller space. This causes them to hit the container walls more frequently and so raise the pressure.

We could say that from state 1 to 3 we have gone through a "packing" operation and thrown away the valuable available shaft energy we used to do this.

In the h-p compressor and aftercooler we do exactly the same thing as in the l-p compressor to increase air pressure further. For an economic compressor $W_H = W_L$ as we learned in Art. 12·9. So at state 5 we have the molecules more tightly packed but have thrown away W_L and W_H as heats Q_L, Q_I, Q_H, and Q_{AC}.

With no net gain in energy in our h-p air we seem to have a useless product on our hands. But let us not forget that pressurized air can exert a force and a moving force can do mechanical work.

When we admit h-p air, 5, into the engine cylinder, the molecules slam into the moving piston, do mechanical work, and rebound with decreased speed and energy. The lowered energy appears as a lowered temperature.

While the pressurized air expands, its temperature drops and heat Q_E from the atmosphere enters the air through the cylinder walls to help produce engine shaft output W_E.

The energy-flow diagram, Fig. 12·11, shows that 51.3 Btu per lb of air joined by 12.3 Btu from the atmosphere produces 63.6 Btu per lb of air engine output. Air exhausted by the engine at state 6 leaves with only 73.5 Btu per lb. It entered the system with 124.8 Btu per lb. As the temperature diagram shows, the air becomes very cold. While its pressure is atmospheric, the density is higher than free air because of the greater number of slower moving molecules.

(*Note:* Figure 12·11 energy flow has been balanced to show 73.5 Btu per lb of air leaving, while Table 12·1 indicates 73.3 Btu. The difference is caused by rounding of figures and slide-rule reading errors.)

Figure 12·12 gives the *P-V* and *T-S* diagrams for the system of Fig. 12·11. In *a*, the gray area to left of process 1–2 measures the l-p compressor work input W_L. The gray area to the left of 3–4 shows the h-p compressor work input W_H. Then the total gray areas measure the total shaft work input to the system. The hatched area to left of 5–6 shows the work output of the engine W_E. Since engine and compressor work areas coincide, the net area 1–2–3–4–5–6 shows mechanical work lost by the system.

In Fig. 12·12b, the gray area under 1–2 shows the heat rejected by the l-p compressor, Q_L. The gray area under 2–3 shows the heat rejected by the intercooler Q_I. The gray area under 3–4 shows the heat rejected by the h-p cylinder, Q_H. The gray area under 4–5 gives the heat rejected by the aftercooler, Q_{AC}.

The hatched area under 5–6 indicates the heat absorbed in the engine cylinder, Q_E. The hatched area under 6–1 measures the amount of heat needed to reheat the exhaust air at 6 back to atmospheric temperature at 1, from U_R to U_A. The net area 1–2–3–4–5–6 shows the net amount of heat rejected by the pressurized air to the atmosphere.

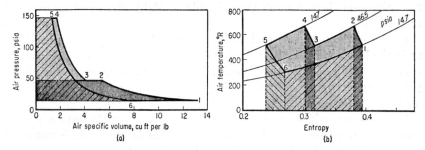

Fig. 12·12 P-V and T-S graphs show state variations of air for the system in Fig. 12·11; net areas 1–2–3–4–5–6 measure net shaft work lost. The compressor and engine use polytropic process with n = 1.3; no pressure loss is assumed in connecting piping or in air coolers.

How are the net areas of Fig. 12·12 related? For this we must go to the circuit diagram of Fig. 12·11 and make a first-law analysis. For steady air flow through the system, no energy accumulates in or drains from the system. So all energy entering the system must equal energy leaving:

$$U_A + W_L + W_H + Q_E = Q_L + Q_I + Q_H + Q_{AC} + W_E + U_R$$

Rearranging terms we get

$$W_L + W_H - W_E = Q_L + Q_I + Q_H + Q_{AC} - Q_E - (U_A - U_R)$$

The left side of the equation is net mechanical work lost by the system; the right side is the net heat lost by the system to the atmosphere. This means that both net areas are equal in Fig. 12·12 and represent net work lost.

The remarkable fact about this system is that the total shaft work we put in is thrown away as heat to the atmosphere. The shaft work we get out as the final product comes from part of the internal energy of the compressed air, which is the same as the internal energy of the free air. This makes the exhaust air very cold.

12·19 Refrigeration. Since we can get cold air from the system of Fig. 12·11, this suggests a scheme for a refrigerating system. In Chap. 9 we talked about the principles of a refrigerating system. Figure 12·13a shows the closed-cycle compressed-air system that can be used to remove energy from a low-temperature region or refrigerator.

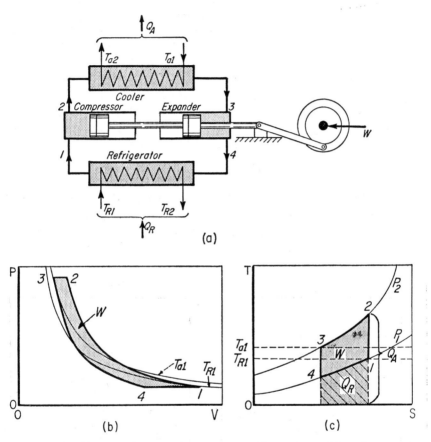

Fig. 12·13 A closed system makes it possible to use cold air exhausted from the expander (air engine) for removing heat energy from refrigerated areas; W activates the cooling system.

Air at atmospheric pressure and state 1 enters the compressor. The compressor pressurizes the air and raises its internal energy and temperature to state 2. Pressurized hot air then enters the cooler, which removes some of the internal energy of the air as Q_A, cooling the air to atmospheric temperature T_{a1}. Cooling water usually does this job. Cooled pressurized air at state 3 then enters the air engine or *expander* and contributes shaft work to drive the compressor. The air expands at constant S, so its pressure drops to atmospheric but its temperature to much lower than atmospheric, to T_4.

The cold air at state 4 then enters the refrigerator and picks up heat Q_R from the medium being cooled. The air leaves at state 1 to reenter the compressor and repeat the cycle.

Figure 12·13b and c shows the state changes on P-V and T-S graphs. The net area 1–2–3–4 measures the difference between compressor work input and the work developed by the expander. This work difference W (equivalent to that lost by a conventional compressed-air system) must be supplied from the outside to make the refrigerating system work.

In Fig. 12·13c the total gray area under 2–3 measures the heat rejected to atmosphere by the cooler, the hatched area under 4–1 measures the heat removed in the refrigerator. The net area 1–2–3–4 measures the work input to the system. By a first-law analysis of Fig. 12·13a we see that energies entering and leaving must balance as

$$W + Q_R = Q_A \quad \text{or} \quad W = Q_A - Q_R$$

12·20 Actual performance. The performance of compressors and engines differs in degree from the ideal reversible processes we have been studying. The ideal performances are easy to calculate and set up the maximum possible goal. We then compare actual performance against the calculated ideals.

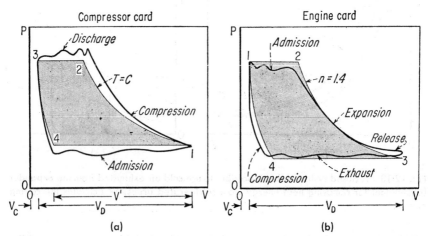

Fig. 12·14 An actual compressor uses more work input than an ideal compressor working between the same pressure limits, while an actual air engine produces less work than an ideal air engine.

Fig. 12·14a compares an actual P-V graph for a compressor against a reversible ideal using constant-T compression. At state 1 the actual pressure is slightly lower than atmospheric. Compression actually follows a path between constant T and constant S. This makes the gas reach the discharge pressure sooner than the constant-T process. The cylinder pressure surges above the discharge pressure because a slightly higher pressure is needed to open the discharge valve automatically.

The varying rate of piston motion and the pressure difference between cylinder air and air in the discharge line set up pressure waves during the discharge stroke. The moving piston also imparts energy to the discharging gas molecules, which raises their pressure, see Art. 12·4. By the time the piston comes to a stop at the end of the discharge stroke, the air pressure does drop to discharge-line pressure at state 3.

In expanding the clearance air, the actual process again lies between constant T and constant S, so the clearance air reaches atmospheric pressure sooner than the ideal constant T. The pressure, however, surges below atmospheric to open the admission valve and allow free air to enter the cylinder. Again we have pressure waves set up in the entering air because of varying piston speed during the stroke, but the pressure level stays below atmospheric during the admission stroke and almost reaches it at the end, state 1.

The net enclosed area of the actual compressor graph or card is larger than the ideal, shaded area. This means that an actual compressor needs more work input than an ideal compressor working between the same pressure limits. The irreversible processes we described contribute to the relative inefficiency of an actual compressor.

Figure 12·14b compares the ideal and actual cards for an air engine. Here the irreversible effects make the net area of the actual engine less than the ideal constant-S engine. This means that an actual engine produces less work than the ideal. The principal adverse effects are the lower admission pressure and higher exhaust pressure.

12·21 Performance factors. In Art. 12·7 we discussed the effect of clearance volume in a general way. To handle the maximum mass of air in a given compressor cylinder we obviously want to make the clearance volume as small as possible. Let us find out what factors affect the amount of air handled by a cylinder.

The conventional volumetric efficiency e_v of a compressor (ideal) is defined as the ratio of compressor capacity to piston displacement. In Fig. 12·14a the compressor capacity is $V' = V_1 - V_4$ and the piston displacement is $V_D = V_1 - V_3$. The per cent clearance for a compressor is defined as $c = V_c/V_D$. Then

$$e_v = \frac{V'}{V_D} = \frac{V_1 - V_4}{V_D}$$

But

$$V_1 = V_D + V_c = V_D + cV_D$$

$$V_4 = V_3 \left(\frac{P_3}{P_4}\right)^{1/n}$$

$$= cV_D \left(\frac{P_2}{P_1}\right)^{1/n}$$

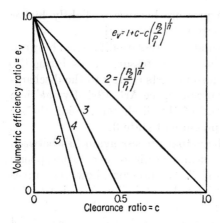

Fig. 12·15 The conventional volumetric efficiency ratio depends on several design factors.

Substituting,

$$e_v = \frac{V_D + cV_D - cV_D(P_2/P_1)^{1/n}}{V_D}$$
$$= 1 + c - c\left(\frac{P_2}{P_1}\right)^{1/n}$$

This means that the conventional volumetric efficiency depends on the per cent clearance and the pressure ratio, as well as the polytropic coefficient n. Figure 12·15 shows how these factors are related.

Power Test Code 9-1954 *Displacement compressors, vacuum pumps and blowers* published by the American Society of Mechanical Engineers covers the method of defining and testing the performance of actual compressors.

REVIEW TOPICS

1. Describe the actions of a simple reciprocating air compressor.
2. Describe the cycle of a reciprocating air compressor with the aid of a P-V graph.
3. Write the equation for each process or action in an air compressor with zero clearance.
4. Write the equation for the net work done on the air in a reciprocating compressor without clearance.
5. Taking account of relative air molecular speed and piston speed, show why the assumption of constant pressure during intake and discharge in a reciprocating compressor introduces an error.
6. What is the advantage of a constant-temperature compressor over a constant-entropy compressor?
7. With what factors does work input to a zero-clearance compressor vary?
8. Describe the actions of an air compressor with clearance with the aid of a P-V graph.
9. Write the equation for work input to an air compressor with clearance.
10. How can the air output of a reciprocating air compressor be varied?
11. Describe the actions of a multistage intercooled reciprocating compressor on both P-V and T-S graphs.
12. Write the optimum work equation for an intercooled multistage air compressor.

13. What are the pressure relations for optimum work input in a two-stage intercooled compressor?
14. What is the function of an aftercooler in a compressed-air system?
15. How do valves for an air compressor and for an air engine differ?
16. Describe the processes of an ideal air engine with clearance.
17. How can the work output per cycle of an air engine be varied?
18. Write the energy equation for a complete compressed-air system.
19. Draw the P-V and T-S graphs for a compressed-air system working on constant-temperature compression and expansion. Interpret the meaning of the areas on each graph.
20. Draw the P-V and T-S graphs for a compressed-air system working with a polytropic reversible compression and expansion. Interpret the meaning of the areas on each graph.
21. Draw the P-V and T-S graphs for a compressed-air system working with a constant-S compression and expansion. Interpret the meaning of the areas on each graph.
22. Describe the energy flow in a simple compressed-air system using a single-stage compressor and single-stage engine.
23. Draw the flow diagram and P-V and T-S graphs of a closed air-cycle refrigerating system. Interpret the meaning of the areas on each graph. How is the temperature of the air lowered?
24. Draw the P-V graph for an ideal and actual compressor with clearance. Describe the causes of the difference in the processes.
25. Draw the P-V graph for an ideal and actual air engine with clearance. Describe the causes of the differences in processes.
26. Define the meaning of volumetric efficiency. Derive the equation for conventional volumetric efficiency of a compressor.

PROBLEMS

1. An ideal zero-clearance compressor pressurizes 80 F air from 14.5 to 90 psia. Calculate the work input per pound of air if the compression process is (a) at constant temperature, (b) at constant entropy, (c) a polytropic with $n = 1.35$.
2. In Prob. 1 find the work of the compression processes alone and compare them with the total work inputs. Discuss.
3. In Prob. 1 calculate the temperature of the air at the end of the compression processes. Will this be the same as the temperature of the air discharged from the compressor?
4. An ideal zero-clearance compressor pressurizes air initially at 60 F from 50 to 120 psia. Calculate the work input per pound of air if the compression process is (a) at constant temperature, (b) at constant entropy, (c) a polytropic with $n = 1.25$. (d) Calculate the discharge temperatures for each process.

5. An ideal zero-clearance compressor pressurizes helium initially at 80 F from 15 to 650 psia. Calculate the work input per pound of helium if the compression process is (a) at constant temperature, (b) at constant entropy, (c) a polytropic with $n = 1.55$. (d) Calculate the temperature of the helium at discharge pressure for (a), (b), and (c).

6. An ideal compressor with clearance inhales 1.2 cu ft of free air per intake stroke with 180 intake strokes per minute and compresses it to 100 psia pressure. When the intake air is at 14.3 psia and 80 F, find the work input per minute if the compression process is (a) at constant temperature, (b) at constant entropy, (c) a polytropic with $n = 1.38$. (d) Find the temperature of the discharged air for (a), (b), and (c).

7. A single-acting compressor with 4% clearance takes in air at 14.5 psia and discharges it at 180 psia. What is its volumetric efficiency if its compression process has an $n = 1.30$?

8. A 12- by 12-in. single-acting compressor with 3% clearance runs at 300 rpm. Pressure and temperature of intake air are 14.3 psia and 65 F. Compression and expansion processes have an $n = 1.33$. For a discharge pressure of 120 psia find (a) the conventional volumetric efficiency, (b) volume and weight of air handled per minute at intake conditions, (c) work input per minute, (d) weight of air in the cylinder at start of compression process.

9. A two-stage intercooled air-compressor-engine cycle, Fig. 12·11, takes in air at 14.2 psia and 90 F. All compression and expansion processes have an $n = 1.25$. The top pressure in the cycle is 250 psia. (a) Find the ideal intermediate pressure of the compressor stages. (b) Find the work input of both l-p and h-p compressors if air entering the h-p cylinder is at the same temperature as the intake air of l-p cylinder. (c) Find the heat rejected by the compressor cylinders and the intercooler. (d) If the aftercooler drops the air temperature to the engine to 90 F, find the work done in the engine, the heat rejected by the aftercooler, and the heat absorbed by the engine. (e) If the aftercooler does not cool the air to the engine, find the work done in the engine and the heat absorbed or rejected by the engine. (f) Find the change in entropy of the air in all components of the system for (a), (b), (c), (d), and (e).

10. In the air refrigerating cycle of Fig. 12·13 the air enters the compressor at 40 F and 18 psia. The ideal compressor pressurizes the air at $n = 1.35$ to 250 psia. The cooler discharges the air at 250 psia and 90 F to the expander. In the expander the air follows an expansion process with $n = 1.35$ and leaves at 18 psia. (a) How much heat per pound of air does it pick up in the refrigerator? (b) How much heat does the cooler reject per pound of air? (c) What is the work input to the cycle per pound of air?

CHAPTER 13

Internal-combustion-engine Cycles

Let us study the heat and work relations of the well-known i-c engines. Numerically and energywise, they are the most important type of prime mover in the United States today. The 75 million motor vehicles on our roads today have a total rated engine capacity of about 7.5 billion hp. They consume about 8×10^{15} Btu of motor fuel yearly. This represents about 20% of our total national fuel consumption for all purposes, such as power and heat.

13·1 Otto engine. Most of our automobiles use four-stroke-cycle spark-ignited i-c engines modeled after the Otto engine first built about 1876. Figure 13·1 shows the essential thermodynamic elements and the processes with which the engine works.

The basic principle depends on heating the working fluid trapped in a cylinder by a piston to raise the fluid pressure. The pressurized fluid then works on the piston to produce engine work output. The cylinder and piston eject spent fluid, inhale a fresh charge, and compress it before heating it for the next cycle. Air mixed with vaporized gasoline forms the working fluid and, when ignited, burns quickly and supplies the heat to raise the pressure.

In Fig. 13·1 and Table 13·1 we trace the actions step by step for an idealized engine. The P-V and T-S graphs show the change in state for the working fluid. The piston drives the constantly rotating flywheel, mounted on the output shaft, through the connecting rod as it oscillates back and forth in the cylinder.

In the first step, state 1, the inlet valve has just opened and the piston is at its head-end or top-dead-center (tdc) position. During the intake stroke 1–2 the fuel-air mixture flows into the cylinder at constant atmospheric pressure. Stored energy in the flywheel moves the piston from 1 to

2, but the entering mixture does a small amount of work on the moving piston, $P_1(V_2 - V_1)$, equal to the area under the intake stroke 1–2 on the P-V graph. This stroke cannot be shown on the T-S graph.

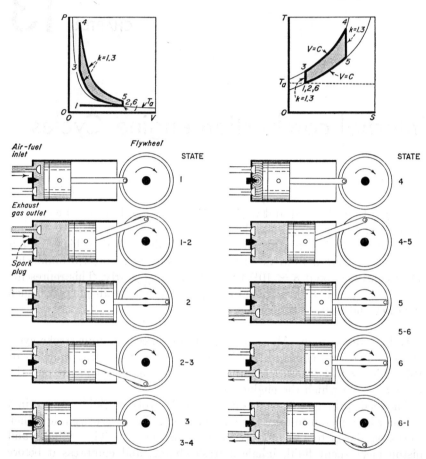

Fig. 13·1 A four-stroke spark-ignited i-c engine working on the Otto cycle uses one stroke out of four to produce net work output. The flywheel stores part of the work output to compress the air-fuel mixture before ignition, exhaust burned gases from the cylinder, and draw in a fresh charge of air-fuel mixture. Burning is assumed to take place instantaneously in the engine cylinder, 3–4. See Table 13·1.

At state 2 the piston passes through its crank end or bottom-dead-center (bdc) position and the working fluid fills the cylinder. The inlet valve closes, and during process 2–3 stored energy in the flywheel moves the piston to compress the fuel-air mixture reversibly at constant S. The area under 2–3 on the P-V graph measures the work of compression. On the T-S graph the process is a vertical line showing that no heat transfer is involved.

A spark ignites the fuel-air mixture at state 3, and it burns quickly, raising its pressure and temperature while the piston passes through its tdc position. This constant-volume heating raises the mixture to state 4. Ideally, heating takes place instantaneously. The vertical line 3–4 on the P-V graph shows that no work is involved, but the area under 3–4 on the T-S graph measures the amount of heat absorbed by the mixture at constant volume. Chemical fuel energy has been released by burning to develop thermal energy of the gas mixture during this process.

Table 13·1 (See Fig. 13·1)

State	Valve position		Action	Work	Heat
	Inlet	Exhaust			
1	Just opened	Just closed			
1–2	Open	Closed	Air-fuel intake	Air to piston	None
2	Just closed	Closed			
2–3	Closed	Closed	Compression	Piston to air	None
3	Closed	Closed	Ignition		
3–4	Closed	Closed	Combustion		Input to gas
4	Closed	Closed	Max T, P		
4–5	Closed	Closed	Expansion	Gas to piston	None
5	Closed	Just opened			
5–6	Closed	Open	Exhaust blowdown	None	Gas to atmosphere
6	Closed	Open	P drop to atmosphere	None	Gas to atmosphere
6–1	Closed	Open	Exhaust	Piston to gas	Gas to atmosphere

From 4 to 5 the hot combustion products expand reversibly and work on the piston at constant S. The area under 4–5 on the P-V graph measures the work developed on the piston during expansion. The vertical line 4–5 on the T-S graph shows that no heat transfer has taken place.

At state 5 the combustion gas is still pressurized with the piston in its bdc position. The exhaust valve opens, and some of the hot gas blows out to the atmosphere till the cylinder pressure drops to atmospheric, 5 to 6. This step cannot be shown directly on the T-S graph.

From 6 to 1 stored energy in the flywheel moves the piston to push out the hot gas to the atmosphere through the exhaust valve. The piston does work $P_1(V_1 - V_2)$ during exhaust. The area under 6–1 on the P-V graph measures exhaust-stroke work done on the gas.

At state 1 the exhaust valve closes and the inlet valve opens and we are ready to repeat the next cycle.

13·2 Otto cycle. We have looked at the idealized engine operations. For easy analysis we can simplify them to form the Otto cycle. The P-V

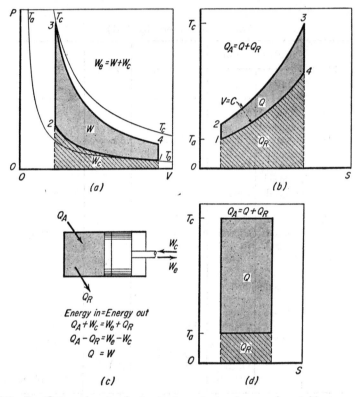

Fig. 13·2 The Otto cycle absorbs heat at constant volume and variable temperature, a and b, making its thermal efficiency lower than that of the Carnot cycle d. The practicality of the Otto engine makes theoretical efficiency of only minor importance.

graph shows that the works of the inlet and exhaust strokes 1–2 and 6–1 are equal but opposite. As far as contributing to the net work output of the engine, we assume they cancel. (See Art. 12·4 about the effect of piston speed on intake and exhaust pressure.)

This leaves the cycle loop 2–3–4–5–2 in Fig. 13·1. We can imagine this as representing 1 lb of working fluid, say air, being compressed, heated and expanded as in the actual engine. But the process 5–2 can be visualized as a constant-volume cooling process, rejecting unavailable energy and returning the air to its initial atmospheric condition.

The area under 5–2 on the T-S graph then measures the amount of heat rejected during the constant-volume cooling. Because of the generally high temperature level of the cycle, we customarily call this the *hot-air standard* and use $k = 1.3$ for the reversible adiabatic processes with $c_p = 0.293$ Btu per lb and $c_v = 0.225$ Btu per lb. This makes the gas constant $R = J(c_p - c_v) = 52.9$.

With the Otto cycle we can study what happens to engine performance as we vary conditions. In Fig. 13·2a we have a cycle working between temperatures T_a and T_c. Expansion work of the hot air W_e is measured by the gray area under 3–4. From this we must deduct the compression work W_c measured as the hatched area under 1–2. This leaves the net area 1–2–3–4 as the net work output W of the cycle.

In Fig. 13·2b we have the total heat added at constant volume Q_A as the gray area under 2–3. From this we deduct the heat rejected during cooling Q_R, the hatched area under 4–1. This leaves the net heat Q used by the cycle as the area 1–2–3–4. Note that we regard the heat added to the cycle as being transmitted through the cylinder wall even though it is released in the cylinder in the actual engine.

Figure 13·2c shows that $Q = W$ by using the first-law analysis.

13·3 Cycle performance. The Carnot cycle taught us that maximum thermal efficiency can be achieved only when heat is added at the highest *constant* temperature and rejected at the lowest *constant* temperature. Figure 13·2b shows that the Otto cycle violates both these conditions, the heats being transferred at *variable* temperatures.

Figure 13·2d shows what happens when the heat Q_A of Fig. 13·2b is added to a Carnot cycle at the highest cycle temperature T_c. Net $Q = W$ for the Carnot cycle is much larger than for the Otto. The much smaller Q_R for the Carnot highlights the better efficiency. Note the smaller entropy change of the Carnot cycle.

13·4 Temperature effect. Knowing the all-important effect of temperature on thermal efficiency, let us study how this applies to the Otto cycle. In Fig. 13·3a and b we have three cycles working between volumes V_1 and V_2. First let us analyze the thermal efficiency of cycle 1–2–3–8 working with a top combustion temperature of 1500°R. Table 13·2 shows the relations for figuring heat transfers and net work in terms of the cycle temperatures. Equation (4) shows that the thermal efficiency can be figured from the temperatures.

But there is a unique relation between these temperatures and the specific volumes as shown by Eqs. (5) and (6) based on the constant-S processes, see Art. 6·9. When we substitute these in Eq. (4) as shown, the temperatures disappear and we find that the thermal efficiency of the Otto cycle depends on the ratio of the specific volumes V_1/V_2 known as the compression ratio, V_r and the ratio of specific heats $k = c_p/c_v$.

176 BASIC THERMODYNAMICS

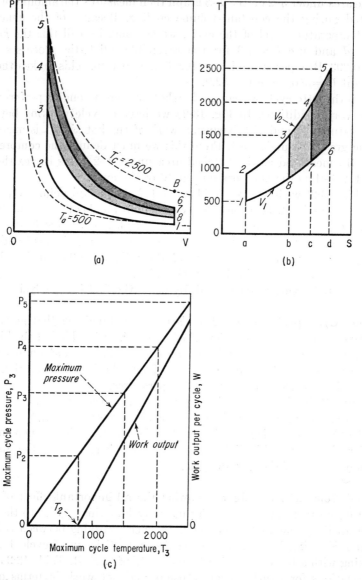

Fig. 13·3 The thermal efficiency of the Otto cycle depends only on the compression ratio V_1/V_2. Varying the combustion temperature T_3 controls the net amount of work output per cycle; the thermal efficiency stays constant. Maximum pressure P_3 varies with T_3.

INTERNAL-COMBUSTION-ENGINE CYCLES

Table 13·2. Otto-cycle thermal efficiency does not depend on temperature

1. $Q_a = E_3 - E_2 = c_v(T_3 - T_2)$ (See Fig. 13·3)
2. $Q_r = E_8 - E_1 = c_v(T_8 - T_1)$
3. $W = Q_a - Q_r = c_v(T_3 - T_2) - c_v(T_8 - T_1)$
4. $e = \dfrac{W}{Q_a} = \dfrac{c_v(T_3 - T_2) - c_v(T_8 - T_1)}{c_v(T_3 - T_2)}$

$$= 1 - \frac{T_8 - T_1}{T_3 - T_2}$$

For the isentropic processes:

5. $T_8 = T_3 \left(\dfrac{V_3}{V_8}\right)^{k-1} = T_3 \left(\dfrac{V_2}{V_1}\right)^{k-1}$

6. $T_1 = T_2 \left(\dfrac{V_2}{V_1}\right)^{k-1}$

Substituting (5) and (6) in (4),

7. $e = 1 - \dfrac{T_3(V_2/V_1)^{k-1} - T_2(V_2/V_1)^{k-1}}{T_3 - T_2}$

$$= 1 - \left(\frac{V_2}{V_1}\right)^{k-1}$$

Letting compression ratio be

8. $V_r = \dfrac{V_1}{V_2}$

9. $e = 1 - \dfrac{1}{V_r{}^{k-1}}$

Thus in Fig. 13·3a and b as we push up the final combustion temperature from T_3 to T_4 and T_5, we can see that the $W = Q$ areas increase while the W_c area remains constant. Q_A also increases to produce these higher temperatures and work outputs, but the thermal efficiency remains constant as long as V_1/V_2 stays constant.

This means that, if we regard each of the increment cycles 8–3–4–7 and 7–4–5–6 as separate cycles, they all run at the same thermal efficiency because they have a common compression ratio $V_r = V_1/V_2$.

13·5 Work variation. Just how is the top temperature and work W related? From Eq. (3) in Table 13·2 we can write

$$W = c_v(T_3 - T_2 - T_8 + T_1)$$

Substituting for T_8 from Eq. (5) we get

$$W = c_v\left[T_3 - T_2 - T_3\left(\frac{V_2}{V_1}\right)^{k-1} + T_1\right]$$

$$= c_v T_3\left[1 - \left(\frac{V_2}{V_1}\right)^{k-1}\right] + c_v(T_1 - T_2)$$

For a given engine with fixed compression ratio, and T_1 the compression temperature, T_2 would also be fixed. The only variables in the work equation are W and the combustion temperature T_3. So lumping together all the constants we get

$$W = CT_3 + K$$

Figure 13·3c shows the plot of this equation of W versus T_3. When the combustion temperature T_3 rises above the compression temperature T_2, the engine develops net work W in direct proportion to the difference $T_3 - T_2$.

An important factor in engine design is the maximum combustion pressure, P_3 developed. Since combustion heating takes place at constant volume, the maximum pressure P_3 varies directly with the absolute combustion temperature T_3.

This tells us that we can vary our engine output simply by varying the amount of heat released in the cylinder and so vary the combustion temperature. To develop high output we must be sure that the cylinder is strong enough to withstand the peak pressure.

In an actual engine we control the heat input by varying the amount of gasoline vapor injected into the air. The mass of air stays essentially con-

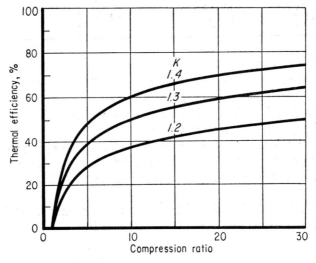

Fig. 13·4 Otto-cycle thermal efficiency rises with increasing compression ratio and specific-heat ratio. It rises fastest at lower ratios.

stant for all outputs. We shall discuss actual engine behavior later. Keep in mind that we are talking about a simplified ideal cycle to sort out the different relations more easily.

Figure 13·4 shows the variation in Otto-cycle thermal efficiency with both compression ratio and specific-heat ratio k. Efficiency rises rapidly up to compression ratio of about 10; then it grows slowly.

13·6 Ratio relations. Peak pressure and temperature have a simple relation to the compression ratio in the Otto cycle. Let us study the overall cycle 1–2–5–6 in Fig. 13·3a. We can relate peak state to the inlet state as follows:

1. Compression ratio = $V_r = \dfrac{V_1}{V_2} = \dfrac{V_B}{V_5}$

2. Temperature ratio = $T_r = \dfrac{T_5}{T_1}$

3. Pressure ratio = $P_r = \dfrac{P_5}{P_1}$
4. From Boyle's law, since $T_B = T_5$
5. $P_B V_B = P_5 V_5$
6. $\dfrac{P_5}{P_B} = \dfrac{V_B}{V_5} = V_r$
7. From Charles' law, since $V_B = V_1$
8. $\dfrac{P_B}{P_1} = \dfrac{T_B}{T_1} = \dfrac{T_5}{T_1} = T_r$

Combining (6) and (8),

9. $\dfrac{P_5}{P_B}\dfrac{P_B}{P_1} = V_r T_r$
10. $P_r = V_r T_r$

13·7 Work variation. We learned that the thermal efficiency of the Otto cycle depends primarily on the compression ratio. Now let us learn how the compression ratio affects the work output of the cycle.

Figure 13·5 shows five cycles running between common top and bottom temperatures T_c and T_a. The P-V graph shows only three of these because the others run off the top of the graph. As drawn, the cycles have the following ratios:

Cycle		V_r	T_r	P_r
1.	1–9–9–1	1	4	4
2.	1–2–8–10	2.5	4	10
3.	1–3–7–11	12	4	48
4.	1–4–6–12	60	4	240
5.	1–5–5–1	102	4	408

In the first cycle we get zero work output and thermal efficiency, since this consists of a constant-V heating followed by a constant-V cooling. But let us assume that at 1 we first compress the gas through a minute increment at constant S, then heat it at constant V to state 9. At 9 the gas can expand through the minute increment at constant S before it is cooled back to 1 at constant V. This would give us net W and Q areas on the graphs thinner than the lines drawn for the process 1–9–9–1. The point is that the work output and also the thermal efficiency grow from zero as we increase the compression ratio above 1.

In the second cycle we get an appreciable growth in work output as the compression ratio rises. The third cycle has the maximum work output of those shown. As we go on to the fourth and fifth cycles we can see the net work areas diminishing to zero. The last cycle simply consists of constant-S compression followed by a constant-S expansion.

Just before getting to this last cycle we can visualize that at state 5 we add a minute amount of heat at constant V before expanding the gas at constant S. At state 1 we would remove a minute amount of heat at constant V before recompressing at constant S. Again we have a net work area slimmer than the thickness of the line for 1–5.

What happens to the thermal efficiency of these cycles? As we go from the first to the fifth cycles in Fig. 13·5 the *average* temperature at which we add heat becomes higher and higher. Similarly the *average* temperature at which we reject unusable heat becomes lower and lower. In the

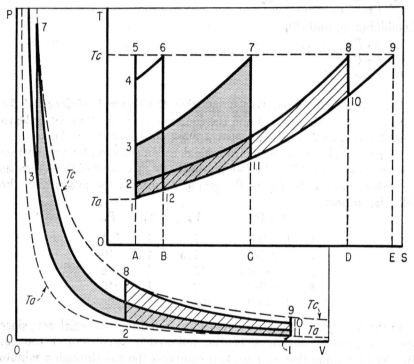

Fig. 13·5 Five Otto cycles working between the same temperature limits but with different compression ratios show that work per cycle grows from zero to a maximum, then diminishes to zero again. Cycle 1–4–6–12 is not shown on the P-V chart to avoid confusion of closely spaced process lines.

first cycle these two average temperatures are equal, and since the work equals zero, the thermal efficiency is zero. In the fifth cycle (with the minute amount of heat added) the average temperatures reach their ultimate limits. So just as the net work vanishes, we approach the Carnot thermal efficiency because we are adding heat at the maximum possible temperature and rejecting it at the lowest possible temperature.

Figure 13·6a and b summarizes the variations in heat transfers, compression work, expansion work, net work, and thermal efficiency for cycles running between 500 and 3000°R. As the compression ratio rises, the heat added Q_A drops quickly at first and then at a decreasing rate. The heat rejected Q_R falls even faster than Q_A and then levels off.

The vertical distance between Q_A and Q_R measures the net heat Q converted to the work output W. This rises to a maximum at a compression ratio of about 20. Practical spark-ignition gasoline-fired engines usually run below compression ratios of 10 to avoid preignition and detonation troubles; more will be said about these later.

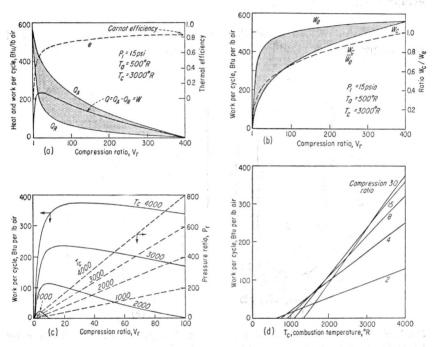

Fig. 13·6 (a) Work per cycle peaks at $V_r = 20$ for given temperatures. (b) Expansion and compression works grow with compression ratio. (c) Rising heat-input temperatures develop more work per cycle and higher pressure ratios. (d) Work per cycle varies directly with the combustion temperature and depends on the compression ratio.

The thermal efficiency of the cycles rises rapidly to a pressure ratio of 20 as does net W and then creeps up slowly to the Carnot efficiency just as W drops to zero at the impractically high V_r of almost 400. Here Carnot efficiency $= (3,000 - 500)/3,000 = 0.833$ or 83.3 per cent.

Figure 13·6b shows that the compression work W_c rises steadily at a diminishing rate as the compression ratio grows. The expansion work W_e grows even more rapidly than W_c. The vertical difference between them equals the net work output of the cycle, plotted in the graph in 13·6a. As V_r grows, the compression work absorbs an increasingly larger part of the expansion work W_e.

Figure 13·6c gives a magnified view of net work variation in the lower range of the compression ratio V_r. It also shows how W varies with the combustion temperature T_c. As T_c rises, the *maximum* W produced occurs at higher V_r. The pressure ratio P_r is directly proportional to temperature T_c and V_r as we learned in Art. 13·6.

Figure 13·6d is simply a cross plot of 13·6c at selected compression ratios. It shows that net work varies linearly with difference of combustion and compression temperatures.

13·8 Water-cooled cycle. So far we have been considering an ideal Otto engine with reversible adiabatic expansions and compressions. The

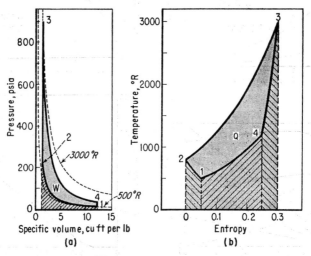

Fig. 13·7 The water-cooled or polytropic Otto cycle rejects heat to the cylinder water jacket during all processes except for the exhaust 4–1.

cylinder of an actual engine would not be able to live under the punishing combustion temperatures and pressures used by the ideal cycle.

To prevent the cylinder from failing (and to preserve the lubricating oil film between cylinder and piston) the cylinder wall is deliberately cooled by a water jacket (or by blowing cooling air over a finned exterior surface). This means that during all processes in the cylinder, the working fluid transfers heat through the cylinder wall. We can study the effect of water cooling by assuming a reversible polytropic cycle as in Fig. 13·7 and Table 13·3.

If you really want to understand how the basic equations are applied, get out your slide rule and check the calculations in the table and the graphs. You will find all the equations explained in Chaps. 6 to 8 for the individual processes, see Appendix F for a summary.

Table 13·3 Performance of water-cooled (polytropic) Otto cycle

1. Cycle specifications; see Fig. 13·7:
   $T_1 = 500°R$ $P_1 = 15$ psia $V_r = 10$ $c_v = 0.225$
   $T_3 = 3000°R$ $k = 1.3$
   Expansion $n = 1.4$
   Compression $n = 1.2$
2. $R = Jc_v(k - 1) = 778 \times 0.225(1.3 - 1) = 52.6$
3. $V_1 = \dfrac{RT_1}{P_1} = \dfrac{52.6 \times 500}{15 \times 144} = 12.18$ cu ft per lb
4. $V_2 = \dfrac{V_1}{V_r} = \dfrac{12.18}{10} = 1.218$ cu ft per lb
5. $P_2 = P_1 \left(\dfrac{V_1}{V_2}\right)^n = 2{,}160 \times 10^{1.2} = 2{,}160 \times 15.86 = 34{,}270$ psfa
6. $T_2 = \dfrac{P_2 V_2}{R} = \dfrac{34{,}270 \times 1.218}{52.6} = 793.2°R$
7. $Q_{12} = c_v \dfrac{k-n}{1-n}(T_2 - T_1) = 0.225 \dfrac{1.3 - 1.2}{1.0 - 1.2}(793.2 - 500) = -33.0$ Btu per lb
8. $W_{12} = \dfrac{P_2 V_2 - P_1 V_1}{1 - n} = \dfrac{34{,}270 \times 1.218 - 2{,}160 \times 12.18}{1.0 - 1.2}$
   $= -77{,}150$ ft-lb per lb, or
   $-77{,}150/778 = -99.1$ Btu per lb
9. $E_2 - E_1 = c_v(T_2 - T_1) = 0.225(793.2 - 500) = 66.0$ Btu per lb
10. Check: $Q_{12} + E_2 - E_1 = 33.0 + 66.0 = 99.0$ Btu per lb $= W_{12}$
11. $Q_{23} = c_v(T_3 - T_2) = 0.225(3{,}000 - 793.2) = 496.4$ Btu per lb
12. Heat transferred to cooling water during process 2–3
    $= Q'_{23} = 0.1\, Q_{23} = 0.1 \times 496.4 = 49.6$ Btu per lb
13. $Q_A = Q_{23} + Q'_{23} = 496.4 + 49.6 = 546.0$ Btu per lb
14. $P_3 = \dfrac{RT_3}{V_3} = 52.6 \times \dfrac{3{,}000}{1.218} = 129{,}600$ psfa
15. $T_4 = T_3 \left(\dfrac{V_3}{V_4}\right)^{n-1} = T_3 \left(\dfrac{V_2}{V_1}\right)^{n-1} = T_3 \left(\dfrac{1}{V_r}\right)^{n-1}$
    $= 3{,}000 \times 0.1^{1.4-1.0} = 3{,}000 \times 0.398 = 1194°R$
16. $P_4 = \dfrac{RT_4}{V_4} = 52.6 \times \dfrac{1{,}194}{12.18} = 5{,}156$ psfa
17. $Q_{34} = c_v \dfrac{k-n}{1-n}(T_4 - T_3) = 0.225 \dfrac{1.3 - 1.4}{1.0 - 1.4}(1194 - 3{,}000) = -101.6$ Btu per lb
18. $W_{34} = \dfrac{P_4 V_4 - P_3 V_3}{1-n} = \dfrac{5{,}156 \times 12.18 - 129{,}600 \times 1.218}{1.0 - 1.4} = 237{,}500$ ft-lb per lb, or
    $\dfrac{237{,}500}{778} = 305.2$ Btu per lb
19. $E_3 - E_4 = c_v(T_3 - T_4) = 0.225(3{,}000 - 1194) = 406.3$ Btu per lb
20. Check: $E_3 - E_4 = Q_{34} + W_{34} = 101.6 + 305.2 = 406.8$ Btu per lb
21. $Q_{41} = c_v(T_1 - T_4) = 0.225(500 - 1194) = -156.2$ Btu per lb
22. $Q = Q_{23} - Q_{12} - Q_{34} - Q_{41} = 496.4 - 33.0 - 101.6 - 156.2 = 205.6$ Btu per lb
23. $W = W_{34} - W_{12} = 305.2 - 99.1 = 206.1$ Btu per lb
24. Heat rejected to cooling water:
    $Q_{cw} = Q_{12} + Q'_{23} + Q_{34} = 33.0 + 49.6 + 101.6 = 184.2$ Btu per lb
25. $W = Q = Q_A - Q_{cw} - Q_{41} = 546.0 - 184.2 - 156.2 = 205.6$ Btu per lb
26. $e = \dfrac{W}{Q_A} = \dfrac{205.6}{546.0} = 0.376$ or 37.6%

(Continued)

Table 13-3 Performance of water-cooled (polytropic) Otto cycle (*Continued*)

27. $\text{Mep} = \dfrac{W}{V_D} = \dfrac{W_{34} - W_{12}}{V_1 - V_2} = \dfrac{237{,}500 - 77{,}150}{12.18 - 1.218} = 14{,}630 \text{ psf}$

$\qquad\qquad\qquad\qquad\text{or } \dfrac{14{,}630}{144} = 101.7 \text{ psi}$

28. $S_2 - S_1 = c_v \dfrac{k - n}{1 - n} \log_e \dfrac{T_2}{T_1} = 0.225 \dfrac{1.3 - 1.2}{1.0 - 1.2} \log_e \dfrac{793.2}{500}$

$\qquad\qquad = -0.1125 \times 0.4613 = -0.05188$

29. $S_3 - S_2 = c_v \log_e \dfrac{T_3}{T_2} = 0.225 \log_e \dfrac{3{,}000}{793.2} = 0.2992$

30. $S_4 - S_3 = c_v \dfrac{k - n}{1 - n} \log_e \dfrac{T_4}{T_3} = 0.225 \dfrac{1.3 - 1.4}{1.0 - 1.4} \log_e \dfrac{1{,}194}{3{,}000}$

$\qquad\qquad = 0.05625(-0.922) = -0.05185$

31. $S_1 - S_4 = c_v \log_e \dfrac{T_1}{T_4} = 0.225 \log_e \dfrac{500}{1194} = -0.1959$

32. Check: $S_3 - S_2 = (S_2 - S_1) + (S_4 - S_3) + (S_1 - S_4)$
$\qquad 0.2992 = 0.05188 + 0.05185 + 0.1959 = 0.2996$

The small difference in the check values between Eqs. (8) and (10), (19) and (20), and (32) is due to slide-rule errors. In Fig. 13·7b the hatched area under 1–2 measures the heat transferred to the cooling water during compression at $n = 1.2$; the hatched area under 3–4 measures heat transferred to cooling water during expansion at $n = 1.4$. The hatched area under 4–1 measures the heat rejected with the exhaust gas at constant V.

Heat rejected during the constant-volume heating 2–3 cannot be shown on the graph, and it is figured as a separate estimate in Eq. (12) in Table 13·3. The gray area under 2–3 in Fig. 13·7b measures only the heat absorbed by the working fluid and not that simultaneously rejected to the cooling water.

The net areas Q and W are equal, measuring the net work output of the engine, since these are all reversible processes. Equation (26) shows that this cycle has a thermal efficiency of 37.6%. In comparison an ideal Otto cycle with constant-S processes and the same $V_r = 10$ and $k = 1.3$ has a theoretical thermal efficiency of 49.8%. To prolong the life of the engine we sacrifice $49.8 - 37.6 = 12.2\%$ in efficiency—not too great a penalty. Actual engine irreversible processes make this efficiency penalty even larger.

Figure 13·8 summarizes the calculations of Table 13·3 in energy-flow form. In Fig. 13·8a, of the 546 Btu in the fuel (per pound of air), 49.6 Btu is rejected to the cooling water. Of the 99.1 Btu returned by the flywheel to compress the air, 33.0 Btu is rejected to the cooling water. The sum of the remaining energies equals 562.5 Btu, which raises the internal energy of the air in the cylinder during the compression and combustion processes.

In Fig. 13·8b, the 562.5 Btu internal energy acquired by the working fluid (combustion products) helps develop the net work output of the engine after satisfying the unavoidable losses. Heat transfer during expansion from the hot working fluid takes 101.6 Btu; the hot exhaust gases take 156.2 Btu with them into the atmosphere. This leaves 305.2 Btu to develop work on the piston during the expansion stroke; of this,

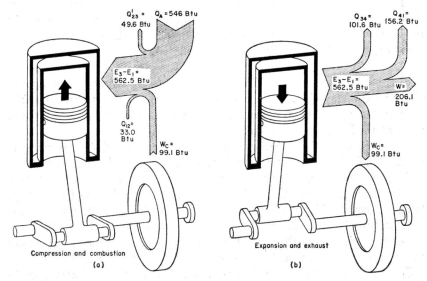

Fig. 13·8 (a) Energy input to the Otto engine during compression and combustion comes from the fuel and the flywheel with some rejected to cooling water. (b) Energy output during expansion and exhaust goes to the jacket water, flywheel, exhaust, and net mechanical work.

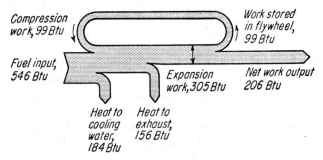

Fig. 13·9 Assuming that the Otto engine has steady energy flow, this diagram indicates that flywheel-compression energy circulates in the engine.

99.1 Btu goes to the flywheel to be stored for compression of the fresh charge during the next cycle. Finally 206.1 Btu leaves the engine as mechanical shaft work—this is the useful product of the original 546 Btu input from the fuel.

Figure 13·8 gives the energy-flow analysis from the standpoint of the individual strokes of the reciprocating engine. Most engines have several cylinders working on the same shaft, developing a practically continuous power output. So we can regard an i-c engine as a continuous-flow machine. With this idea we can develop the energy-flow diagram in Fig. 13·9. Here we see the 546 Btu fuel input distributed to cooling water, exhaust heat, and net work output. The compression work merely circulates between the cylinder and flywheel without contributing directly to the net work output. We know, of course, that the engine could not work without this internally circulating energy.

13·9 Salvaging energy. We know by now that to convert a *maximum* of heat to work, *all* the heat must enter a cycle at the highest possible temperature; also, *all* the unavailable heat must be rejected from the cycle at the lowest possible temperature.

The Otto cycle, Fig. 13·10a, cannot meet these conditions because it absorbs and rejects heat at constant volume, that is, variable temperature. This makes the *average* temperature at which Q_a enters much lower than the top temperature of the cycle T_3 (see the T-S graph) and the average temperature at which it rejects unused heat much higher than the lowest temperature T_1. Because of its simple, practical nature we use the Otto cycle even though it throws away a considerable amount of available energy measured by area 1–4–A–1 on the T-S graph of Fig. 13·10a.

Since $P = C$ is the atmospheric-pressure line on the T-S graph, state A is at a pressure much below atmospheric. There is no practical way of expanding the exhaust gas at state 4 to state A to salvage all the unused available energy as work. But state 5 is at atmospheric pressure, so if we can find a way to make the gas in the cylinder do work in expanding from state 4 down to atmospheric pressure at state 5, we can get the extra work output measured by the area 1–4–5–1 on both the T-S and P-V graphs.

Increasing the piston stroke to V_5 is impractical, but we can pass the exhaust gas through a small turbine and make it do work like wind on a windmill. We have not studied steady-flow processes yet, so you will have to take our word that this can be done. This arrangement gives us a compound engine, developing main work output through the piston-driven main engine shaft, and generating a smaller "byproduct" work output through the turbine shaft.

There is one difficulty with this setup—it is not practical to couple the shafts together to drive a common load because the turbine usually turns much faster than the main engine shaft. Also, we seldom find two loads in the right proportion that vary in step so they can use the main engine and turbine outputs individually. So how can we use the turbine power?

13·10 Supercharging. The extra work developed by the exhaust turbine makes it possible to increase the total work produced in the engine

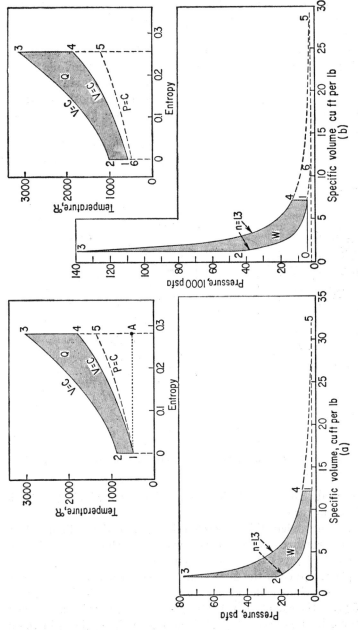

Fig. 13·10 (a) The basic Otto cycle wastes theoretical available energy measured by area 1–4–A on the T-S graph. The part that could be recovered by expansion of the exhaust is 1–4–5. (b) The supercharged Otto cycle raises peak pressure and temperature and produces more work.

cylinder. Figure 13·11 shows a four-cylinder Otto engine using a supercharger which consists of an exhaust turbine driving a centrifugal compressor.

The four cylinders are displaced in timing by a quarter cycle, so the crankshaft gets a power stroke every half revolution. The four cylinders discharge a pulsating flow of exhaust gas into the exhaust manifold, which acts as a receiver to damp pressure pulsations. This smooths out the flow of gas to a steady stream through the turbine, so it generates a steady shaft-work output.

The turbine shaft drives the centrifugal compressor, which inhales a steady flow of air and pressurizes it to double atmospheric or more. The

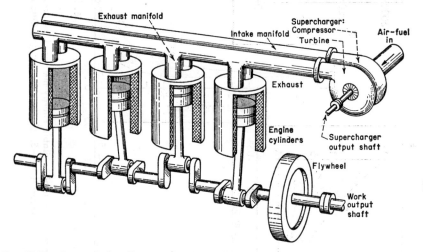

Fig. 13·11 Four-cylinder Otto engine has individual cycles phased so the crankshaft gets a power stroke every half revolution. Exhaust gas flows through a turbine that generates shaft work to drive the centrifugal compressor that pressurizes the air input to the engine cylinders.

pressurized intake air enters the intake manifold from which the cylinders draw their air in turn.

Let us see just what this supercharging device does. Figure 13·10a shows the performance of an Otto cycle working with a compression ratio of 6.0; Table 13·4 gives the computations for this cycle. Figure 13·10b shows the performance of the same engine fitted with a supercharger that pressurizes the cylinder intake air to double atmospheric at state 1. Table 13·5 gives the calculations for the supercharged engine and assumes that the same heat *per pound of air* is added to the cycle as in the unsupercharged engine in Fig. 13·10a.

In Table 13·4 we figure the additional work that can be generated by an exhaust turbine shown as area 1–4–5 in Fig. 13·10a. Item 11 figures the gross work generated by expanding the exhaust to atmospheric pressure

INTERNAL-COMBUSTION-ENGINE CYCLES   189

at constant S, from state 4 to 5. But, remember, in the equivalent closed cycle the gas must be returned to its original starting state. So after expanding to P_5 (atmospheric), the gas must be cooled back to state 1 at constant pressure. To do this we must do work on the gas during the constant-pressure cooling from V_5 to V_1 in compressing it to the smaller volume. This is figured in Item 14.

Item 15 gives the net work w that could be generated by an exhaust turbine. For the conditions specified this equals 21.9% of the net work W generated by the main engine as shown in Item 17.

Table 13·4 Performance of Otto cycle without supercharging

1. Cycle specifications, see Fig. 13·10a:
   $T_1 = 500°R$ $T_3 = 3000°R$ $T_r = 3000/500 = 6.0$
   $V_r = 6.0$ $k = 1.3$ $c_v = 0.225$
   $Q_A = 482.3$ Btu per lb $R = 52.6$
   $V_1 = 12.18$ cu ft per lb $P_1 = 2160$ psfa

2. $V_2 = \dfrac{V_1}{V_r} = \dfrac{12.18}{6.0} = 2.03$ cu ft per lb

3. $P_2 = P_1 V_r^k = 2160 \times 6^{1.3} = 2160 \times 10.28 = 22{,}200$ psfa

4. $T_2 = \dfrac{P_2 V_2}{R} = 22{,}200 \times \dfrac{2.03}{52.6} = 856°R$

5. $Q_A = c_v(T_3 - T_2) = 0.225(3000 - 856) = 482.3$ Btu per lb

6. $P_3 = P_2 \dfrac{T_3}{T_2} = 22{,}000 \dfrac{3000}{856} = 77{,}800$ psfa

7. $P_r = \dfrac{P_3}{P_1} = V_r T_r = \dfrac{77{,}800}{2160} = 6 \times 6 = 36$

8. $P_4 = P_3 \left(\dfrac{1}{V_r}\right)^k = 77{,}800 \left(\dfrac{1}{6}\right)^{1.3} = 7565$ psfa

9. $T_4 = \dfrac{P_4 V_4}{R} = 7565 \times \dfrac{12.18}{52.6} = 1752°R$

10. $T_5 = T_3 \left(\dfrac{P_5}{P_3}\right)^{(k-1)/k} = 3000 \left(\dfrac{2160}{77{,}800}\right)^{0.3/1.3}$
    $= 3000(1/36)^{0.2308} = 1313°R$

11. $Q_{45} = c_v(T_4 - T_5) = 0.225(1752 - 1313) = 98.8$ Btu per lb $= W_{45}$

12. $Q_{41} = c_v(T_4 - T_1) = 0.225(1752 - 500) = 281.1$ Btu per lb $= Q_R$

13. $V_5 = \dfrac{RT_5}{P_5} = 52.6 \times \dfrac{1313}{2160} = 31.97$ cu ft per lb

14. $W_{15} = P_1(V_5 - V_1)/J = 2160 \dfrac{31.97 - 12.18}{778} = 54.95$ Btu per lb

15. $w = W_{45} - W_{15} = 98.8 - 55.0 = 43.8$ Btu per lb available for supercharging

16. $W = Q = Q_A - Q_R = 482.3 - 281.8 = 200.5$ Btu per lb

17. $\dfrac{w}{W} = \dfrac{43.8}{200.5} = 0.219$ or 21.9%

18. $e = \dfrac{Q}{Q_A} = \dfrac{200.5}{482.3} = 0.4155$ or 41.55%

When we apply this byproduct power output in a supercharger, we find the states of the working fluid completely changed throughout the cycle as in Fig. 13·10b, and Table 13·5. The cycle really starts at state 6,

Table 13·5 Performance of supercharged Otto cycle

1. Cycle specification, see Fig. 13·10b:
   Supercharging pressure ratio $P_1/P_6 = 2.0$
   $Q_A = 482.3$ Btu per lb (same as uncharged cycle)
   $V_r = 6$ $k = 1.3$ $c_v = 0.225$ Btu per lb
   $R = 52.6$ $P_6 = P_5 = 2{,}160$ psfa
   $V_c = 12.18$ cu ft cylinder (same as V_1 for uncharged cycle numerically, note V_1 is specific volume of gas, but V_c is space volume in cylinder)

2. $T_1 = T_6 \left(\dfrac{P_1}{P_6}\right)^{(k-1)/k} = 500 \times 2^{0.2308} = 586.8°R$

3. $M = \dfrac{P_1 V_c}{R T_1} = \dfrac{4320 \times 12.18}{52.6 \times 586.8} = 1.705$ lb of working fluid per cycle

4. $P_2 = P_1 \left(\dfrac{V_1}{V_2}\right)^k = 4320 \times 6^{1.3} = 44{,}400$ psfa

5. $V_1 = \dfrac{V_c}{M} = \dfrac{12.18}{1.705} = 7.14$ cu ft per lb

6. $V_2 = \dfrac{V_1}{V_r} = \dfrac{7.14}{6} = 1.19$ cu ft per lb

7. $T_2 = \dfrac{P_2 V_2}{R} = 44{,}400 \times \dfrac{1.19}{52.6} = 1005°R$

8. $T_3 = \dfrac{Q_A + c_v T_2}{c_v} = \dfrac{482.3 + 0.225 \times 1005}{0.225} = 3148°R$

9. $P_3 = P_2 \dfrac{T_3}{T_2} = 44{,}400 \dfrac{3148}{1005} = 139{,}000$ psfa

10. $P_4 = P_3 \left(\dfrac{V_3}{V_4}\right)^k = 139{,}000 \left(\dfrac{1}{6}\right)^{1.3} = 13{,}520$ psfa

11. $T_4 = \dfrac{P_4 V_4}{R} = 13{,}520 \times \dfrac{7.14}{52.6} = 1835°R$

12. $Q_R = Q_{41} = c_v(T_4 - T_1) = 0.225(1835 - 587) = 280.8$ Btu per lb

13. $Q = Q_A - Q_R = 482.3 - 280.8 = 201.5$ Btu per lb $= W$

14. $\Sigma Q = QM = 201.5 \times 1.705 = 343.6$ Btu per cycle

15. $T_r = \dfrac{T_3}{T_1} = \dfrac{3148}{586.8} = 5.365$

16. $P_r = \dfrac{P_3}{P_1} = T_r V_r = \dfrac{139{,}000}{4320} = 5.36 \times 6 = 32.16$

17. $T_5 = T_3 \left(\dfrac{P_3}{P_5}\right)^{(1-k)/k} = 3148 \left(\dfrac{139{,}000}{2160}\right)^{-0.2308} = 1204°R$

18. $Q_{45} = c_v(T_4 - T_5) = 0.225(1835 - 1204) = 142.0$ Btu per lb

19. $\Sigma Q_{45} = Q_{45} M = 142.0 \times 1.705 = 242.2$ Btu per cycle

20. $V_5 = \dfrac{R T_5}{P_5} = 52.6 \times \dfrac{1204}{2160} = 29.3$ cu ft per lb

21. $\Sigma V_5 = M V_5 = 1.705 \times 29.3 = 50.0$ cu ft per cycle

22. $W_{56} = P_5 \dfrac{V_5 - V_6}{J} = 2160 \dfrac{29.3 - 12.18}{778} = 47.6$ Btu per lb

23. $W_{61} = c_v(T_1 - T_6) = 0.225(586.8 - 500) = 19.5$ Btu per lb

24. $w = W_{45} - W_{56} - W_{61} = 142.0 - 47.6 - 19.5 = 74.9$ Btu per lb

25. $\Sigma w = Mw = 1.705 \times 74.9 = 127.7$ Btu per cycle

26. $\dfrac{\Sigma w}{\Sigma Q} = \dfrac{127.7}{343.6} = 0.372$ or 37.2%

27. $\dfrac{\Sigma Q}{Q_1} = \dfrac{343.6}{200.5} = 1.713$ or 71.3% more work output available at turbine shaft than *unsupercharged* cycle

28. $e = \dfrac{Q}{Q_A} = \dfrac{201.5}{482.3} = 0.4175$ or 41.75%

where the air enters the centrifugal compressor at atmospheric conditions, the same as in Fig. 13·10a. The compressor delivers the air at state 1 into the cylinder at just double atmospheric pressure; that is, $P_1/P_6 = 2$. We assume that 6–1 is a constant-S compression.

Since we are using the same engine as in Fig. 13·10a, the cylinder has the same volume. But at a higher pressure it can hold a greater mass of air because the air has a smaller specific volume. Item 3 in Table 13·5 figures the mass in pounds per cycle. Note that it is not double, even though the pressure is double, because the constant-S compression raises the temperature of the air forced into the cylinder.

Throughout the supercharged cycle we find that pressures and temperatures are higher and the specific volumes smaller. Primarily because of the greater mass of air passing through the engine we find that it produces a greater work output per cycle as shown in Item 14, Table 13·5, compared with Item 16, Table 13·4. There is a slight gain in W per pound (201.5 versus 200.5 Btu) because of the higher average temperature at which heat Q_A enters the cycle; this makes the supercharged cycle slightly more efficient—compare Item 18, Table 13·4, with Item 28, Table 13·5.

Studying the T-S graph of Fig. 13·10b we find that the net work output that could be produced by the supercharger (beyond what is used by the compressor) is measured by area 6–1–4–5. This is considerably larger than that produced by the turbine in the unsupercharged engine in Fig. 13·10a, measured by area 1–4–5.

Item 26, Table 13·5, shows that the supercharger can produce byproduct power equal to 37.2 per cent of the work produced per cycle. Compared with the unsupercharged engine it produces 71.3% more work, Item 27. So we find that supercharging increases not only the work output per cycle of an engine but also the amount of byproduct work it could produce in the exhaust turbine.

Figure 13·12 summarizes the changes in Otto-cycle states, as numbered in Fig. 13·10, for supercharging pressure ratios ranging from 1.0 to 6.0. This assumes, as in Tables 13·4 and 13·5, that the heat input *per pound of air* stays constant. Figure 13·12a shows moderate changes in temperature with increasing supercharging. The top temperature of the cycle, T_3, rises by 532 F over the base cycle (supercharging ratio = 1), less than 18% of the base 3000°R. But since T_1 also rises with supercharging ratio, we find that this works out to only a minor rise in cycle thermal efficiency, from 41.6 to 42.3% over the range shown.

Highly supercharged engine cylinders and pistons would have to be made of metals that could withstand the higher intermittent top temperatures. This would not be too knotty a problem. But the rise in working fluid, to 3.7 lb per cycle at the supercharging ratio of 6, proves to be the key to increased work output per cylinder or per cycle—the basic advantage of supercharging.

Figure 13·12b shows a tremendous rise in top pressure, P_3, with supercharging, about 4.3 times the base cycle at a supercharging ratio of 6. This means that we would have to build the cylinder wall much heavier and use generally heavier construction with supercharging but the engine dimensions would be about the same. The rise in exhaust pressure P_4 contributes to the rise in byproduct power of an exhaust turbine working with a supercharged engine, but more pounds of air per cycle is the controlling factor.

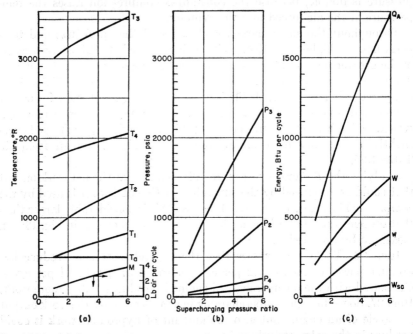

Fig. 13·12 (a) Supercharging pressure ratio rise causes all cycle temperatures to rise moderately. The biggest change is in pounds of air passing through the cylinder. (b) Pressures rise sharply with supercharging. (c) Supercharging boosts heat input and work output.

Figure 13·12c shows the tremendous rise in heat added per cycle Q_A caused by the greater amount of working fluid passing through the cylinder of fixed dimensions and compression ratio. This causes a corresponding rise of 370% in main work output W. Concurrently, we have a rise in byproduct work w that could be generated by an exhaust turbine. This jumps from 21.5% of W for the base cycle to 51% for the supercharging ratio of 6.

Despite this great rise in availability of byproduct power, the work of supercharging needed by the compressor W_{sc} is only 9.3% of the main work W or 18% of the byproduct work w at the supercharging ratio of 6.

The supercharger uses only a small part of the work that could be developed by the turbine.

This waste of available energy by supercharged cycles has led to the development of the free-piston compressor and the free-piston gas generator which we shall discuss later.

13·11 Diesel cycle. The cycle which we will study now has many similarities to the Otto cycle, differing primarily in the way it receives heat energy.

In Fig. 13·13 the induction stroke 0–1 inhales a charge of air into the cylinder. From 1 to 2 the piston compresses the air at constant-S, so

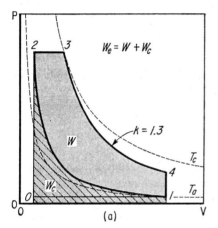
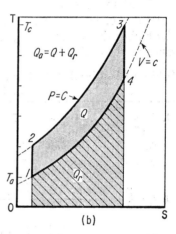

Fig. 13·13 Diesel-engine cycle resembles the Otto cycle but differs in the heat-adding process by absorbing energy at constant pressure from state 2 to state 3 while doing shaft work.

its temperature rises to a level high enough to ignite the distillate oil that fuels the engine. From 2 to 3 oil injected into the hot compressed air ignites and burns as it enters; at the same time the air and combustion products expand so their pressure stays constant. Since this is a constant-pressure process, the internal energy of the mixture of air and combustion products rises and simultaneously does work on piston.

During 3–4 the hot gases expand at constant S doing work on the piston. At 4 the exhaust valve opens and most of the hot gases blow off to atmosphere until cylinder pressure drops to atmospheric at 1. During the exhaust stroke 1–0 the piston pushes the remaining gases out of the cylinder; then the engine starts its next cycle.

As with the Otto cycle we can analyze the ideal diesel cycle by assuming that 1 lb of air stays in the cylinder at all times. Then 2–3 is a simple constant-pressure heating process and 4–1 a constant-volume cooling process.

Heat Q_a is added to the cycle during 2–3, and heat Q_r is rejected by the cycle during 4–1. By the first law the net shaft work output

$$W = Q = Q_a - Q_r$$

By a parallel analysis the net shaft work $W = W_e - W_c$, the difference of the expansion work done on the piston and the compression work done by the piston. Figure 13·13 shows these areas.

13·12 Thermal efficiency. The thermal efficiency of the diesel cycle is one of our prime interests, so let us see what basic relationships we can develop. In Table 13·6 we derive Eq. (4) for the thermal efficiency in terms of the end process temperatures. As always, the average temperatures of heat addition and rejection control the thermal efficiency; the higher the upper average and the lower the lower average, the more efficient the cycle.

Table 13·6 Diesel-cycle thermal efficiency depends on compression ratio and fuel-cutoff ratio

1. $Q_a = E_3 - E_2 + W_{23} = c_p(T_3 - T_2)$
   (See Fig. 13·13)
2. $Q_r = E_4 - E_1 = c_v(T_4 - T_1)$
3. $W = Q_a - Q_r = c_p(T_3 - T_2) - c_v(T_4 - T_1)$
4. $e = \dfrac{W}{Q_a} = \dfrac{c_p(T_3 - T_2) - c_v(T_4 - T_1)}{c_p(T_3 - T_2)}$

   $= 1 - \dfrac{T_4 - T_1}{k(T_3 - T_2)}$

For the constant-S processes:

5. $T_2 = T_1\left(\dfrac{V_1}{V_2}\right)^{k-1} = T_1 V_r^{k-1}$
6. $T_4 = T_3\left(\dfrac{V_3}{V_4}\right)^{k-1}$

For the constant-P process:

7. $\dfrac{V_3}{V_2} = \dfrac{T_3}{T_2} = V_c$ (fuel-cutoff ratio by definition)
8. $T_3 = T_2 V_c$

Substituting from (5),

9. $T_3 = T_1 V_r^{k-1} V_c$

Substituting in (6),

10. $T_4 = T_1 V_r^{k-1} V_c \left(\dfrac{V_3}{V_4}\right)^{k-1}$

But from (7)

11. $V_3 = V_2 V_c$

Substituting in (10) and noting that $V_4 = V_1$,

12. $T_4 = T_1 V_r^{k-1} V_c \left(\dfrac{V_2 V_c}{V_1}\right)^{k-1}$

   $= T_1 V_r^{k-1} V_c \left(\dfrac{V_c}{V_r}\right)^{k-1} = T_1 V_c^k$

Substituting for T_2, T_3, and T_4 in (4) from (5), (9), and (12),

13. $e = 1 - \dfrac{T_1 V_c^k - T_1}{k(T_1 V_r^{k-1} V_c - T_1 V_r^{k-1})}$

   $= 1 - \dfrac{V_c^k - 1}{k V_r^{k-1}(V_c - 1)}$

   $= 1 - \dfrac{1}{V_r^{k-1}}\left[\dfrac{V_c^k - 1}{k(V_c - 1)}\right]$

But as in the Otto cycle, these temperatures are controlled by the mechanical dimensions and ratios of the cylinder volume and piston displacement. Two ratios control in the diesel cycle: the compression ratio $V_r = V_1/V_2$ and the fuel-cutoff ratio $V_c = V_3/V_2$, Fig. 13·13.

The derivation in Table 13·6 works up relations for all the temperatures T_2, T_3, and T_4 in terms of T_1 and the two volume ratios. This is done in Eqs. (5), (9), and (12). Making these substitutions we find Eq. (13) which shows that thermal efficiency of the theoretical diesel cycle depends only on the compression ratio, fuel-cutoff ratio, and the ratio of specific heats k.

When the quantity in the bracket equals 1, the diesel efficiency equals the Otto efficiency. This happens when we have a cycle where at state 2 a minute amount of Q_a is added. Then, expanding the gas to state 1, we reject a minute amount of heat Q_r. The amount of work developed is practically zero, but it is produced at the best efficiency (Otto cycle) for the given compression ratio and practically unity fuel-cutoff ratio.

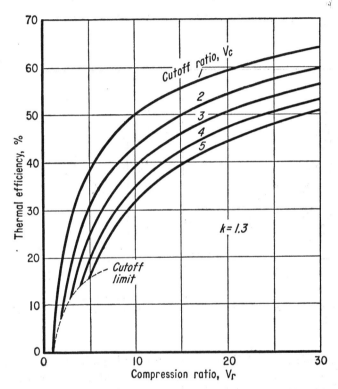

Fig. 13·14 Diesel-cycle thermal efficiency equals the Otto-cycle efficiency at unity cutoff ratio but drops with increasing cutoff ratio.

But as the fuel-cutoff ratio grows, the efficiency at a given compression ratio drops. So we see that the diesel cycle has a lower efficiency than the Otto cycle for the same compression ratio. Figure 13·14 shows how the thermal efficiency of the diesel cycle varies when $k = 1.3$. The thermal efficiency rises rapidly at low compression ratios but above about 20 the rate of rise tapers off. For increasing fuel-cutoff ratios the increment drop in efficiency decreases. In practical diesel engines the maximum cutoff ratios run about 2.5.

In Fig. 13·13a we can define another ratio called the expansion ratio, $V_e = V_4/V_3 = V_1/V_3$. The three ratios are then related as $V_e V_c = V_r$, since $(V_1/V_3)(V_3/V_2) = V_1/V_2$. To maintain high efficiency in a diesel engine the expansion ratio should be kept high and fuel-cutoff ratio low; this relation is automatic, since their product equals the compression ratio, a design factor of every engine.

13·13 Load control. Load control for the diesel cycle obviously depends on the fuel-cutoff ratio which depends on the amount of fuel injected. Figure 13·15 shows how the net work output and top temperature varies with cutoff.

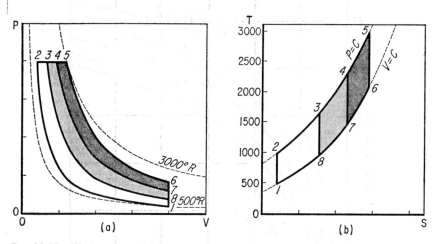

Fig. 13·15 Work output of the diesel cycle varies by controlling the amount of heat injection because a constant amount of air passes through the cylinder regardless of the engine load.

With cutoff at state 3, the net work produced and heat converted are the areas 1–2–3–8. For cutoff at state 4, work and heat increase to areas 1–2–4–7. Cutoff at state 5 raises work and heat to areas 1–2–5–6. But as Fig. 13·14 shows, getting more work per cycle is done at the expense of lower thermal efficiency, since the compression ratio is constant in a given engine.

Delaying cutoff to obtain higher output also raises the top cycle temperature, as shown in Fig. 13·15b. This would seem to point to the possibility of higher thermal efficiency because this raises the *average* temperature at which Q_a is injected into the cycle. But, unfortunately, the *average* temperature at which Q_r is rejected from the cycle also rises and at a faster rate than the temperature for Q_a. So we end up with an overall drop in cycle thermal efficiency.

INTERNAL-COMBUSTION-ENGINE CYCLES

13·14 Supercharging. We can supercharge a diesel-engine cylinder in the same way we do an Otto engine. As in the Otto engine we do not raise thermal efficiency by salvaging some of the available energy in the exhaust, but we do raise the work output yielded by a given cylinder and piston. Theoretically, we can get a large byproduct output of shaft work through an exhaust turbine for high supercharging ratios and raise the thermal efficiency in this way.

13·15 Preignition. The smooth addition of Q_a in the theoretical diesel and Otto cycles is a greatly simplified assumption of the actual way that heat develops in an engine cylinder. The spark-ignited Otto engine inhales a mixture of air and gasoline vapor and compresses it.

To prevent preignition of the fuel-air mixture the compression temperature must be kept below the ignition temperature. An electric spark ignites the mixture at or near the end of the compression stroke. To prevent preignition the compression ratio of the engine must be kept below about 11 or 12. Modern automobile engines have compression ratios ranging from about 6 to 10; this limits the maximum efficiency at which an Otto engine can work.

In contrast to the Otto engine, the diesel engine depends on compressing the air enough to heat it to the ignition temperature of the fuel. Ideally, the fuel starts burning as soon as it mixes with the compressed air at the beginning of the constant-pressure process. To achieve ignition temperatures, diesel engines must have compression ratios of at least 12. Practical diesel-engine compression ratios range from about 12 to 22.

Because of the fuel-ignition limitations of the Otto engine and the low compression ratios that must be used, we find *actual* diesel *engines* more efficient than automotive Otto *engines*, despite the Otto *cycle* being more efficient than the diesel. The practical compression ratios for the engines explain the paradox.

13·16 Otto detonation. Spark-ignited gasoline-fired engines must be designed to avoid detonation or knocking—a phenomenon not anticipated in the simple Otto cycle. Figure 13·16a to g shows schematically what takes place in the head end of the cylinder after compression when the fuel vapor burns in the compressed air.

In *a*, the spark ignites the gasoline-air mixture surrounding it. In *b* through *e*, the flame front spreads more or less smoothly through the fuel-air mixture, which burns rapidly as the flame front passes through. The sharp temperature rise of the combustion products quickly raises the pressure of both the burned and unburned parts of the fuel-air mixture.

This rise in pressure boosts the temperature of the unburned portion of the fuel-air mixture. This may be enough to ignite it before the flame front passes through. This causes an explosive rise in pressure (detonation) that strains the engine, causes audible knocking, and produces local hot spots on the cylinder wall and piston face.

Figure 13·16h to n shows the changes in volume of the different parts of the air-fuel mixture and assumes that the total volume stays constant.

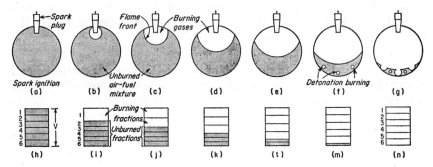

Fig. 13·16 Looking down into an Otto-engine cylinder during burning, we see the flame front moving uniformly through the fuel-air mixture, a to g. Block graphs h to n show that the last part of the charge to burn (6) is highly compressed into a small volume and may cause knocking.

As part 1 burns, it expands and squeezes the other five unburned parts into a smaller volume while the total volume stays constant. This raises both pressure and temperature of the unburned parts. As additional parts burn, the remaining unburned fractions continue rising in temperature and pressure. The last fraction (6) experiences the highest pressure and temperature rise of all. As we said, if temperature exceeds ignition level, this fraction burns explosively.

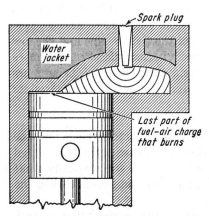

Fig. 13·17 Cylinder-head shape helps to cool the last part of the fuel-air charge to burn.

This shows that detonation in a gasoline engine develops from the uncontrolled burning of the last part of the fuel-air charge. We can prevent this if we find ways of keeping the compression temperature of the last unburned part somewhere below its ignition temperature.

We use two ways. One consists of designing the cylinder head so the last part of the charge spreads itself thinly over a cool heat-transfer surface, Fig. 13·17. This cools the last fraction enough to prevent it from detonating and allows the flame front to move through it smoothly so that the cylinder pressure rises steadily without abrupt surge.

The other method consists of treating gasoline to raise its ignition level.

This characteristic is measured by an arbitrary scale of *octane rating*. Higher octane-number gasolines are less likely to detonate. Both methods are ordinarily used to control detonation. Higher compression engines need higher octane gasolines.

13·17 Diesel detonation. Diesel engines also have knocking troubles, but for a reason different from Otto engines. Figure 13·18a to f shows schematically the burning procedure at constant pressure in the cylinder head as we look down into it. Pistons and cylinder heads of diesel engines usually have special shapes to promote turbulent motion of the air when it is compressed—not shown in Fig. 13·18.

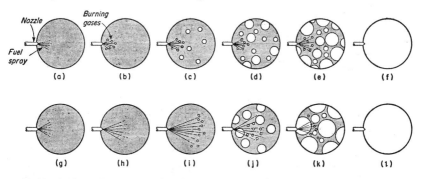

Fig. 13·18 Looking down into a diesel-engine cylinder during burning, a to f, we see the fuel spray igniting right after it mixes with turbulent air to burn at uniform rate. In g to l delayed ignition of the fuel causes an explosive pressure rise because of accumulated fuel in the cylinder.

In Fig. 13·18a the fuel has just begun to spray into the turbulent high-temperature h-p air; in b some fuel-air mixture "cells" start burning. In c the turbulence has carried these growing cells into other parts of the cylinder, but new burning cells keep forming at the nozzle as more fuel sprays into the cylinder.

This action produces a uniform burning rate at constant pressure until all the fuel is consumed at f. The action depends on prompt ignition of the fuel as it enters the hot air; fuel must have low enough ignition temperature.

When the ignition temperature of the fuel is too high, quite a bit of fuel enters during Fig. 13·18g to i before the first part to enter ignites. Once it does and starts burning, the temperature of the later fuel already in the cylinder rises rapidly. This fuel then ignites and burns at an explosive rate, causing detonation or knocking. Pressure cannot be held constant with this kind of burning, so we get a knock.

So we see that late ignition of the first part of the fuel to enter the cylinder causes diesel detonation. The fuel must have a relatively low ignition temperature to avoid knocking.

This characteristic is measured by an arbitrary scale of *cetane rating*. Higher cetane-number diesel or distillate oils are less likely to detonate. Lower compression diesel engines generally run satisfactorily on lower cetane-number oils.

13·18 Rotary engines. Reciprocating engines have proved themselves overwhelmingly as shown by the millions used world-wide today. Despite this good record there have always been efforts to design engines

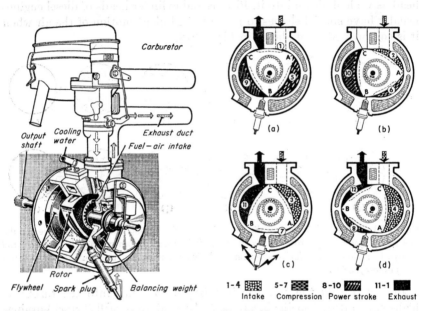

Fig. 13·19 Rotary engine has two moving parts, rotor and output shaft, and can be made to work on the diesel cycle by using fuel injection.

Fig. 13·20 Engine rotor turns through a planetary path guided by the casing meshing with the output shaft through the internal gear and pinion.

that would minimize the wide swing in forces that i-c engines must handle. Also, i-c engines have many moving parts, all needing adjustment and maintenance at times.

Efforts to make gas engines that do not reciprocate date as far back as 1699. James Watt tried to build a rotary steam engine between 1782 and 1786 but was defeated by the sealing problem. Many unsuccessful tries have been made since then, but the new NSU-Wankel engine now appears to be on the verge of feasibility. Felix Wankel of Germany invented the engine.

Figure 13·19 shows a cutaway view of the engine with carburetor and exhaust line on top. Figure 13·20 shows the internal arrangement of rotor, casing, ports, and output shaft.

First let us study arrangement of the engine, Fig. 13·19. The engine handles three separate charges of working fluid simultaneously, all undergoing different processes at any instant. The feature of the engine is only two moving parts: (1) a triangular-shaped rotor having an internal gear and convex sides and (2) an output shaft geared to the rotor. The center of the output shaft is fixed, but the rotor turns through a planetary path and has no fixed center.

The three apexes of the rotor always rub over the interior of the casing and fix the path of the rotor. Technically the outline of the casing is a double epitrochoid. Sealing strips at the apexes divide the space between the rotor and casing into three compartments.

As the rotor turns, these compartments vary in volume, being at a minimum when any one of the sides parallels the bottom or top of the casing and at a maximum when any of the sides stands about vertical.

Since the center of gravity of the rotor turns about the fixed output shaft, the rotor is unbalanced. A balancing weight on the shaft counterbalances the centrifugal force of the rotor. Like a reciprocating engine, this engine develops only intermittent and varying turning forces, so it needs a flywheel to store energy for brief rapid bursts to be used during compression.

The gear ratio makes the output shaft turn three times as fast as the rotor, helping minimize friction losses between rotor and casing. The engine in Fig. 13·19 has a carburetor to prepare the fuel-air mixture for entry to the engine through the inlet port, at top. The exhaust port, to the right of the inlet port, carries off the burned expanded gas. The spark plug in the lower part of the casing does its usual ignition duty. A cooling-water jacket surrounds the entire casing.

Elimination of a crankshaft and piston rods makes this a compact engine, reducing the weight to about 1 lb per hp; designers expect to cut this to $\frac{1}{2}$ lb per hp eventually.

Engine temperature runs as low as 300 F, and the engine develops 1 bhphr (brake horsepower-hour) for 0.45-lb gasoline. NSU says that special treatment of the casing-wall metal resists heat developed in the ignition area and helps maintain a good seal between the three compartments formed by the rotor and casing.

13·19 Processes. Any one charge of air-fuel going through the engine passes through the four processes of the cycle during one revolution of the rotor. In contrast, the crankshaft of a reciprocating engine makes two complete revolutions for the four processes.

Figure 13·20 gives the details of engine operation. The four diagrams shown cover only about 120° rotation of the rotor but show all processes taking place as three separate but simultaneous series.

Let us trace one charge passing through the engine. In Fig. 13·20a the charge 1 is just beginning to enter the minimum upper space between the rotor and casing. The tail end of a burned previous charge is leaving the same space through the exhaust port, to the left. In b the fuel-air charge continues entering the expanding volume at 2, while seal C is just starting to close the exhaust port from this space. In c, since the volume continues expanding into 3, fuel and air keep flowing in; at C the exhaust port has been completely cut off.

In d the intake volume has expanded to its maximum at 4 and C starts to close off the space from the intake port. To keep tracing this charge of fuel and air we skip back to a, but note that seal C is now seal A. Keeping this transition in lettering in mind we see that our fuel-air charge is now being compressed in the diminishing volume 5, seal A having cut it off completely from the inlet port.

In b the fuel-air charge continues being compressed by the shrinking volume of the space at 6. In c the charge has now been squeezed into the minimum volume at 7. Shortly after the rotor face AB passes the midpoint of its lower position, the spark ignites the fuel-air charge. The heated gas now exerts a sharply rising pressure on the rotor and casing. Since the rotor is in an unbalanced position about the output shaft, it exerts a net force on the rotor face AB to turn the rotor in a clockwise direction.

In d the expanding volume at 8 allows the gas pressure to change as it does work on the rotor face AB. In this position the gas has more leverage, so it exerts greater turning torque on the rotor face.

To continue tracing the processes we again go back to a. Again we get a letter transition, and seal A now becomes seal B. In a the hot gas continues working on the turning rotor as it expands to volume 9. Gas has stopped burning, and the pressure starts dropping because of increasing volume and no heat input—in fact, water cooling contributes to pressure drop as in a reciprocating engine.

In b gas expands into maximum volume at 10 and seal C is about to uncover the exhaust port. In c the seal C has completely uncovered the exhaust port and the gas has quickly blown down to about atmospheric pressure as it leaves through the exhaust duct.

In d the diminishing volume at 12 forces out the remaining burned-gas charge and seal C is beginning to uncover the inlet port. Again returning to a we have just completed one revolution of the rotor and find that seal C is again seal A (to accommodate the letter transition). The minimum volume above face CA has now forced the tail end of the burned gas out of the casing at about atmospheric pressure, and this space is now receiving its next charge of fuel and air, ready to repeat the cycle of four processes during the next revolution of the rotor.

To summarize, the engine carries on three separate series of processes or cycles, one at each rotor face. Since all four processes of each cycle take place during one revolution of the rotor, the rotor gets three power impulses per revolution, each impulse varying in intensity and lasting for almost one-third of a revolution. In one sense this rotary engine equals a six-cylinder four-cycle reciprocating engine, in terms of power impulses per revolution of the output shaft.

13·20 Engine shape. The symmetrical appearance of Fig. 13·20 makes it hard to judge the movement of the rotary engine parts. Figure 13·21 reproduces parts of Fig. 13·20 by themselves so we can judge their shape

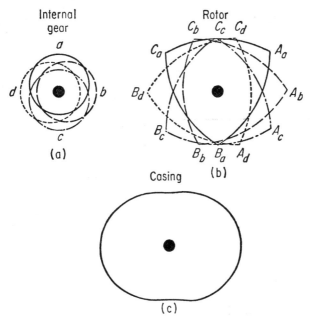

Fig. 13·21 Separate drawings of internal gear and rotor outline emphasize eccentric path of the rotor about center of the output shaft.

and motion. In Fig. 13·21a four positions of the internal gear of the rotor are reproduced in relation to the output-shaft center, which remains stationary. The letters correspond to positions shown in Fig. 13·20a, b, c, and d.

In Fig. 13·21b we see the corresponding rotor positions. Notice the difference in travel of the seal points. Seal C has a much shorter travel than seal A for this particular part of one revolution (about one-third). Travel from C_a to C_d is much slower than travel from A_a to A_d, since they each take place in the same time interval. For constant-output shaft speed the speed of the linear travel of the seals will be the same as each traverses the same part of the casing wall. This shows the complex planetary path the rotor takes during each revolution.

Figure 13·21c shows the symmetrical layout of the casing interior wall. This shape, called a double epitrochoid, must be machined closely to maintain contact with the seals on the rotor. The critical question of success of the engine depends on how well the casing wall can maintain its shape under the stress of changing gas pressure and temperature.

Original engines using this principle have been lubricated by mixing the oil with the fuel, as in a two-cycle engine. But experience shows this makes it impossible to have a clear exhaust from the engine. Air-pollution agitation makes it inadvisable to promote an engine with this fault. NSU Company report that they are developing a forced-lubrication system to deliver oil to the seals at the point of friction, keeping the exhaust clear.

13·21 Engine cycle. Figure 13·22 shows how the engine would run on an ideal cycle on P-V coordinates. For reciprocating engines we could draw the equivalent cylinder volume and corresponding piston position below the volume axis. We cannot do this for the rotary engine, but we can correlate it by position numbers.

The numbers on the cycle of Fig. 13·22 correspond to the numbered volumes and positions shown in Fig. 13·20. Fuel-air induction takes place through 1-2-3-4; the next compression of the fuel-air charge through 4-5-6-7. The ignition of the mixture takes place at 7 for a sharp pressure rise to A (not shown on Fig. 13·20). Then expansion and work take place through A-8-9-10. Most of the gas blows out of the cylinder as the exhaust port is uncovered after position 10. Then the rotor squeezes the rest of the spent gas from the compartment through 10-11-12-1. An actual engine cycle, of course, differs from this ideal because of irreversibilities and heat transfer.

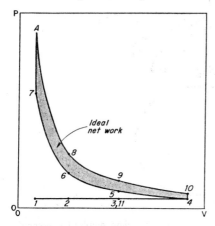

Fig. 13·22 Ideal P-V diagram shows conventional thermodynamic cycle engine uses; numbers key the cycle to the rotor position in Fig. 13·20.

REVIEW TOPICS

1. With the aid of a piston-and-cylinder sketch, P-V graph, and T-S graph, describe the operation of an ideal Otto engine.

2. With the aid of P-V and T-S graphs, describe the operation of an ideal Otto cycle. What do the areas mean on the graphs?

INTERNAL-COMBUSTION-ENGINE CYCLES 205

3. Why cannot the Otto-cycle thermal efficiency be as high as that of the Carnot cycle?
4. What factors control the thermal efficiency of the ideal Otto cycle?
5. How is the work output of an ideal Otto engine varied?
6. Does the top combustion temperature of an Otto cycle control the thermal efficiency?
7. On a graph show the trend of Otto-cycle thermal efficiency with increasing compression ratio.
8. What is the relation between pressure ratio and temperature ratio for an ideal Otto cycle?
9. How does work, heat added, heat rejected, and thermal efficiency of an ideal Otto cycle vary with compression ratio?
10. Describe the action of an ideal water-cooled Otto cycle on a T-S graph.
11. Show the energy that can be salvaged in an Otto cycle on P-V and T-S graphs.
12. With the aid of T-S and P-V graphs describe the actions in an Otto cycle caused by supercharging.
13. With the aid of P-V and T-S graphs describe the processes of the diesel engine and cycle. What do the areas mean on the graphs?
14. What are the compression ratio and fuel-cutoff ratio in a diesel engine?
15. Write the equation defining the thermal efficiency of the diesel cycle.
16. How are expansion ratio, compression ratio, and fuel-cutoff ratio related in a diesel cycle?
17. How is the output of a diesel engine controlled?
18. Can a diesel engine be supercharged?
19. How do Otto and diesel cycles compare in efficiency for comparable compression ratios and the same specific heat ratio?
20. How do the thermal efficiencies of Otto and diesel engines usually compare in actual practice? Why?
21. Explain the phenomenon of detonation in an Otto engine. How can this knocking be minimized or eliminated?
22. Explain the phenomenon of detonation in a diesel engine. How can this knocking be minimized?
23. What is meant by octane rating and cetane rating of a fuel?
24. Describe the operation of a Wankel rotary engine.

PROBLEMS

1. Find the ideal thermal efficiency of an Otto cycle with a compression ratio of 5 and a specific-heat ratio of 1.4.
2. Find the ideal thermal efficiency of an Otto cycle with a compression ratio of 10 and a specific-heat ratio of 1.4.

3. Find the thermal efficiency of an Otto cycle with a compression ratio of 5 and a specific-heat ratio of 1.3.

4. Find the thermal efficiency of an Otto cycle with a compression ratio of 10 and a specific heat ratio of 1.3.

5. Make the necessary calculations to reproduce the curves of Fig. 13-4.

6. If maximum combustion temperature in an Otto engine is 2500 F and the engine has a compression ratio of 12, what will be the maximum cylinder pressure? Assume an air temperature of 80 F.

7. Compute the (a) heat added, (b) heat rejected, and (c) mechanical work produced in an ideal Otto cycle using air initially at 14.7 psia and 80 F. The maximum combustion temperature is 2500 F, and the compression ratio is 12. Assume that $n = k = 1.3$ for all constant-entropy processes; $c_v = 0.225$.

8. What is the mean effective pressure of the cycle in Prob. 7?

9. What is the thermal efficiency of the cycle in Prob. 7?

10. What is the maximum change in entropy of the working fluid in Prob. 7?

11. An Otto cycle takes in air at 550°R and compresses it through a compression ratio of 15 in a polytropic process with $n = 1.25$. A heating process raises the air temperature to 3200°R with $c_v = 0.225$ Btu per lb, F, and $k = 1.28$. Heat transfer to cooling water during heating is equal to 15 per cent of the energy absorbed by the air. The heated air expands polytropically at $n = 1.35$. Calculate (a) heat rejected to cooling water during compression, (b) heat rejected to cooling water during heating (combustion), (c) heat rejected in the exhaust, (d) heat added to the cycle, (e) work output of the cycle, (f) thermal efficiency of the cycle, (g) mean effective pressure, (h) change in entropy of the air in each process of the cycle.

12. Recalculate Table 13-4, assuming the compression ratio as 10, all other factors remaining the same.

13. Recalculate Table 13-5, assuming a compression ratio of 10.

14. Make the necessary calculations to reproduce the curves of Fig. 13-14.

15. A diesel cycle takes in air at 100 F and 14.7 psia and works with a compression ratio of 20 and cutoff ratio of 3. Assuming that $k = 1.3$ and $c_v = 0.225$, calculate the state of the air at the end of each process and plot the cycle on P-V and T-S graphs. Calculate (a) work output per pound, (b) heat rejected per pound, (c) overall thermal efficiency of the cycle, (d) the mean effective pressure of the cycle.

CHAPTER 14

Free-piston Gas Generator

One of the newest heat engines is the free-piston gas generator supplying a gas turbine. The generator, Fig. 14-1, has undergone a long period of development since about 1922. Its principal advantage lies in its light weight and good thermal efficiency. While it is a reciprocating engine it needs no crankshaft or flywheel. There are about 350 gas generators operating or under construction for both stationary and transportation applications.

14·1 Operation. Figure 14·1 and Table 14·1 show the processes and principal parts of a free-piston gas generator. Two piston assemblies move in opposite directions centered on a common *diesel-engine cylinder*.

Expanding gas in the engine cylinder pushes the piston assemblies outward. This makes the outer faces of the large pistons compress air trapped in the outer *bounce cylinders*. The rising bounce-air pressure decelerates the piston assemblies and brings them to a stop.

The expanding bounce air then pushes the piston assemblies toward each other. This compresses air trapped in the engine cylinder, which again decelerates the pistons and brings them to a stop in the inner position. Fuel injected into the engine cylinder then burns, and this energy helps to push the pistons apart during the next cycle.

The inner faces of the large pistons work in a pair of air-compressor cylinders that take air from the atmosphere as working fluid for the cycle. Let us trace the processes of the working fluid in Fig. 14.1 with the aid of the graphs of Fig. 14-2.

14·2 Cylinder actions. In the outer positions, atmospheric air fills the compressor cylinders. During the inward stroke, under the influence of the expanding bounce air, the air compresses to an intermediate pressure

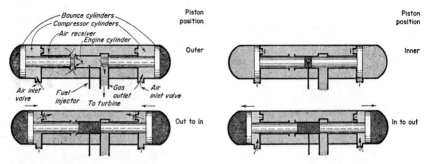

Fig. 14·1 Free-piston gas generator takes atmospheric air into the compressor cylinders for pressurizing, then delivers it to the engine cylinder through the air receiver. Fuel burned in the engine supplies compressor work; gas exhausting from engine is hot and pressurized, capable of developing shaft work in a turbine. (See Table 14·1.)

Table 14·1 (See Fig. 14·1)

Piston position	Compressor action		Engine action		Bounce-cylinder pressure	Work transfer	Heat or energy transfer
	State	Process	State	Process			
Outer	1	End air intake	5–2 2	Scavenging: Gas exhaust to turbine. Air intake from receiver	Maximum	None	Engine exhaust to gas turbine
Out to in	1–2	Compression and air discharge to receiver	2–3	Compression	Intermediate	Bounce piston to compressor and engine pistons	None
Inner	a–b	End air discharge to receiver. Start air intake	3	Fuel injection and ignition	Minimum	None	Fuel heating energy to engine air
In to out	b–1	Air intake	3–4 4–5	Combustion and expansion	Intermediate	Engine and compressor pistons to bounce pistons	Combustion in engine cylinder

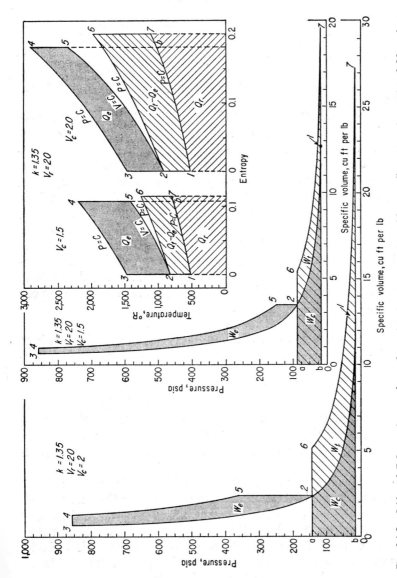

Fig. 14·2 P-V and T-S graphs are for a free-piston gas generator with overall compression ratio of 20 and at fuel cutoff ratios of 1.5 and 2.0. This air-standard analysis shows that air-receiver pressure at a rises with increasing output measured by area a-6-7-b.

and then discharges to the air receiver surrounding the engine cylinder. During the next outward stroke, the air remains in the receiver at this pressure, but toward the end of the stroke, air-inlet ports in the engine cylinder uncover to admit pressurized air into the cylinder. This *scavenges* or pushes the remaining combustion gases out of the cylinder into the gas outlet of the cylinder.

Ideally, we assume that the fresh air charge just displaces the spent gas scavenged from the cylinder. (Actually, some of the scavenge air passes out with the combustion gas and flows on to the turbine.) During the following inward stroke, the pistons cover the gas-outlet and air-inlet ports, trapping the air charge and compressing it in the engine cylinder.

At the end of the stroke, fuel injected into the cylinder burns to keep engine pressure constant during the early part of the next outward stroke. The burned gases then expand while doing work on the pistons. As the pistons move outward, the gas-outlet ports are first uncovered and the gases throttle to a lower pressure, usually equal to the air-receiver pressure. By the time the gas throttles to the outlet pressure, the air-inlet ports uncover to admit pressurized air that pushes the remaining gas out of the cylinder.

14·3 Processes. Figure 14·2 shows the processes, work, and heat transfers in a gas generator. On the P-V graph, 1–2 is the constant-S compression process in the air compressors; the gray area to the left of the curve measures total compressor work W_c. Line b–1 represents atmospheric air entering the compressor cylinder as on an indicator card. The area under the line measures the work done on the piston by the entering air. Line 2–a shows the pressurized air discharging from the compressors into the receiver. We assume zero compressor clearance. The receiver air is at state 2.

Process 2–3 compresses the air in the engine cylinder at constant S; 3–4 is the constant-P heat addition Q_a to the cycle by the burning fuel. From 4 to 5 the gases expand at constant S as they work on the engine piston faces. Process 5–2 shows the pressure drop of the gases as they throttle out of the cylinder through the gas-outlet ports. This variable-flow throttling (an irreversible process) ends up with the gas in an average state 6 at the pressure of the receiver air.

In an equivalent ideal cycle, we assume that 5–2 represents a constant-V cooling of the working fluid in the engine cylinder. The energy removed during this cooling transfers irreversibly to heat an equal amount of air initially in state 2 at constant P to the final state 6.

14·4 Net cycle output. The free-piston generator takes air at state 1 and energizes it to state 6. The unit does not produce any mechanical work output. To get this work we must pass the gas flow through a gas turbine. Even though the gas discharges in pulses, the usual operating rate of 1,000 cycles per min for these free-piston units gives an effective steady flow. Since we have not studied steady-flow processes, let us assume that the air does work in an ideal complete-expansion air engine. This will give the same result.

In Fig. 14·2 we assume that the gas enters the engine during a–6; it then expands at constant S to state 7 where its pressure has dropped to atmospheric. The spent gas exhausts to atmosphere during 7–b. Total hatched area a–6–7–b measures the work output of the engine which is also the cycle net work output, W_t.

14·5 Work and heat transfers. Between the air and gases in the generator there is considerable transfer of mechanical work, but the unit produces no net work. Let us take a close look at what happens.

All the gross engine work produced during the working stroke 3–4–5 is stored as internal energy in the bounce air. Air entering the compressor cylinders during this stroke also contributes a small amount of energy to the bounce air; this is measured by the area under b–1.

When the bounce air expands on the following stroke, it does work in compressing the entering air and the air charge in the engine cylinder. Since no net work leaves the gas generator and no net work is stored in the bounce cylinders, the net work done by the free-piston engine must equal the net work absorbed by the compressors, $W_e = W_c$.

This means, on the P-V graphs of Fig. 14·2, that area 2–3–4–5 equals area a–2–1–b, the net work transfer between engine and compressor cylinders. On the T-S graphs, we find constant-S processes 1–2 and 2–3 in the air compressor and engine cylinders. Heat Q_a is added to the cycle during 3–4 at constant P. The total area under 3–4 measures Q_a.

From 4 to 5 the engine gas expands at constant S. From 5 to 2 the engine working fluid cools at constant V. Then area 2–3–4–5 measures the net heat used by the engine cylinder Q_e. By the first law we see that $Q_e = W_e$, but remember that $W_e = W_c$, too.

We have assumed that the heat rejected by the engine heats air at constant P; then the total area, under 2–5 equals the area under 2–6, that is, $c_v(T_5 - T_2) = c_p(T_6 - T_2)$. This assumption enables us to calculate state 6, note increase in entropy.

The air expands at constant S from 6 to 7 in the air engine. Assuming an equivalent closed cycle, the air cools at constant P from 7 to 1 while it rejects unusable heat Q_r from the cycle. Since all the processes we assumed are internally reversible, the net heat Q_t converted to work W_t must be $Q_t = Q_a - Q_r$. Since the last two heat areas are not coincident because of irreversible heat transfer from 5 to 6, we have no single area to show Q_t on the graph.

But note on the P-V graph that area 1–2–6–7 represents the difference of W_t and W_c and furthermore $W_c = Q_c$. Then, on the T-S graph, this area must represent $Q_t - Q_e$. The total net cycle work output then equals the sum of the two areas 2-3-4-5 and 1-2-6-7 or $Q_e + (Q_t - Q_e) = Q_t$, remembering that $Q_e = Q_c$. But Q_c cannot be shown on the T-S graph because no heat transfer is involved in air-compressor processes.

14·6 Variable output. Graphs in Fig. 14·2 drawn for two different work outputs show that the air-receiver pressure varies. The center P-V and T-S graphs are drawn for the overall compression ratio of compressors

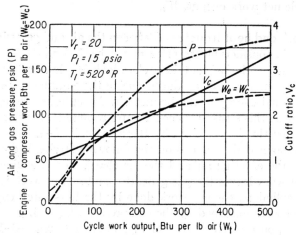

Fig. 14·3 Air-receiver and exhaust-gas pressures rise with increasing cycle output as well as with cutoff ratio and compressor work.

and engine of $V_1/V_3 = V_r = 20$ and a cutoff ratio of $V_4/V_3 = V_c = 1.5$. At this output the air-receiver pressure is about 87 psia, and maximum gas temperature in the engine is about 2200°R.

For the higher output developed at a cutoff ratio of 2.0 shown in the outer P-V and T-S graphs of Fig. 14·2, the air-receiver pressure rises to 143 psia and the maximum gas temperature rises to 2960°R. Figure 14·3 shows how receiver pressure, engine and compressor works, and cutoff ratio vary with output.

When we assume constant-pressure heat addition to the cycle, we find that the theoretical thermal efficiency is the same as for the diesel cycle, as shown in Table 14·2. Figure 14·4 shows how thermal efficiency varies with compression and cutoff ratios when $k = 1.35$. The cutoff has a much lower upper limit than in a diesel engine. This follows from the two-stage compression used and the much smaller expansion ratio in the free-piston engine cylinder.

14·7 Energy flow. This can be shown as in Fig. 14·5 for an ideal free-piston gas-generator turbine cycle. For the ideal condition, all the energy input to the engine cylinder (cycle) appears in the exhaust gas. A constant amount of energy circulates in all other parts of the gas generator, apparently not contributing to cycle work output. To make the cycle run, however, this energy circulating among the engine, bounce cylinders, air compressors, and air receiver is essential.

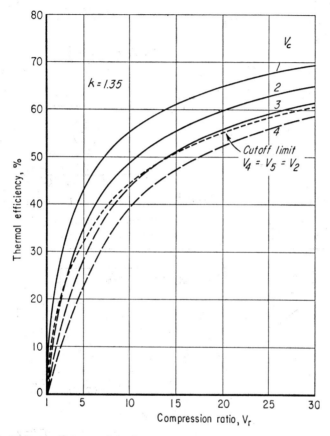

Fig. 14·4 Thermal efficiency of the free-piston cycle is the same as that of the diesel cycle except that the cutoff ratio is much more limited.

The bounce cylinders perform the same duty as a flywheel in the Otto and diesel engines. They store some of the gross engine work intermittently to compress the engine air before firing. In the free-piston gas generator this compression is done in two stages: first, in the air compressors and, finally, in the engine cylinder.

Many papers have been written on estimating the performance of actual free-piston gas generators. We believe that this chapter is the first study published on setting up an air standard for the ideal free-piston unit comparable to that used for many years for the Otto and diesel cycles.

Table 14·2 Free-piston gas-generator power cycle has same efficiency as diesel cycle

1. $e = 1 - \dfrac{Q_r}{Q_a} = 1 - \dfrac{c_p(T_7 - T_1)}{c_p(T_4 - T_3)}$
   (see Fig. 14.2)
2. $c_p(T_6 - T_2) = c_v(T_5 - T_2)$
3. $T_6 = \dfrac{T_5 + T_2(k-1)}{k}$
4. $\dfrac{P_1}{P_2} = \left(\dfrac{V_2}{V_1}\right)^k = \left(\dfrac{V_6}{V_7}\right)^k = \left(\dfrac{1}{V_{rc}}\right)^k$
   where V_{rc} = compressor compression ratio
5. $T_2 = T_1\left(\dfrac{V_1}{V_2}\right)^{k-1} = T_1 V_{rc}^{k-1}$
6. $T_3 = T_1\left(\dfrac{V_1}{V_3}\right)^{k-1} = T_1 V_r^{k-1}$
7. $T_4 = T_3\left(\dfrac{V_4}{V_3}\right) = T_3 V_c = T_1 V_c V_r^{k-1}$
8. $\dfrac{V_3}{V_2} = \dfrac{V_3}{V_1}\dfrac{V_1}{V_2} = \dfrac{V_{rc}}{V_r}$

9. $T_5 = T_4\left(\dfrac{V_4}{V_5}\right)^{k-1}$
   $= T_1 V_c V_r^{k-1}\left(\dfrac{V_c V_3}{V_2}\right)^{k-1}$
   $= T_1 V_c^k V_{rc}^{k-1}$
10. $T_7 = T_6\left(\dfrac{V_6}{V_7}\right)^{k-1}$
    $= \dfrac{T_5 + T_2(k-1)}{k}\left(\dfrac{1}{V_{rc}}\right)^{k-1}$
    $= \dfrac{T_1 V_c^k V_{rc}^{k-1} + T_1 V_{rc}^{k-1}(k-1)}{k}$
    $\left(\dfrac{1}{V_{rc}}\right)^{k-1} = \dfrac{T_1(V_c^k + k - 1)}{k}$

Substituting in (1) and after canceling T_1's

11. $e = 1 - \dfrac{(V_c^k + k - 1)/k - 1}{V_r^{k-1}(V_c - 1)}$
    $= 1 - \dfrac{1}{V_r^{k-1}}\left[\dfrac{V_c^k - 1}{k(V_c - 1)}\right]$

Compare with diesel-cycle thermal efficiency, Table 13·6.

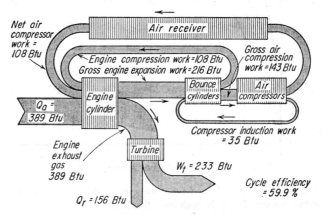

Fig. 14·5 Energy-flow diagram for a free-piston cycle shows that mechanical work circulates internally in the unit among elements.

REVIEW TOPICS

1. With the aid of a cross-section sketch describe the actions in a free-piston gas generator including the diesel cylinder, bounce cylinders, and compressed-air cylinders.

2. Describe the free-piston compressor cycle with the aid of P-V and T-S graphs, assuming equivalent closed air cycle where necessary.

CHAPTER 15

Actual Engine Cycles

Before we leave the topic of reciprocating i-c engines, let us compare their actual operation with the theoretical with which we have been dealing. This will bring out the limitations of elementary thermodynamic analysis. While thermodynamics outlines the broad theory of engine operation, it cannot always anticipate *all* the problems raised by actual conditions.

15·1 Actual P-V changes. Chapter 13 outlined the theoretical operation of Otto engines in detail. Figure 15·1 shows the theoretical P-V diagram (dashed) of a four-stroke Otto or gasoline engine. Air enters the cylinder and gas leaves at pressure P_a, so the works of induction and exhaust, measured by the area under the dashed horizontal process line, are equal and opposite. They cancel each other as far as contributing to net work output.

The piston compresses the air charge during process a–b; heat addition raises the pressure during process b–c; the gas expands and does work on the piston during process c–d; the spent gas throttles to atmosphere during d–a. The area a–b–c–d measures the net mechanical work output of the piston during a complete cycle of four strokes.

The actual changes in pressure of a gasoline-fired four-stroke-cycle engine are shown by the solid lines in Fig. 15·1. At 0 the gases have just finished exhausting from the cylinder at a pressure higher than atmosphere P_a. To make a gas flow from one region to another past the restriction of valves and pipe or ducting, its pressure at the source must be higher than in the discharge region.

Before the piston starts its induction stroke, the inlet valve opens and the gas in the clearance volume begins expanding from 0 to f. After the gas pressure drops below P_a, air starts flowing into the cylinder. During the entire intake stroke the air pressure in the cylinder stays below P_a.

After the piston starts its compression stroke, the intake valve closes—more about this later. The actual compression process is lower than the theoretical for several reasons: (1) The pressure at 1 is lower than at a. (2) During the early part of the stroke the cooler gas absorbs heat from the hotter cylinder walls, but during the later part of the stroke the hotter gas gives up heat to the relatively cooler walls. (3) The actual specific heat varies during the stroke and the process is far from adiabatic. (4) Air escapes from the cylinder by leaking past the piston rings and through closed valves. (5) Pure air contains the residual combustion gases left in the clearance volume, and vaporized gasoline.

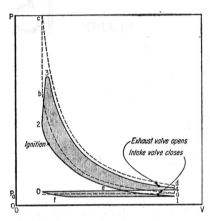

Fig. 15·1 Actual gasoline-fired engine P-V changes lie below theoretical with less work output reduced by the negative intake-exhaust loop.

The actual cycle starts its heat addition before the end of compression because the combustion process needs time to run to completion. From the ignition point to state 3 the fuel-air mixture burns and produces heat energy. This happens during the late part of compression and the early part of expansion. The maximum cycle pressure at 3 is much lower than c. Several conditions cause this: (1) The high combustion temperature transfers considerable heat to the walls and cooling water. (2) The expanding volume reduces the pressure. (3) The moving piston removes energy as work. (4) Incomplete combustion does not develop full heat release. (5) The gas mixture reabsorbs some of the heat by dissociation.

The actual expansion stroke 3–4 lies below the theoretical stroke c–d because: (1) Maximum pressure at 3 is lower. (2) The expansion is not adiabatic, since the gas transmits heat to the cooling water. (3) Some gas leaks past the piston rings and closed valves.

The exhaust valve opens before the end of the expansion stroke and allows some of the gas to leave the cylinder early. During the exhaust stroke 4–0 the piston pushes most of the gas out of the cylinder to the atmosphere. During this stroke the gas pressure in the cylinder stays higher than the atmosphere to overcome the frictional resistance of the exhaust valve opening, exhaust manifold, muffler, and piping.

15·2 Work output. As we know, the dashed loop a–b–c–d measures the theoretical work output of an engine. But Fig. 15·1 shows two loops marked in gray for the actual engine—what do these mean? In the upper loop 4–e–2–3, during the expansion process 2–3–4, the gas works *on* the piston, so total area underneath to $P = 0$ measures the gross work output.

But during the first part of the exhaust stroke 4–e the piston does work on the gas, and during the last part of the compression stroke e–2 the piston does work on the air-fuel mixture. The areas underneath these curves measure work done *by* the piston. The difference of the coincident areas gives us the net loop 4–e–2–3, which measures the net work *output* of the piston.

Now let us examine the events around the lower loop 1–e–0–f. From 1 to e, the piston works on the air-fuel mixture; from e to 0, the piston works on the exhaust gas. The area underneath measures the gross work done *by* the piston. During 0–f–1 the expanding clearance gas and incoming air work *on* the piston; the area underneath measures the work. The difference between the two work areas gives us the net lower loop, which measures the net work done *by* the piston on the gas. To get the net work output of the engine, we must subtract the area of the lower loop from the upper. The difference is considerably less than the theoretical work.

15·3 Lower outputs. Figure 15·1 shows the P-V changes for an engine running at full load. To vary the output of a gasoline-fueled engine, we vary the mass of air-fuel mix placed in the cylinder during the induction or intake stroke. A butterfly valve ahead of the carburetor (which injects a fuel spray into the incoming air) varies the resistance to air flow entering the cylinder.

By closing down on the butterfly-valve opening, we increase the air pressure drop across it and lower the pressure of the fuel-air mix in the cylinder. This, of course, lowers the mass of working fluid in the cylinder and so lowers the work output for the cycle. Figure 15·2 shows the lower pressure at 1 compared with 15·1. This decrease reflects throughout the entire cycle by lowering cylinder pressures at 2, 3, and 4. The upper loop e–2–3–4 is smaller in Fig. 15·2 than in Fig. 15·1, while the lower loop e–0–1 is larger. So we have a "double" effect that lowers the net output of the engine, which is the difference of the two loops.

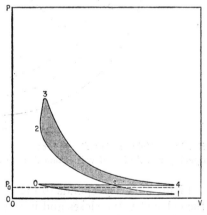

Fig. 15·2 Throttling the intake of a gasoline-fired engine lowers the initial compression pressure at 1 to increase the negative loop work area and reduce net work output.

15·4 Valve timing. Figure 15·1 indicated that the valves and ignition operated at times different from the theoretical. These can be easily studied with the aid of a timing diagram, Fig. 15·3. This shows the position of the crank throw when the different events take place. TDC means *top dead center* and corresponds to the piston in its full-in position, that is, at the end of its compression or exhaust strokes. BDC means *bottom dead center*, corresponding to the full-out position of the piston in the cylinder.

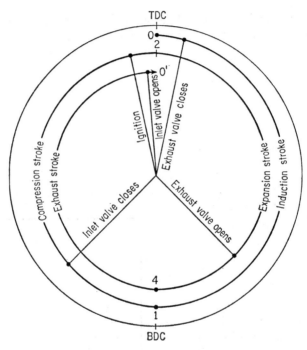

Fig. 15·3 Valve-timing diagram compares crankshaft position for theoretical and actual opening and closing of valves and timing of ignition.

Figure 15·3 corresponds to Figs. 15·1 and 15·2. For the theoretical cycle the exhaust valve closes and inlet valve opens at 0; the inlet valve closes at 1; ignition takes place at 2; the exhaust valve opens at 4 and closes at 0'. As we see, the actual events take place at appreciably different times.

For the actual cycle the inlet valve opens before the beginning of the cycle and the exhaust valve remains open till about 10° after the beginning of the cycle. The early inlet opening gives extra time for accelerating the air, which is important in our modern high-speed engines. The late outlet closing gives extra time for getting out more of the exhaust gas, which lowers the dilution of the fresh air-fuel mixture.

The inlet valve does not close until about 45° after BDC. The air coming in at considerable speed can pack more of itself into the cylinder even when the piston has already moved through one-quarter of its compression stroke. The pressure wave set up inside the cylinder by the piston does not get to the inlet valve until about that time.

Ignition takes place at various angles before TDC depending on the load and speed. The exhaust valve opens as much as 45° before BDC. This blows down some of the pressurized gas before the end of the expansion stroke and causes a small loss in work output. The advantage lies in getting the gases moving earlier and lowering the pressure during the exhaust stroke.

15·5 Pressure-time diagram. Figures 15·1 and 15·2 are indicator diagrams that can be taken directly off the engine if its speed is not too high. They prove valuable in analyzing engine performance and measuring internal work output. For high-speed engines the mechanical parts of the indicator mechanism have too much inertia and it becomes difficult to get an accurate card.

To overcome this limitation an electric oscillograph measures cylinder pressures and relates them to time or crank angle. Figure 15·4 shows an

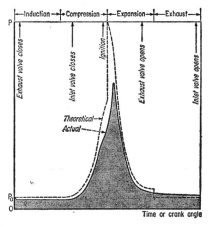

Fig. 15·4 Pressure-time diagram of theoretical (dashed) and actual gasoline-fired engine can be converted to a *P-V* graph for analysis.

actual cycle in solid line and the theoretical in dashed line. Both relate directly to Fig. 15·1. By suitable measurement, Fig. 15·4 can be translated into the *P-V* diagram of Fig. 15·1. The strokes are indicated at the top of the graph and the time of valve operations and ignition.

15·6 Two-stroke-cycle engines. In Chap. 14, we studied the free-piston gas generator which uses a two-stroke-cycle diesel cylinder. Let us study a more conventional two-stroke diesel engine, Fig. 15·5. This engine offers opportunities for almost doubling work output in a given cylinder as compared with a four-stroke-cycle.

The major problem in this engine centers on getting the exhaust gas out of the cylinder after the expansion stroke and the fresh air in before the immediately following compression stroke. Figure 15·5 shows the *P-V* graph for the engine cylinder. Just before the compression stroke is started at 1, both intake and exhaust ports are open so, during the first part, the piston pushes some air back out the intake and exhaust ports. When the piston skirt closes these ports, the pressure rises. Before the end of the compression stroke, fuel is forced in and ignites to raise the pressure to 3 after starting the expansion stroke.

Near the end of the expansion stroke at 4 the piston uncovers the exhaust port, allowing part of the gas to blow down to the atmosphere.

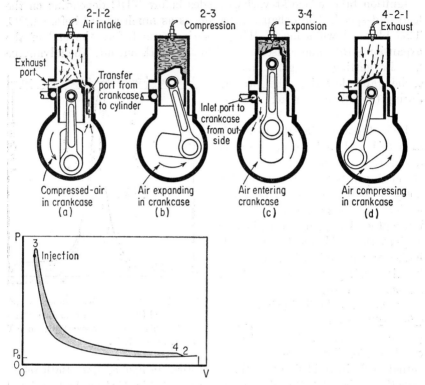

Fig. 15·5 Two-stroke-cycle diesel engine develops 50 to 80% more work per cylinder than a four-stroke engine. This engine uses the crankcase to precompress intake air a few pounds before it enters the cylinder. (a) Spent exhaust gas leaves through the port; incoming pressurized air sweeps out the residual gas. (b) Rising piston closes the inlet and exhaust ports and compresses the air charge; it also raises a vacuum in the crankcase. (c) Injected fuel ignites before the expansion stroke; uncovered crankcase inlet port admits air. (d) The expanding gas works on the piston, and starts exhausting as soon as the exhaust port is uncovered by the piston; descending piston compresses air trapped in the crankcase.

From 4 to 1 to 2 fresh air forced into the cylinder through the intake port sweeps out the burned gas through the exhaust port.

Let us see how the air is compressed for this cycle. In (a) air which has been compressed in the crankcase leaves through the cylinder intake or transfer port. In the cylinder the shape of the piston face directs the air toward the cylinder top, so it scavenges out the gas in the cylinder. As the piston rises in (b), the skirt closes off the cylinder transfer port and the air trapped in the crankcase expands below atmospheric pressure.

In (c) the cylinder skirt has uncovered the crankcase intake or inlet port, admitting atmospheric air into the crankcase. In (d) the descending cylinder compresses the air in the crankcase as it closes the inlet port. The compressed air can then force its way into the cylinder as in (a).

The net work to precompress the air and the part of the piston stroke devoted to scavenging the cylinder makes the two-stroke cylinder do less than twice the work of a four-stroke cylinder.

15·7 Stirling engine research. (Arts. 15·7 to 15·11 from Paul N. Garay, *Power*, September, 1960, pp. 73–75.) Robert Stirling patented and

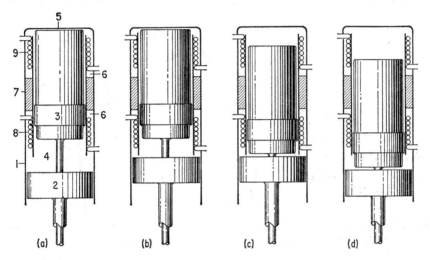

Fig. 15·6 Stirling engine uses pistons 2 and 3 in cylinder 1. The annular space 6 houses the regenerator 7, heating coil 9, and cooling coil 8. Operating sequence of the engine is: (a) Displacer piston 3 moves upward, displacing spent hot working gas from hot space 5 to cold space 4 between two pistons. Gas gives up recycling heat to regenerator 7 and the unavailable energy to cooling coil 8. (b) Power piston 2 compresses the cooled gas in cold space 4 between pistons 2 and 3. (c) Displacer piston 3 moves downward, displacing compressed gas into hot space 5; gas picks up heat from regenerator 7 and gets energy input from heating coil 9. (d) Expanding hot gas in 5 forces down both pistons during working stroke of engine.

built his two-piston closed-cycle external-combustion engine in 1827. Many Stirling engines were built later to compete with the steam engine of that time. But as the nineteenth century drew to a close, engines using the Rankine, Otto, and diesel cycles supplanted the Stirling, which ran at low efficiency.

The large difference between theoretical and actual engine performance interested N. V. Philips Gloeilampenfabrieken, Netherlands. They started a research program in 1938; by 1948 they had built several successful engines. A one-cylinder engine was built to develop several horsepower, and it proved that the potential could be realized. A four-cylinder engine using a swash-plate drive was 19 in. long, developed 15 hp at 3,000 rpm, and ran as quietly as a sewing machine.

A 1954 engine with 20-cu-in. displacement used hydrogen as the working fluid and delivered 40 hp at 36% thermal efficiency with a maximum cycle temperature of 1300 F. It has run since then as a development unit. Improvements raise efficiency to over 38%; correcting for combustion losses, the thermal efficiency is 45%. Larger engines show the same performance as the basic research unit.

15·8 Advantages. The Stirling engine has no valves, timing mechanisms, injection or ignition systems. It uses a closed cycle—heat comes

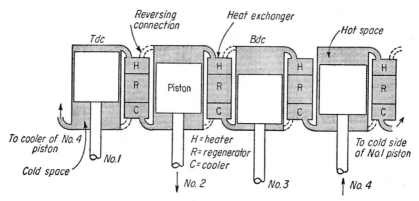

Fig. 15·7 The four-cylinder Stirling engine has pistons connected to the crankshaft at 90° intervals of rotation. External heat-exchange equipment connects between the cylinders.

from external combustion. Figure 15·6 shows the basic elements of the cylinder and the sequence of operation. Figure 15·7 shows a four-cylinder arrangement using double-acting pistons and external heat exchangers. Figure 15·8 shows another arrangement using a pair of cylinders, single acting, with external heat exchangers; in this way the hot coolant from a nuclear reactor might drive it.

In Fig. 15·8 the two pistons attached to a common crank throw have a rotational phase difference of 90°. The fixed relative motions of the pistons completely govern engine operation. The working fluid completely fills the volume of hot space, cold space, heat exchangers, and connecting ducts. The relative position of the two pistons determines the varying volume of the fluid; the "hot" piston leads the "cold" piston.

Figure 15·9a is the ideal P-V diagram for the Stirling cycle, and Fig. 15·9b shows sequence of piston positions. The output of the engine depends on the mass of fluid in the cycle. By supercharging the engine, a high output relative to size can be obtained.

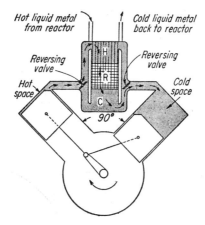

Fig. 15·8 Stirling engine with displacer and power pistons in separate cylinders could use heat-carrying fluid from a nuclear- or fossil-fuel-fired furnace.

15·9 Cycle. In Fig. 15·9 the ideal cycle is made up of four phases:

A—uniform motion of both pistons transfers expanded gas through the regenerator and cooler into cold space; gas cools at constant volume from state 3 to 4.

B—isothermal compression of the cold gas from state 4 to 1.

C—uniform motion of both pistons transfers compressed gas through the regenerator and heater into hot space; gas heats at constant volume from state 1 to 2.

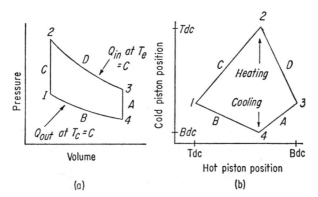

Fig. 15·9 (a) Ideal P-V chart of a Stirling engine uses constant-temperature heating and cooling. (b) The chart plots piston positions relative to one another during the cycle.

D—the pistons move relatively, so the system volume grows and gas expands isothermally from state 2 to 3 doing work on pistons.

The ideal cycle, being reversible, has the same efficiency as the Carnot cycle: the maximum possible.

15·10 Actual cycle. Figure 15·10 shows actual processes for P-V relations and piston position. The P-V chart varies considerably from the ideal, but the piston-position chart has clearly defined positions of the top and bottom dead centers as in the ideal.

Between temperature limits of 1250 and 200 F, T_e and T_c of Fig. 15·9a, the ideal cycle has an efficiency of 70%. Actual engine-shaft thermal efficiency with modern design including all friction and process losses may be anticipated as about 60% of the ideal. The regenerator contributes

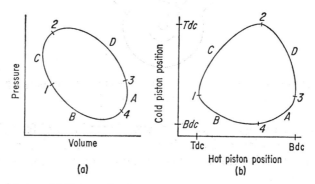

Fig. 15·10 (a) Actual P-V chart of a Stirling engine shows the effects of variable temperatures and frictional losses. (b) An actual piston chart also becomes rounded.

greatly to high efficiency. Extensive research has been done on regenerators and they show an effectiveness of 98% as a possibility.

15·11 Controls. Changing the pressure of the working fluid within the engine volume regulates the shaft output. The heat input produces constant temperature of the working fluid regardless of engine load.

Reversing the direction of gas flow through the heat exchanger reverses the engine rotation. Interconnected reversing valves in Fig. 15·8 control the gas flow to regulate the rotation.

The excellent properties of hydrogen make it a good working fluid for the engine. Helium works with a lower efficiency—a drop of about 12%.

Aluminum chloride in gaseous form offers excellent possibilities. It exists as the monomer $AlCl_3$ at high temperatures and as the dimer Al_2Cl_6 at low temperatures and may be an exceptionally good heat-transfer medium. It dissociates with increasing effective specific heat and thermal conductivity. Lowering the temperature makes the gas not only yield heat given off as if the gas composition were "frozen" but also give off chemical heat of association of some of the monomer molecules resulting from the lower temperature. The effect on the efficiency and size of the regenerator should be significant.

The ideal Stirling engine runs more efficiently than any steam or steam-mercury cycle operating within the same temperature limits. Table 15·1 gives results of studies made by the General Motors Research Laboratories of a Stirling cycle using a 47-cu-in. cylinder compared with other cycles. The study concluded:

1. For given temperature limits the Stirling engine is the only practical engine cycle to attain maximum thermodynamic efficiency.

2. At the same *minimum* cycle pressure, the specific output of a Stirling engine is 85% of a Brayton engine of the same displacement and one-third of a four-stroke Otto engine of the same displacement.

3. At the same *maximum* cycle pressure, the specific output of a Stirling engine is more than double that of the same size four-stroke Otto engine.

Table 15·1 Ideal Stirling air-cycle data

Cycle →	Otto	Brayton	Carnot	Stirling
Efficiency, %	59.5	59.5	59.5	59.5
Min cycle temperature, °R	540	540	540	540
Max cycle temperature, °R	3418	2100	1330	1330
Corresponding Carnot-cycle efficiency, %	84.1	74.3	59.5	59.5
Net work output, Btu per lb	211	110	37.7	37.7
Max cylinder volume, cu in.	47	47	47	47
Working-fluid mass, lb	0.0020	0.0008	0.0020	0.0020
Net work per cycle, Btu	0.422	0.088	0.075	0.075
Hp at 3,000 cycles per min	29.9	6.2	5.3	5.3
Min cylinder volume, cu in.	4.94	6.10	2.48	23.5
Volume ratio	9.5	7.8	19.1	2.0
Piston displacement, cu in.	42.06	40.90	44.52	23.50
Hp per cu in.	0.71	0.15	0.12	0.23
Min cylinder pressure, psia	14.7	14.7	14.7	14.7
Max cylinder pressure, psia	884.0	72.4	688.0	72.4
Pressure ratio	60.0	4.94	46.7	4.94

Inherently the Stirling engine has high efficiency, but it must handle large masses of working fluid to attain competitive specific outputs.

Table 15·2 compares actual engine types. The Stirling data are from the General Motors research tests on the original two engines sent here by Philips. The Stirling is a single-cylinder engine, the diesel six-cylinder, and the automotive eight-cylinder. The Stirling is similar to the diesel in efficiency and specific weight, but both are more efficient and heavier than the automotive.

One disadvantage of the Stirling is the radiator—it is about 2½ times the size needed by a diesel engine. The Stirling rejects all its heat through the radiator (cooling coil) except for the small part going up the stack from the furnace. But in waste-heat-recovery arrangements this can be economically advantageous.

Table 15·2 Actual engine performance

Engine → Ideal cycle for engine → Engine fuel →	Stirling Stirling Diesel 1	Diesel 6-cylinder Diesel Diesel 1	Automotive V-8 Otto Gasoline
Fuel heating value, Btu per lb	18,200	18,200	18,900
Cylinder bore, in.	3.47	4.25	4.06
Piston stroke, in.	2.37	5.00	3.56
Cylinder displacement, cu in.	20.08	71	46.15
Total engine weight, lb	450	2190	678
Max economy data:			
Min bsfc, lb per bhphr	0.358	0.40	0.415
Brake output, hp	30	181	199
Shaft speed, rpm	1500	1700	3000
Maximum thermal efficiency, %	39.0	34.8	32.4
Full-load data:			
Total hp output	40	210	242
Output per cylinder, hp	40	35	30.2
Shaft speed, rpm	2500	2100	4600
Bsfc, lb per bhphr	0.418	0.41	0.468
Thermal efficiency, %	33.3	34.0	28.7
Bmep, psi	317	93	113
Specific weight, lb per hp	11	10.1	2.80
Specific output, hp per cu in.	2.00	0.49	0.65

A Stirling engine may cost more than a diesel engine because of the heat exchangers it needs. But eliminating high peak pressures, valves and valve gear, injection and timing systems may more than counterbalance cost of heat exchangers.

The improved Stirling engine may compete with or displace steam and gas turbines up to 5000-hp capacity. For large installations, weight and efficiency will be at least as good as those of high-economy diesel engines, and advantages will include quiet operation; invisible, odorless exhaust; adaptability to special fuels or heat sources; and maximum possibilities for efficient waste-heat recovery.

In small engine installations, 10 hp or less, the Stirling engine can be used where its significant efficiency and low noise level compensate for extra weight and cost. Immunity to ambient temperature and pressure conditions, insensitivity to type of fuel, and absence of wear and corrosion caused by dust, moisture, or gases in the air are advantages which cannot be matched by other existing engines now being widely used.

REVIEW TOPICS

1. With the aid of a P-V graph describe the differences between an ideal Otto engine and an actual spark-fired gasoline engine.

2. Why does the indicator card of an actual gasoline engine have a double loop?

3. How is the output varied in a gasoline-fired engine?

4. Compare the theoretical and actual valve actions of an Otto and gasoline-fired engine on a valve-timing diagram.

5. With the aid of sketches show the relation between an indicator card and a pressure-time diagram for a gasoline-fired engine.

6. Using a cross-section sketch of the piston and cylinder and a P-V graph show the operation of a two-cycle diesel engine.

7. With the aid of cross-section sketches describe the operation of an actual Stirling engine.

8. On P-V and T-S graphs show how theoretical and actual Stirling engines compare in operation.

CHAPTER 16

Steady Flow Energy Equation

We have studied the nonflow processes of gases based on certain simple ideas that we talked about in Chap. 6. From these we built up an analysis of the operation of reciprocating machinery such as air compressors and i-c engines. Now we are ready to deal with steady-flow processes.

16·1 Flow work. Machines such as turbines, compressors, furnaces, pumps, fans, and heat exchangers handle a steady flow of fluid. This usually enters and leaves the machine through a pipe or duct, Fig. 16·1.

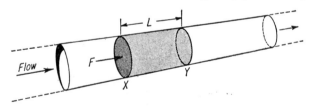

Fig. 16·1 Net force acting on a fluid flowing in a pipeline transmits energy that resides in the fluid when it moves. This is flow work energy.

Ordinarily the fluid moves into the entrance pipeline because the space ahead of it is being opened by the machine constantly removing the fluid from the entering pipeline at the machine impeller. The fluid moves through the discharge pipeline because the impeller forces it into the discharge opening and so forces fluid ahead of it. The fluid can move only if the fluid ahead has space to move into.

For some machines the impelling force may simply be the initial pressure of the fluid itself in the machine. No matter how this flow is created, the fluid moves because there is more pressure behind it than ahead of it. As a result there is a net force that moves it through the pipeline.

228

In Fig. 16·1 we have a continuous pipeline filled with fluid and for study choose a 1-lb slug of it occupying the length L between planes X and Y. A net force F moves this slug (and all the slugs ahead of it). But as this force F acts on the slug, it moves through a distance, and a force moving through a distance acting on a body does mechanical work. The fluid possesses this energy, since the work is done on it, and it can transfer this work to any object in its path. This energy is called *flow work*. As soon as the fluid stops flowing, it loses any flow work it had.

Now let us see how we can measure it. The work done on the slug in Fig. 16·1 is the flow work in foot-pounds:

$$W_f = FL \tag{16·1}$$

But the total net force acting is PA where A is the cross-sectional area of the fluid in square feet. Then

$$W_f = PAL \tag{16·2}$$

But AL is the volume of the slug of fluid V, so

$$W_f = PV \tag{16·3}$$

Then the flow work of any pound of moving fluid is simply the product of its absolute pressure in pounds per square foot absolute and its specific volume in cubic feet per pound. In one sense we can regard flow work as the ability of a fluid to transmit energy from the source of pressure or force. This term, of course, does not apply to any of the nonflow processes.

16·2 Kinetic energy. Never forget that flow work is associated with the force transmitted through the fluid. A moving fluid, however, also has kinetic energy because of its motion. Work had to be done in moving it from rest, and it can give up this energy by doing work on something else that brings it to rest (see Chap. 1).

The kinetic energy of a mass of fluid is given by

$$E_k = \frac{wv^2}{2g} \tag{16·4}$$

where E_k = kinetic energy, ft-lb
w = weight of fluid, lb
v = fluid velocity, fps
g = acceleration of gravity, ft per sec²

(See Art. 5·7 for derivation). This energy is separate and distinct from flow work, so be careful not to confuse the two energy forms.

16·3 Potential energy. In Chap. 1 we also discussed potential energy. It is measured by the mechanical work a body can do in falling from a high elevation to a lower elevation. We figure it as

$$E_p = wD \tag{16·5}$$

where E_p = potential energy, ft-lb
w = weight of fluid, lb
D = vertical distance between elevations, ft

This also applies in nonflow processes, but the change in elevation usually is so small that it can be neglected. In most steady-flow processes we shall also find that E_p is so small compared with the other energy forms that we can neglect it without significant error.

16·4 Mass flow. Figure 16·2 shows a steady-flow machine receiving a constant flow of working fluid and exhausting an equal amount through

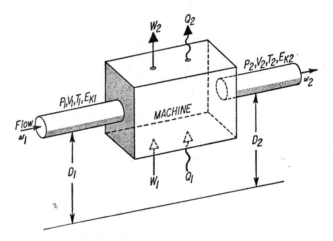

Fig. 16·2 Steady flow energy equation centers on a machine that receives and discharges energy in many forms; w_1 always equals w_2, the mass of working fluid passing through the machine.

the discharge line. The inlet and discharge lines may be of different areas A_1 and A_2; the inlet and outlet velocities will probably be different, v_1 and v_2; and the inlet and outlet states as measured by the specific volume will also differ, V_1 and V_2. Then for constant mass flow rate through a machine, in pounds per second,

$$M = \frac{A_1 v_1}{V_1} = \frac{A_2 v_2}{V_2} \tag{16·6}$$

If inflow and outflow are different in mass or weight, then, of course, Eq. 16·6 does not apply.

16·5 Steady flow energy equation. With the energy forms we have used for nonflow processes and the new ones we have learned or reviewed, we can list those to be applied to steady-flow processes:

1. Potential energy, E_p
2. Kinetic energy, E_k
3. Flow work, W_f
4. Internal energy, E
5. Mechanical or shaft work, W
6. Heat, Q

In Fig. 16·2 we show a machine, schematically, that receives and discharges a fluid at constant mass rate. To satisfy the First Law of Thermodynamics the energy that enters the machine must equal the energy that leaves it, providing that no energy accumulates in the machine and no stored energy in the machine drains off. This is true of most thermodynamic machines in practice once they carry a steady load.

We can express this first law in the steady flow energy equation:

$$E_{p1} + E_{k1} + W_{f1} + E_1 + W_1 + Q_1 \\ = E_{p2} + E_{k2} + W_{f2} + E_2 + W_2 + Q_2 \quad (16\cdot7)$$

All these energy quantities must be in the same units, for example, Btu per pound, so when we evaluate the individual terms, they must be alike. In this complete equation, if we know 11 terms we can figure the twelfth unknown. For our purposes we can immediately simplify Eq. (16·7) by noting that in any machine we deal with the change in E_p can be ignored, so we shall drop it from further discussion.

Substituting the property terms in (16·7) we get, in terms of Btu per pound,

$$\frac{wv_1^2}{2gJ} + \frac{P_1V_1}{J} + E_1 + W_1 + Q_1 = \frac{wv_2^2}{2gJ} + \frac{P_2V_2}{J} + E_2 + W_2 + Q_2 \quad (16\cdot8)$$

This outlines the information we must have to analyze the operation of a machine.

16·6 Enthalpy. The two properties internal energy E and flow work PV/J appear in combination so often that we find it convenient to combine them into one. So *by definition*

$$H = E + \frac{PV}{J} \quad (16\cdot9)$$

where H is the *enthalpy* in Btu per lb, a combination property of a fluid, and it applies only when the fluid *moves*, with some exceptions [see Eq. (6·12)]. Then our steady-flow equation reduces to

$$\frac{wv_1^2}{2gJ} + H_1 + W_1 + Q_1 = \frac{wv_2^2}{2gJ} + H_2 + W_2 + Q_2 \quad (16\cdot10)$$

In addition to the simple schematic sketch of Fig. 16·2, we can visualize the energy flow by a scaled diagram as in Fig. 16·3. We can identify the various forms of the energy entering and leaving by this manner, but it is impractical to show just how much of each converts to the other

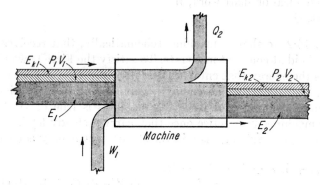

Fig. 16·3 Energy-flow diagram helps visualize the quantities involved and emphasizes the exact balance between energy in and out.

forms inside the device. But the total width of flow bands entering must be the same as the width of flow bands leaving. We have used this scheme in studying air compressors and i-c engines, which are essentially nonflow machines, but it applies best of all to steady-flow machines. This helps us imagine rivers of energy with all their incoming and outgoing branches.

Now that we have shaken down this equation let us study some simple cases and learn how to use it.

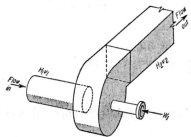

Fig. 16·4 Mechanical work input to the fan shaft adds energy to the air or gas flowing through the unit by raising the enthalpy and velocity of the fluid.

16·7 Fan. To apply Eq. (16·10) we must be able to identify all forms of energy entering a device, such as a fan, and all the forms leaving. The air, Fig. 16·4, enters with a given enthalpy H_1 and velocity v_1; it leaves with enthalpy H_2 and velocity v_2. We need work input W_1 to turn the fan rotor, and of course, it has no shaft work output. For most fans we have no heat transfer into or out of the fan, so we can write

$$\text{energy entering} = \text{energy leaving}$$

or
$$\frac{wv_1^2}{2gJ} + H_1 + W_1 = \frac{wv_2^2}{2gJ} + H_2 \qquad (16·11)$$

We can then solve for

$$W_1 = H_2 - H_1 + \frac{w}{2gJ}(v_2^2 - v_1^2) \qquad (16\cdot12)$$

This gives us directly the amount of work input we need to run the fan to produce given enthalpy and kinetic-energy changes. Sometimes the changes in kinetic energy are very small and can be ignored; then the energy equation reduces to simply $W_1 = H_2 - H_1$, a form we shall see quite frequently.

16·8 Gas turbine. The gas turbine (Fig. 16.5) converts energy in the gas flow to shaft work, so again, using our basic premise

energy entering = energy leaving

or

$$\frac{wv_1^2}{2gJ} + H_1 = \frac{wv_2^2}{2gJ} + H_2 + W_2$$

$$(16\cdot13)$$

Then if we know the state of the gas entering and leaving, we can figure the amount of shaft work output:

$$W_2 = H_1 - H_2 + \frac{w}{2gJ}(v_1^2 - v_2^2)$$

$$(16\cdot14)$$

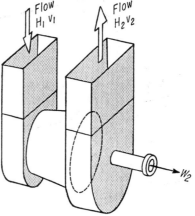

Fig. 16·5 Gas turbine develops mechanical shaft work output from the energy given up by the working fluid.

If the change in kinetic energy is small, as it usually is, the work output of the turbine is $W_2 = H_1 - H_2$. Compare this with the work input to the fan, above. Note how the subscripts arrange themselves always to give a positive answer for work input or work output.

16·9 Nozzle. Turbines use nozzles to direct a jet of working fluid over the blades and so convert some of the internal energy of the fluid into mechanical work. Let us study the energy balance around a nozzle. In Fig. 16·6 we see the entering h-p, P_1, fluid approach at a relatively low velocity v_1 and with an enthalpy of H_1. A jet of the working fluid leaves the nozzle in a region with lower pressure P_2; the jet has a high velocity v_2 and an enthalpy H_2. There is no heat transfer or mechanical work involved, so we can write

$$\frac{wv_1^2}{2gJ} + H_1 = \frac{wv_2^2}{2gJ} + H_2 \qquad (16\cdot15)$$

234   BASIC THERMODYNAMICS

The kinetic energy of the jet is our point of interest, so

$$\frac{w}{2gJ}(v_2^2 - v_1^2) = H_1 - H_2 \tag{16·16}$$

Often the entering velocity is negligible, so we can consider v_1 equal to zero and for 1 lb per sec flow write

$$v_2^2 = 2gJ(H_1 - H_2) \tag{16·17}$$

This gives a direct method of calculating the velocity of the jet. No W and Q terms appear because there is no work or heat transfer.

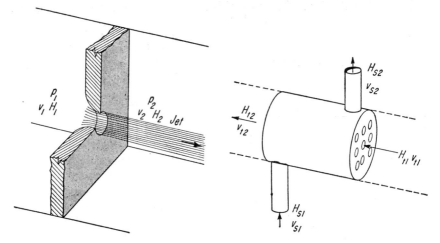

Fig. 16·6 Nozzle reorganizes motion in a pressurized fluid to produce a jet with high leaving speed.

Fig. 16·7 Heat exchanger brings two fluids in close contact to transfer internal energy from one to the other.

16·10 Heat exchanger. In the shell-and-tube exchanger of Fig. 16·7 we have two fluids to consider. By the first law we can write for the shell fluid

$$\frac{w_s v_{s1}^2}{2gJ} + H_{s1} = \frac{w_s v_{s2}^2}{2gJ} + H_{s2} + Q_{s2} \tag{16·18}$$

assuming that the fluid is cooled in passing through the exchanger. Since the velocity change is negligible in all cases, this simplifies to

$$H_{s1} = H_{s2} + Q_{s2} \tag{16·19}$$

and
$$Q_{s2} = H_{s1} - H_{s2} \tag{16·20}$$

We can write the same relations for the tube fluid, and since velocity changes are negligible and the fluid is heated,

$$H_{t1} + Q_{t1} = H_{t2} \tag{16·21}$$

and
$$Q_{t1} = H_{t2} - H_{t1} \tag{16·22}$$

But the heat given up by the shell fluid is the heat acquired by the tube fluid, so

$$Q_{t1} = Q_{s2} \tag{16.23}$$
and
$$H_{t2} - H_{t1} = H_{s1} - H_{s2} \tag{16.24}$$

16·11 Throttling. We throttle the flow of a gas or vapor primarily to control its rate of flow. This is done by making it pass through an opening smaller than the pipeline through which the fluid flows; a nozzle does essentially the same thing to produce a high-speed jet of fluid.

With a throttle we are not trying to reorganize the flow. Figure 16·8 shows a simple needle valve that can provide various sized annular openings for the fluid to pass through. Let us analyze what happens.

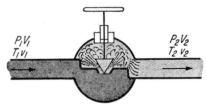

Fig. 16·8 Throttling a gas or vapor flow reduces the downstream pressure and controls the rate of flow through the system.

The fluid enters at a pressure P_1, a temperature of T_1, and a velocity of v_1. It leaves with a lower pressure P_2, a temperature of T_2, and a velocity of v_2. The change in velocity normally is small and can be ignored; there is no work or heat transferred during the process, so we have no W or Q terms to deal with.

So shaking down our steady flow energy equation,

$$\frac{wv_1^2}{2gJ} + H_1 + W_1 + Q_1 = \frac{wv_2^2}{2gJ} + H_2 + W_2 + Q_2 \tag{16.25}$$

for the throttling process; since $v_1 = v_2$ in most cases and both W and Q are zero, we have

$$H_1 = H_2 \tag{16.26}$$

Then if the fluid is a perfect gas with a constant specific heat, we find that

$$c_p T_1 = c_p T_2 \tag{16.27}$$

which reduces to

$$T_1 = T_2 \tag{16.28}$$

This means that the temperature of a perfect gas does not change when throttled to a lower pressure. Actual gases do drop somewhat, depending on pressure and temperature level; we shall study this later. We shall deal only with perfect gases when theorizing.

Constant temperature during throttling makes sense when we study the molecular kinetic theory. The rms average of the individual molecular velocities determines the temperature of the gas mass, (see Chap. 5). In passing through the throttle opening, the molecules do not give up or acquire energy, so their average velocity stays constant. This means that the gas temperature must stay constant.

Remember, in a perfect gas we assume the molecules so far apart that their mutual gravitational attraction can be ignored. But at certain pressures they are close enough together so that, when the gas expands, part of the kinetic energy of the molecules does work against these attractive forces. This slows them down, meaning that their temperature drops. The disappearing kinetic energy changes to potential energy. This explains the behavior of actual gases.

In the throttling process the gas pressure largely depends on the density D of the molecules in a given volume, where $D = 1/V$. Then $P = RDT$, and since both R and T stay constant during throttling, the pressure depends only on gas density. In essence, the throttle holds back some of the molecules in the flow to vary the density.

16·12 Nozzle flow. Figure 16·9 shows a schematic arrangement for testing the behavior of a gas flowing through a nozzle. A compressor

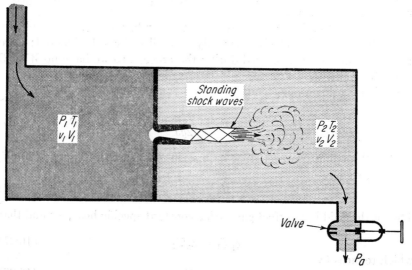

Fig. 16·9 Test arrangement for studying nozzle behavior receives gas at constant pressure P_1. Gas approaches the nozzle at a slow rate but speeds up in passing through the opening to the lower pressure region at P_2.

(not shown) supplies air at *constant* pressure P_1 to a large chamber feeding the air to the nozzle under test. Air flowing through the nozzle discharges into a second large chamber held at a pressure P_2. From here the air flows to the atmosphere through an outlet valve.

Closing the valve stops the air flow (assuming the supply compressor also stops) and the discharge pressure $P_2 = P_1$. As we gradually open the outlet valve, the air flow increases through the nozzle. By keeping P_1 constant, we find that P_2 will drop. By opening the valve more and more, we increase the air flow as P_2 drops. But beyond a certain level of P_2 the air flow reaches a maximum and stays there no matter how far P_2 drops.

Figure 16·10 shows how the air flow varies as the ratio of P_2/P_1 changes. At about a *critical pressure ratio* of 0.53 the air flow reaches its peak. As the pressure ratio drops below critical (by dropping P_2), the flow does not increase. Instead, stationary *shock waves* appear in the air flowing through the nozzle. Across these shock waves the air suddenly rises in pressure and density, changes temperature, and drops in velocity.

This action seems to contradict expectations that flow ought to vary directly with pressure difference across the nozzle. We begin to approach this expectation when we put a vacuum pump at the test tank outlet and bleed air into the tank at the inlet at constant P_1. Figure 16·10 shows that the flow at very low pressures is nearly proportional to the pressure ratio. At molecular flows where the molecules do not collide with one another, the flow is exactly proportional to pressure ratio. These l-p flows, however, do not ordinarily concern us in energy equipment, so let us return to the pressure level we shall work with in practical applications.

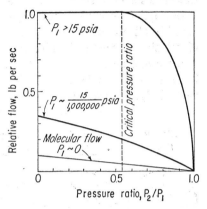

Fig. 16·10 Rate of gas flow through the nozzle depends on the upstream pressure and the ratio of downstream and upstream pressures.

Extreme crowding of the molecules in ordinary working gases makes the flows act in an unexpected fashion. Let us attempt to understand what happens to the molecules by a little speculation.

16·13 Molecular velocities. For future work we shall find the velocity of sound through quiet air of great importance. Many experiments and theory have shown it to be

$$v_s = \sqrt{gkRT} \quad \text{fps} \tag{16·29}$$

At 59 F or 519°R this works out to

$$v_s = \sqrt{32.17 \times 1.4 \times 53.3 \times 519}$$
$$= 1{,}115 \text{ fps}$$

By comparison, the molecular kinetic theory shows the rms average speed at 519°R to be 1,635 fps for the individual molecules of air. A sound wave propagates through a mass of air at a speed less than that of the average molecular velocity, the ratio being

$$\frac{v_{\text{mol}}}{v_s} = \frac{1{,}635}{1{,}115} = 1.465$$

In other words, the average speed of molecules is only 46.5% more than the speed of sound—a wave disturbance in the air.

In a quiet mass of gas we expect the direction of molecule travel to be equal in all directions, on the average. We could visualize this by a star of velocity vectors as in Fig. 16·11.

If we can now set a mass of quiet air moving into a duct by using an impeller, each molecule will have the overall velocity v added to it vectorially. Figure 16·11b shows the velocity distribution when the entire mass of gas moves with a subsonic velocity of v. This diagram suggests

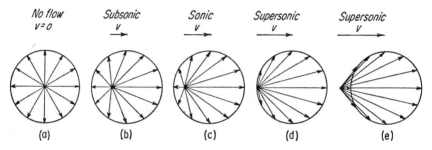

Fig. 16·11 In a stationary gas the molecular velocities radiate equally in all directions, but when gas as a whole moves with speed v, the molecular velocities become unequal.

that about one-third of the molecules "backflow" against the general mass flow direction.

When we step up the gas mass to the sonic velocity in Fig. 16·11c, about 25% of the molecules still backflow. At some supersonic velocity as in Fig. 16·11d or e all the molecules flow in the generally forward direction. This rough analysis assumes that all the molecules have the rms speed—the true situation is not that simple. Remember that molecular speeds in any mass of quiet gas vary from zero to more than three times the average (see Art. 5·11). This means that the spread of speed vectors is much broader in a gas flow than suggested by Fig. 16·11.

16·14 Nozzle molecular flow. In Fig. 16·12 we imagine the molecules of a gas near the mouth of the nozzle moving in all directions as suggested by the velocity vector star. For certain conditions only those molecules having a speed component in the direction of the nozzle mouth enter it and flow through to the l-p region.

For the condition assumed, this could mean that the gas density in the nozzle is less than in the upstream chamber. The mass flow could be supersonic in speed—at some point in the nozzle or just beyond it the speeding molecules pile up abruptly forming a shock wave. The molecules still have their individual speeds that they exchange billions of times a second with one another, but there are more of them, and the mass as a whole moves more slowly out of the nozzle into the lower-pressure area. Depending on the pressure ratio and temperature there may be several shock waves in the jet until it balances out to the discharge pressure P_2 (see Fig. 16·9).

16·15 Vortex flow. The most important use of fluid flow through nozzles is in gas and steam turbines. The nozzles organize some of the

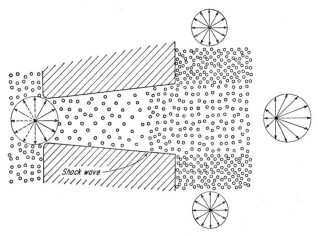

Fig. 16·12 Molecules racing through a nozzle at supersonic speeds catch up with molecules ahead grouped in higher density to form a shock wave that makes the overall flow slow up to slower leaving speed.

random motion of the molecules into powerful jets that can exert mechanical forces.

Figure 16·13 shows a simplified arrangement of a ring of nozzles in a stationary diaphragm receiving air flowing in an axial direction from the left. The nozzles direct this air flow into a series of jets that merge and spiral away toward the rotor buckets (not shown) in the direction of the arrows. This forms a *vortex* of air, shown separated to the right.

Whenever a fluid is forced to flow in a vortex, the circumferential speed of the fluid varies—faster on the inner edge and slower on the outer edge. The basic law for vortex flow states that the product of the circumferential velocity and its radius r, from the center, must be constant throughout the ring of flow, or

$$v_1 r_1 = v_2 r_2 = v_3 r_3 = \text{constant} \tag{16·30}$$

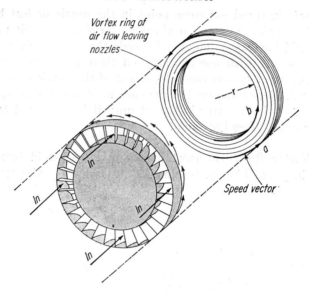

Fig. 16·13 Gas flowing through a ring of nozzles in the turbine leaves as a vortex before entering the blades; the inner edge of the vortex spins faster than the outer edge.

In Fig. 16·13, the arced vector a shows the distance a particle on the outer edge would travel while a particle on the inner edge would travel the longer distance b in the same time.

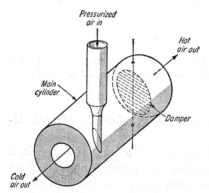

Fig. 16·14 In a Ranque-Hilsch tube the air jet enters the main cylinder tangentially to form a vortex that splits into hot and cold air jets and can be separated by controlling the damper position.

16·16 Ranque-Hilsch tube. The complex motion and energy relations of air are demonstrated by the Ranque-Hilsch tube. A Frenchman, Georges Ranque, received a basic U.S. patent in March, 1934, for the idea, and Rudolph Hilsch developed the tube further during World War II in Germany. Others are now developing it.

Figure 16·14 shows the simple arrangement of such a tube. A main cylinder receives a jet of air injected tangentially so the air stream is forced into a vortex flow. The vortex divides into two flows, so a stream of cold air can be withdrawn through a centered hole in a diaphragm in one end and a stream of hot air withdrawn from the other end past an adjustable damper.

The vortex flow takes place at slightly above atmospheric pressure in the cylinder. The damper seems to act primarily as a flow separator; both streams can leave from one end of the tube (as in the original Ranque tube), but the arrangement of Fig. 16·14 makes it easier to separate the flows.

Considerable experimentation has been and is going on in developing these tubes as sources of cold air, primarily for laboratory use. With pressurized air at 90 psia entering the nozzle, about half the air can be taken out of the main cylinder at about 60 F colder and the other half at about 40 F hotter than entering air.

These few examples of actual behavior of gas flows show that a steady-flow energy analysis will not necessarily predict all the results. To make a complete prediction or calculation we must use aerodynamic principles as well, but we shall not cover these in this book.

16·17 Blowdown cooling. Anyone who has blown down a compressed-air tank learns to expect the fast drop in temperature of the air staying in the tank. Let us see why the air should cool.

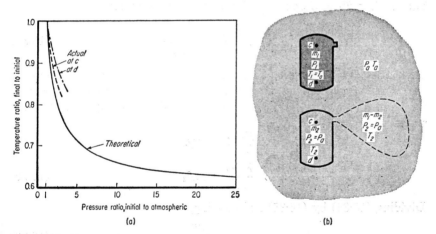

Fig. 16·15 Blowing down a pressurized air tank with the air at room temperature initially makes the air remaining in the tank cool off sharply, the drop depending on initial pressure. Thermodynamics assumes that tank air cools because it does work on the air pushed out to the atmosphere.

In Fig. 16·15b we have a tank of compressed air with a mass m_1 at a temperature $T_1 = T_a$ (the temperature of the atmosphere) and pressure P_1. The atmosphere surrounds the tank with pressure P_a.

When we open the valve, the tank air blows out at considerable speed. If m_2 lb of air stays in the tank, $(m_1 - m_2)$ lb leaves the tank and pushes back the atmosphere at constant pressure P_a. Let us concentrate on what happens to m_1.

Actually the discharged air and atmosphere mix, but in Fig. 16·15b we assume that the expanded air is separated by a tight, weightless, expandable envelope that exerts no pressure of its own. The air that has left the tank, $(m_1 - m_2)$ lb, has done flow work $P_a V_a/J$ Btu per lb on the atmosphere. The flow work has come from the original internal energy of the m_1 lb of pressurized tank air. Experience tells us that T_2 is lower than T_1. Let us balance out the energy before and after the blowdown.

Initially all we have is the internal energy of m_1, which is $m_1 E_1$. After the process has ended we have $m_1 E_2$ internal energy, and during the process we did flow work on the atmosphere of $(m_1 - m_2) P_a V_a/J$. From the first law we can then write

$$m_1 E_1 = m_1 E_2 + (m_1 - m_2) \frac{P_a V_a}{J} \qquad (16\cdot31)$$

Substituting, we get

$$m_1 c_v T_1 = m_1 c_v T_2 + (m_1 - m_2) \frac{R T_a}{J} \qquad (16\cdot32)$$

Substituting $T_1 = T_a$ and clearing this equation we get

$$\frac{T_2}{T_1} = 1 - \left(1 - \frac{m_2}{m_1}\right) \frac{R}{c_v J} \qquad (16\cdot33)$$

With a fixed tank volume V_t cu ft, the tank pressure depends on both the mass of gas in the tank and its temperature. Before expansion we have $V_1 = V_t/m_1$ and after expansion $V_2 = V_t/m_2$; then for both states

$$\frac{P_1 V_t}{m_1} = R T_1 \qquad (16\cdot34)$$

$$\frac{P_2 V_t}{m_2} = R T_2 \qquad (16\cdot35)$$

Dividing (16·34) by (16·35) and clearing,

$$\frac{P_1}{P_2} = \frac{m_1 T_1}{m_2 T_2} \qquad (16\cdot36)$$

Substituting symbols for the pressure and temperature ratios,

$$P_r = \frac{m_1}{m_2} \frac{1}{T_r} \qquad (16\cdot37)$$

$$\frac{m_2}{m_1} = \frac{1}{P_r T_r} \qquad (16\cdot38)$$

Substituting (16·38) in (16·33) and remembering that $R/c_v J = k - 1$, we have

$$T_r = 1 - \left(1 - \frac{1}{P_r T_r}\right)(k - 1) \qquad (16\cdot39)$$

Clearing this equation we find that

$$P_r = \frac{k-1}{T_r(T_r + k - 2)} \qquad (16\cdot 40)$$

This equation is plotted in Fig. 16·15a as the theoretical curve. Curves c and d show the temperatures actually found by experiment at point c near the tank discharge and at point d remote from the discharge. One experimenter found that the gas temperature near the tank outlet dropped as soon as the valve opened, as at c in Fig. 16·15b. But the air temperature remote from the outlet, as at d, did not start dropping until about 1 min later. The tank held about 20 cu ft.

Our thermodynamic analysis, making simplifying assumptions, also assumes that the gas acts homogeneously throughout its mass. So it is not surprising that the actual P-T behavior differs from the theoretical, since our assumptions do not match actual conditions.

On a molecular basis we can offer another explanation than the one assuming that the air in the tank works on the air leaving. Both tank and atmospheric air are at the same temperature, so the tank pressure is solely due to higher molecular density. Immediately after we open the outlet valve, molecules flow out of the tank because there are more molecules per cubic foot in the tank than in the atmosphere.

But molecular velocities of the tank air range from zero to over three times the rms average (Maxwell distribution). It seems that the faster molecules leave the tank first because of their higher speed. The lower average speed of the slower molecules left behind in the tank means that the air remaining will have a lower temperature.

The lag in temperature drop in remote parts of the tank shows that the diffusion of the higher velocities through the mass of air takes time; the air does not act homogeneously.

The separation of molecules by speed differences can take place in transient flows such as this where there is a limited source of air or gas. In steady flow from unlimited sources of gas there is a constant distribution of slow to fast molecules, so we do not find similar temperature drops in the source as for transient flows.

16·18 Air induction. Admitting atmospheric air to an evacuated flask is the converse process to blowing down a tank of pressurized air. Experiments show that, when we suddenly admit a burst of air into an evacuated flask, Fig. 16·16b, and immediately recork the flask, the trapped air has become very hot inside the flask.

It is practically impossible to get a perfect vacuum in the flask; the flask initially contains a small amount of air m_t at a state P_t, V_t, T_1. This mass gets compressed by the incoming air m_a and has a common final state with it of P_1, V_2, T_2. Conventional thermodynamic theory assumes that the atmosphere does flow work on the mass m_a entering the tank; this mass takes with it the flow-work energy and its own original internal energy E_{a1} when it enters the flask.

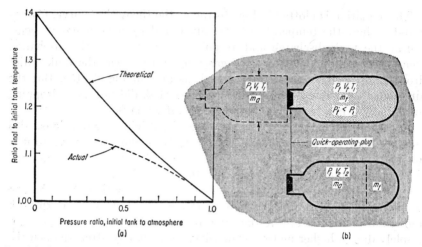

Fig. 16·16 Split-second opening and closing of a bottle originally holding a vacuum makes the entering air trapped in the bottle very hot. Thermodynamic analysis assumes work done on incoming air by the atmosphere gives extra energy to the trapped air and raises its temperature sharply.

Then accounting for all the energy before and after the process, which must be equal according to the first law,

$$m_a E_{a1} + m_t E_{t1} + \frac{m_a P_1 V_{a1}}{J} = m_a E_{a2} + m_t E_{t2} \qquad (16 \cdot 41)$$

Substituting $PV = RT$ and $E = c_v T$, we get

$$m_a \frac{RT_1}{J} = (m_a + m_t) c_v (T_2 - T_1) \qquad (16 \cdot 42)$$

Substituting $R/Jc_v = k - 1$ and clearing,

$$\frac{m_a}{m_a + m_t} = \frac{1}{k-1} \left(\frac{T_2}{T_1} - 1 \right) \qquad (16 \cdot 43)$$

Now to eliminate the masses in favor of the pressures they exert, let S_t be the flask volume in cubic feet; then the specific volumes of m_t and $m_t + m_a$ are

$$V_t = \frac{S_t}{m_t} = \frac{RT_1}{P_t} \qquad (16 \cdot 44)$$

$$V_{t+a} = \frac{S_t}{m_a + m_t} = \frac{RT_2}{P_1} \qquad (16 \cdot 45)$$

STEADY FLOW ENERGY EQUATION

Then from (16·44) and (16·45),

$$S_t = m_t \frac{RT_1}{P_t} = (m_a + m_t) \frac{RT_2}{P_1} \tag{16·46}$$

Clearing and transposing, we have

$$\frac{m_a}{m_a + m_t} = 1 - \frac{T_2 P_t}{T_1 P_1} \tag{16·47}$$

Equating (16·43) and (16·45),

$$1 - \frac{T_2 P_t}{T_1 P_1} = \frac{1}{k-1}\left(\frac{T_2}{T_1} - 1\right) \tag{16·48}$$

Substituting pressure- and temperature-ratio symbols,

$$1 - T_r P_r = \frac{1}{k-1}(T_r - 1) \tag{16·49}$$

$$T_r = \frac{k}{1 + P_r(k-1)} \tag{16·50}$$

This equation gives us the temperature rise in the flask for given initial pressure ratios in the flask before admitting air. Figure 16·16a shows the corresponding curve, marked *theoretical*.

The *actual* curve in Fig. 16·16a shows the experimental results. Final trapped-air temperature never becomes quite as hot as predicted. Again predictions may be too high because of the thermodynamic assumptions made of homogeneous action of the atmosphere on the air entering.

Offering an alternative molecular explanation, we again turn to the Maxwell velocity distribution as a possibility. Before the flask is opened, molecules with a wide range of speeds collide with the plug. When we momentarily remove the plug, the fastest molecules get in first; if we replug quickly enough, we keep out the slower molecules. This raises the average temperature of the molecules trapped in the flask higher than that of the atmosphere where they originated.

As we start with lower pressures, that is, less m_t in the flask, there is less diluting effect on the entering fast molecules and so we get higher flask temperatures. With zero initial flask pressure and making the entering molecules travel a longer path, we theoretically could get better separation between fast and slow molecules. The experiment suggested below could cause such a separation and confirm this hypothesis if it is correct.

16·19 Suggested experiment. This experiment using a length of pipe and two quick-closing valves, Fig. 16·17, might demonstrate the separation of molecules during transient flow because of the Maxwell velocity distribution. This is an extension of the vacuum-bottle experiment in Fig. 16·16.

A length of pipe closed at one end has a quick-opening and -closing inlet valve at the open end and a quick-closing sectionalizing valve near the closed end to isolate a relatively small chamber.

To start the experiment, close the inlet valve and keep the sectionalizing valve open as in step 1 in Fig. 16·17, and pump the pipe down to as near a perfect vacuum as possible.

In step 2 open the inlet valve instantaneously and immediately slam shut both inlet and sectionalizing valves. This operation must be done in a fraction of a second—the shorter the pipe length, the shorter the time.

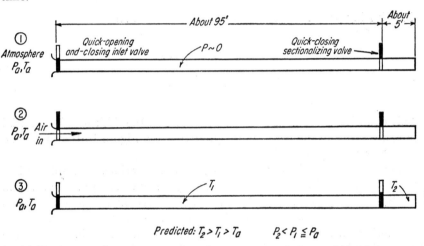

Fig. 16·17 Suggested experiment to show that an evacuated pipe would hold air at two higher temperatures than atmospheric when a short burst of air is allowed to enter and be trapped.

In step 3 we expect to find the temperature in the small chamber higher than the temperature in the large section and both higher than atmospheric temperature, where the air originated.

Theoretically, the fastest molecules should reach the closed pipe end first, leaving slower molecules behind in the first section, separated by the sectionalizing valve. The end chamber trapping the fastest molecules should be hottest, while the main chamber holding a mixture of fast and slower molecules should not be so hot.

The author would like to know if anyone has performed this or a similar experiment.

REVIEW TOPICS

1. What is flow work of a moving fluid?
2. Does a stationary fluid have flow work?

STEADY FLOW ENERGY EQUATION   247

3. Derive the *P-V* formula for flow work of a fluid.
4. How do we evaluate the kinetic energy of a fluid?
5. How is potential energy of a fluid figured?
6. Show how inlet and outlet areas, fluid velocities, and specific volumes are related when the mass flow through a thermodynamic machine remains constant.
7. Write the complete steady flow energy equation. Explain each term.
8. What is the enthalpy of a fluid? What is its physical significance?
9. Derive the steady flow energy equation for a fan or blower.
10. Derive the steady flow energy equation for a gas turbine.
11. Derive the steady flow energy equation for a gas flowing through a nozzle.
12. Derive the steady flow energy equation for a shell-and-tube heat exchanger relating the two fluids passing through the unit.
13. Derive the steady flow energy equation for the throttling process of a perfect gas.
14. What is the fundamental characteristic of a throttling process?
15. With the aid of a graph explain how gas flow through a nozzle behaves as the exhaust pressure drops while the inlet pressure stays constant.
16. How are sonic velocity and molecular velocity of a gas related?
17. How does air behave when in vortex flow? What law governs?
18. What is the function of the Ranque-Hilsch tube?
19. Describe the events of blowing down a tank of compressed air to atmospheric pressure. What molecular behavior is involved?
20. Describe the experiment of admitting a brief burst of atmospheric air into an evacuated flask. What molecular behavior is involved?

PROBLEMS

1. Water weighing 62.4 lb per cu ft flows through a pipe under a pressure of 200 psig. What is its flow work?
2. Air at 120 psig pressure and 150 F flows through a pipe. Calculate its flow work.
3. If the water in Prob. 1 moves through the pipe at 15 mph, what is its kinetic energy per pound?
4. The air in Prob. 2 flows at 75 mph through the pipe. Calculate its kinetic energy per pound.
5. The water in Prob. 1 flows through a vertical pipe upward for 50-ft rise. What is its gain in potential energy per pound?
6. The air in Prob. 2 flows through a vertical pipe upward for a 300-ft rise. Calculate its gain in potential energy per pound.
7. A machine receives a flow of gas of 150 lb per hr through an open-

ing of 2-sq-in. area. The specific volume of the incoming gas is 3.5 cu ft per lb. The gas leaves the machine through an opening of 0.1 sq in. with a specific volume of 0.8 cu ft per lb. Find the entering and leaving velocities of the gas.

8. A gas at 30 psig pressure with a specific volume of 7.2 cu ft per lb has an internal energy of 20 Btu per lb. What is its enthalpy?

9. A fan takes in air at a speed of 5 fpm, and the air has an enthalpy of 20 Btu per lb. The fan discharges the air with an enthalpy of 25 Btu per lb at a speed of 30 fpm. How much work is done by the fan per pound of air? Give the answer in Btu and foot-pounds.

10. A gas turbine receives gas with an enthalpy of 400 Btu per lb entering at 3,000 fpm. The gas leaves with an enthalpy of 250 Btu per lb and a speed of 4,500 fpm. How much work has been done by each pound of the gas in the turbine? How much has been contributed by the change from entering to leaving speed?

11. Gas enters a nozzle with an enthalpy of 350 Btu per lb at a mass flow rate of 2 lb per sec. The gas leaves with an enthalpy of 225 Btu per lb. If the entering velocity is close to zero (negligible), what is the leaving velocity of the gas? What must the entering gas velocity be to raise the leaving velocity by 10%?

12. A shell-and-tube exchanger transfers heat from a gas to a liquid. The gas has an entering enthalpy of 260 Btu per lb and a leaving enthalpy of 180 Btu per lb; the liquid an entering enthalpy of 90 Btu per lb. If there is a flow of 10 lb of gas for every pound of liquid through the exchanger, what is the leaving enthalpy of the liquid?

13. A valve throttles a gas with an entering enthalpy of 210 Btu per lb to half its entering pressure. What is the leaving enthalpy of the gas? What is the leaving temperature relative to the entering temperature?

14. A gas enters a nozzle at a pressure of 150 psig. At what exhaust pressure will the gas reach its maximum flow through the nozzle?

15. What is the velocity of sound in air at 0 F? At 120 F? What is the velocity of sound in hydrogen at −30 F? At 200 F?

16. A gas-turbine nozzle ring as in Fig. 16·13 has an outer diameter of 4 ft 6 in. and an inner diameter of 4 ft. If the inner edge of the gas vortex leaving the ring has a velocity of 3,500 fpm, what is the velocity of the outer edge?

17. Make a series of calculations to reproduce the theoretical curve in Fig. 16·15 for air. Calculate the same graph for hydrogen.

18. Make a series of calculations to reproduce the curve in Fig. 16·16 for air. Calculate the same graph for helium.

CHAPTER 17

Gas-turbine Cycles

17·1 Gas-turbine engine. Now that we know the basic thermodynamics of steady-flow working fluids, we are ready to study the well-known gas-turbine engine, Fig. 17·1. The simple open-cycle engine has three main parts: (1) rotary compressor C, (2) combustor or furnace F,

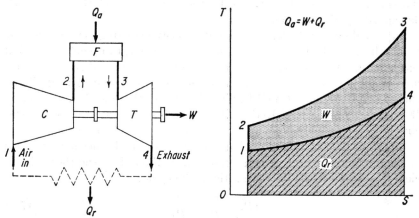

Fig. 17·1 Simple open-cycle gas-turbine unit, left, has three major elements: C, compressor; F, combustor; T, turbine. Compressed air heated in the combustor drives the turbine, which in turn drives the compressor and also produces net cycle work output.

and (3) gas turbine T. The unit gets its name from the turbine which takes its energy input from hot pressurized gas flowing through it.

Compressor C draws air in steady flow from the atmosphere and pressurizes it adiabatically. The air then flows into the furnace, or combustor. Fuel injected into the air flowing through the furnace burns to raise the temperature of both the air and combustion gases, at constant pressure.

The heated pressurized air-gas mixture then flows into the turbine and does mechanical work on its rotating shaft. The air-gas mixture expands and cools as it gives up energy to do this work and then exhausts to the atmosphere. The turbine develops all the mechanical work of the cycle, part being used to drive the compressor; the remainder is net output of the cycle.

17·2 Cycle analysis. To simplify analysis we shall assume that the working fluid is air throughout the cycle, that is, we shall use an air standard. We shall regard the heat of combustion as heat energy transferred to the working fluid, Q_a. In Fig. 17·2 we analyze the operation of each element in terms of the steady flow energy equation and assume that the kinetic energy of the air stays constant throughout the cycle.

The lower three sketches of Fig. 17·2 give the P-V graphs for the compressor, turbine, and total engine. The corresponding T-S graph for the cycle is in Fig. 17·1. Process 1–2 raises the air pressure and temperature at $S = C$ by absorbing work input W_c. Process 2–3 in the combustor raises the air temperature at $P = C$ by adding the cycle energy input Q_a. Process 3–4 in the turbine develops the gross mechanical work W_t by expanding the air at $S = C$. The WORK OUTPUT graph of Fig. 17·2 shows that the compressor absorbs the major part of the gross work developed by the turbine.

An actual engine works with a continuously changing mass of air, but to simplify, we shall assume that the turbine exhaust air cools in flowing through the atmosphere, giving up rejected heat Q_r, and reenters compressor intake to form an equivalent closed cycle. Summarizing all the energy transfers under CYCLE, Fig. 17·2, we see that ideally and actually $W = Q_a - Q_r$. The ENERGY FLOW sketch shows that the compressor circulates a relatively large amount of energy through the cycle.

Equation (22) of Fig. 17·2 gives the cycle thermal efficiency in terms of the air temperatures. A more practical method gives the thermal efficiency in terms of the pressure ratio of the cycle:

$$P_r = \frac{P_2}{P_1} = \frac{P_3}{P_4} \tag{17·1}$$

For the isentropic processes (Art. 6·9), temperatures and pressures are related by

$$\frac{T_2}{T_1} = \left(\frac{P_2}{P_1}\right)^{(k-1)/k} = P_r^{(k-1)/k} \tag{17·2}$$

$$\frac{T_3}{T_4} = \left(\frac{P_3}{P_4}\right)^{(k-1)/k} = P_r^{(k-1)/k} = \frac{T_2}{T_1} \tag{17·3}$$

Transposing the temperature ratios,

$$\frac{T_4}{T_1} = \frac{T_3}{T_2} \tag{17·4}$$

GAS-TURBINE CYCLES

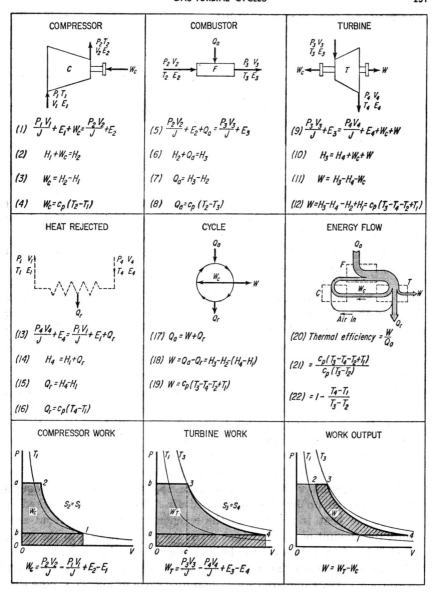

Fig. 17·2 Steady-flow energy analysis of each element of the gas-turbine engine shows how they fit together to develop net work output from heat input. P-V graphs for the compressor, turbine, and total unit correlate with the T-S graph in Fig. 17·1 and show the work distribution.

Subtracting 1 from each side we get

$$\frac{T_4}{T_1} - 1 = \frac{T_3}{T_2} - 1 \tag{17.5}$$

$$\frac{T_4 - T_1}{T_1} = \frac{T_3 - T_2}{T_2} \tag{17.6}$$

$$\frac{T_4 - T_1}{T_3 - T_2} = \frac{T_1}{T_2} = \frac{1}{P_r^{(k-1)/k}} \tag{17.7}$$

Substituting in Eq. (22) of Fig. 17·2 we get

$$\text{Thermal eff} = 1 - \frac{1}{P_r^{(k-1)/k}} \tag{17.8}$$

Processwise the gas-turbine cycle is the same as the Joule or Brayton cycle (Art. 11·4) developed for reciprocating engines. Figure 11·4 shows how the thermal efficiency of the cycle varies with pressure ratio. It also shows that the work output of the cycle reaches a maximum at a certain pressure and temperature ratio.

17·3 Regenerative cycle. The T-S graph, Fig. 17·1, shows that heat is added and rejected from the gas-turbine cycle at temperatures varying over a wide range. The Carnot principle shows that to achieve high efficiency we must absorb Q_a at the highest possible level and reject Q_r at the lowest possible level.

We can do something about this by noting that most practical gas-turbine engines exhaust their gas at temperatures higher than the air discharged by the compressor before it enters the combustor, Fig. 17·1. By passing the air circuit through a heat exchanger called a *regenerator*, we can heat the air to the exhaust temperature before it enters the combustor, Fig. 17·3. Ideally, the turbine exhaust cools to the exhaust temperature of the compressor. This makes $T_3 = T_5$ and $T_6 = T_2$.

Figure 17·3c shows that the heat given up by the exhaust gas equals the heat picked up by the compressed air. The two areas marked Q_t measure the heat transferred in the regenerator. The right-hand area shows the amount of rejected heat saved by recirculating it in the cycle. The left-hand area shows the reduction in cycle heat input Q_a.

Figure 17·3b shows the heat added and rejected in the regenerative gas-turbine cycle. Compared with the simple cycle in Fig. 17·1, note that the average temperature of heat addition from T_3 to T_4 has been raised in Fig. 17·3b and the average temperature of heat rejection from T_6 to T_1 has been lowered. This raises the cycle thermal efficiency.

The net work of the cycle $W = $ area 1-2-3-4-5-6 even though the areas for Q_a and Q_r are not coincident. This follows from

$$\begin{aligned} W &= Q_a + Q_t - (Q_r + Q_t) \\ &= Q_a - Q_r \end{aligned} \tag{17.9}$$

GAS-TURBINE CYCLES

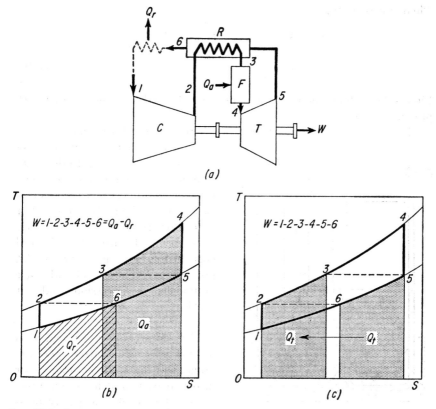

Fig. 17·3 Regenerative gas-turbine cycle passes the turbine exhaust through heat exchanger R to heat pressurized air entering the combustor. The circulated energy Q_t of graph (c) helps reduce both heat added and rejected to improve the overall thermal efficiency of the cycle.

Next let us find the thermal efficiency for this improved cycle:

$$e_t = \frac{W}{Q_a} = \frac{Q_a - Q_r}{Q_a} = 1 - \frac{Q_r}{Q_a}$$

$$= 1 - \frac{c_p(T_6 - T_1)}{c_p(T_4 - T_3)}$$

$$= 1 - \frac{T_2 - T_1}{T_4 - T_5}$$

Multiplying the numerator of the fraction by T_1/T_1 and the denominator by T_4/T_4 we get

$$e_t = 1 - \frac{T_1}{T_4} \frac{(T_2/T_1) - 1}{1 - (T_5/T_4)}$$

But

$$\frac{T_2}{T_1} = \frac{T_4}{T_5} = P_r^{(k-1)/k} = A$$

Then substituting,

$$e_t = 1 - \frac{T_1}{T_4}\frac{A-1}{1-(1/A)} = 1 - \frac{T_1}{T_4}A$$

$$= 1 - \frac{T_1}{T_4}P_r^{(k-1)/k} \tag{17·10}$$

Figure 11·4 shows how the regenerative thermal efficiency varies with pressure ratio.

Figure 17·4 shows the energy-flow diagram for an ideal regenerative cycle. The recirculated energy Q_t is of the same order of magnitude as Q_a.

When irreversibilities are taken into account, cycle thermal efficiencies are considerably lower. For practical performances see "Electric Generation—Hydro, Diesel and Gas-turbine Stations" by Skrotzki, pp. 407–417. These cycles also use intercooled compression and reheating.

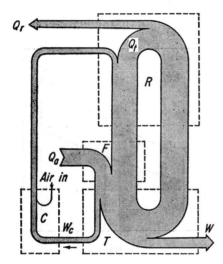

Fig. 17·4 Energy-flow diagram for the regenerative gas-turbine cycle has a large quantity of energy recirculating within the cycle.

17·4 Intercooling. In Fig. 17·5 we use the regenerative-cycle arrangement but split the compressor into two parts and insert an intercooler. This cools the partly pressurized air back to atmospheric at state 1, before completing compression to the top pressure at 4 and sending air to the regenerator.

17·5 Reheating. We split the turbine in two parts, letting the hot gas expand partly in the first section. Then we reheat the partly expanded gas in a second combustor F_2 back to initial temperature so $T_6 = T_8$. The l-p gas then completes its expansion to atmospheric at 9.

17·6 Heat transfer. If we again assume this as an equivalent closed air-standard cycle, we can apply the first law and write

$$W + Q_{r1} + Q_{r2} = Q_{a1} + Q_{a2} \tag{17·11}$$

and so

$$W = (Q_{a1} + Q_{a2}) - (Q_{r1} + Q_{r2}) \tag{17·12}$$

In effect this cycle splits the heat additions and rejections into two parts compared with the simple cycle (see Fig. 17·1). Figure 17·6 shows the

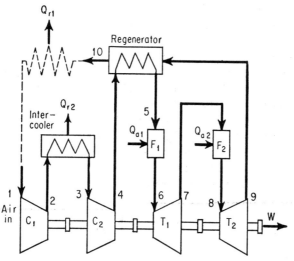

Fig. 17·5 Regenerative cycle with intercooling and reheating gives most opportunity for raising efficiency at cost of more equipment.

T-S graph for this cycle under ideal conditions. In this graph the states correspond to the points in Fig. 17·5.

According to Eq. (17·12) the net area of the cycle loop W is the net work output of the cycle. We can show this in Fig. 17·6 by remembering that the total area under 9–10 is the heat given up by the turbine exhaust in the regenerator, Q_t. This heat transfers to the pressurized air and equals the area under process 4–5. Then

$$W = Q_t + Q_{a1} + Q_{a2} \\ - (Q_t + Q_{r1} + Q_{r2})$$

Canceling Q_t's gives us Eq. (17·12).

To understand what the intercooling and reheating does for the cycle, we have to refer back to the basic regenerative cycle, Fig. 17·7. Here we do all the compressing in one step or process 1–2 and all the expansion in one process 4–5 instead of breaking them up as we have done in Fig. 17·6.

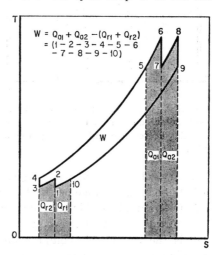

Fig. 17·6 The ideal intercooled and reheated regenerative cycle receives heat input at the highest average temperature and rejects unused energy at the lowest temperature.

The net loop W measures the work output of the cycle, and of course, $W = Q_a - Q_r$. Remember that heat transferred in regenerator Q_t equals the areas under 2–3 and 5–6. This cycle has the same pressure ratio, compressor-inlet state, and turbine-inlet temperature as the cycle of Fig. 17·6, yet it has a lower overall thermal efficiency. The W output is less by the two "ears" shown on each end of the cycle. But most significantly the average temperature of heat addition during 3–4 is lower than the heat addition during 5–6, 7–8 of Fig. 17·6.

At the other end of the cycle we find a like comparison. The average temperature of heat rejection during 6–1 in Fig. 17·7 is higher than the average during 10–1, 2–3 of Fig. 17·6. So we see that intercooling lowers the average temperature of heat rejection and reheating raises the average temperature of heat addition to raise the overall cycle thermal efficiency in accordance with the basic Carnot principles.

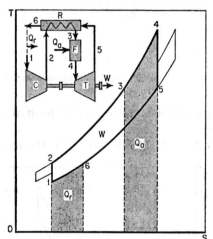

Fig. 17·7 Regenerative cycle has lower work output and thermal efficiency when not using intercooling and reheating of Fig. 17·6.

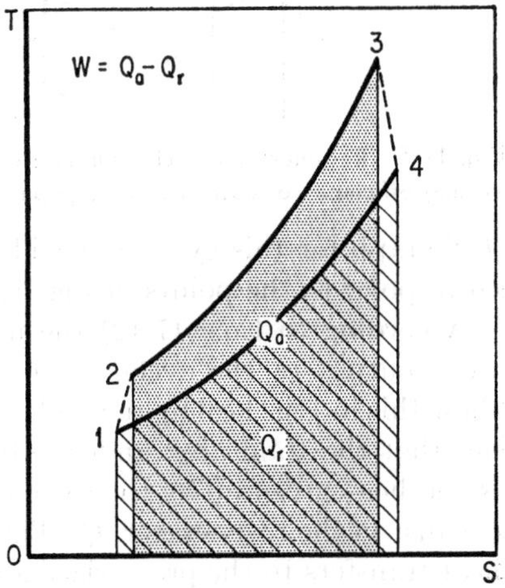

Fig. 17·8 The simple-cycle gas turbine with 80% efficient compressor and turbine has net work output less than net area of loop.

17·7 Actual cycles. Studying ideal cycles always helps to separate the effects of basic arrangements. This we have done; now let us study the effects of actual machines in the gas-turbine cycle.

Figure 17·8 shows the T-S graph for a simple-cycle gas turbine. Process 1–2 is the irreversible compression of the incoming air in a compressor working at 80% efficiency. Process 2–3 is the constant-pressure heating in the combustor. Process 3–4 is the irreversible expansion of the heated pressurized air in the turbine at 80% efficiency. Process 4–1 is the constant-pressure heat rejection to the atmosphere between the turbine exhaust and compressor air inlet.

This does not show all the irreversibilities possible. Others that can be important would be pressure drops between atmosphere and compressor

GAS-TURBINE CYCLES

first stage, between compressor last stage and combustor, in the combustor, between combustor and turbine first stage, between turbine last stage and atmosphere. Radiation from all parts of the cycle would also lower efficiency. But let us keep this simple and just study the most important effects—the efficiency of the compressor and the turbine.

We define compressor efficiency e_c as

$$e_c = \frac{\text{ideal work}}{\text{actual work}} = \frac{h_1 - h_{2s}}{h_1 - h_2} \qquad (17\cdot13)$$

This shows that the actual compressor work is more than the ideal at constant entropy. The irreversibility, of course, makes the entropy increase at state 2 in Fig. 17·8. We draw the process 1–2 dotted because no heat is transferred during the adiabatic compression and the area below has no meaning.

We assume during the heating 2–3 that the pressure stays constant and the process is reversible. So the total gray area under the process measures the heat added to the cycle, Q_a.

We define turbine engine efficiency e_e as

$$e_e = \frac{\text{actual work}}{\text{ideal work}} = \frac{h_3 - h_4}{h_3 - h_{4s}} \qquad (17\cdot14)$$

During its irreversible adiabatic expansion the gas entropy increases in the turbine. Again because the area under the process has no meaning, we draw the process dotted.

During the constant-pressure cooling 1–4 the cycle rejects the unusable heat Q_r to atmosphere. Even for an irreversible cycle the first law applies and $W = Q_a - Q_r$. The net work is the difference of the shaded and crosshatched areas. But these areas are not coincident, so no net area measures W. Note that the loop of processes 1–2–3–4 does *not* measure W. This always holds true when we have irreversible processes in a cycle.

In Fig. 17·8 the less efficient the compressor and turbine, the greater the increase in the gas entropy changes. The units can become so inefficient that Q_r will equal Q_a and make $W = 0$. On the graph this would show by Q_a becoming narrower and Q_r becoming wider as efficiencies drop.

17·8 Cycle efficiency. Figure 17·9a shows the variation in simple-cycle thermal efficiency with machine efficiency. By machine efficiency we mean that for the 90% curve, both compressor and turbine efficiency are each 90%, and so on. In Art. 17·2 we learned that the ideal simple-cycle efficiency depended only on the overall pressure ratio—the 100% curve defines this relation.

For irreversible cycles the machine efficiencies, air-inlet temperature, and turbine-inlet temperature have a marked effect on cycle efficiency.

In Fig. 17·9a the curves apply to a cycle with 60 F air inlet, 1500 F turbine inlet, 100% combustion efficiency, and no pressure losses in the

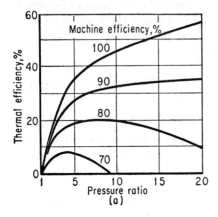
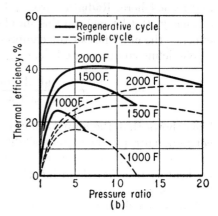

Fig. 17·9 (a) Compressor, turbine efficiencies, and pressure ratios control simple-cycle thermal efficiency. (b) Turbine inlet temperature and pressure ratio control efficiencies of simple and regenerative cycles. Beyond the intersection of simple- and regenerative-cycle efficiency curves, the latter becomes less than the former.

cycle. Practical machine efficiencies lie in the range of 80 to 90%. The curves show that cycle efficiency peaks at a certain pressure ratio; the optimum pressure ratio drops as the machine efficiency drops. When machine efficiency drops to 60% for these conditions, the cycle cannot even run itself because of the irreversible losses.

In contrast the 100% curve has no peak but continues rising and is asymptotic to the Carnot-cycle efficiency, the ideal for the cycle.

17·9 Regenerative cycle. Figure 17·10 shows the T-S graph for a practical cycle layout as shown in Fig. 17·7. The cycle T-S loop is a duplicate of that for the simple cycle as in Fig. 17·8, but the regenerator fixes certain states during the heating and cooling processes that affect cycle efficiency.

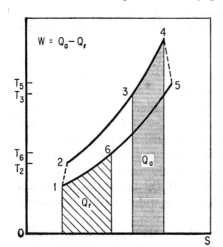

Fig. 17·10 A regenerator improves actual gas-turbine performance even though there must be temperature drops in the heat exchanger.

In an ideal regenerator $T_3 = T_5$ and $T_2 = T_6$, but since an actual heat transfer needs a temperature difference, we find $T_3 < T_5$ and $T_2 < T_6$.

GAS-TURBINE CYCLES 259

The degree to which these temperatures can be made equal in a regenerator is measured by its effectiveness, which depends largely on the amount of heat-transfer surface. We measure regenerator effectiveness e_r as

$$e_r = \frac{\text{actual heat transfer}}{\text{potential heat transfer}}$$
$$= \frac{c_p(T_3 - T_2)}{c_p(T_5 - T_2)} \tag{17.15}$$

Regardless of the actual effectiveness, the heat absorbed by the pressurized air equals the heat given up by the turbine exhaust in the regenerator. This means then that the area under 2–3 equals the area under 5–6. The area under the latter will be narrower because of the higher average temperature. As the effectiveness decreases, the widths of Q_a and Q_r increase, showing that the overall cycle efficiency drops, because the average temperatures of heat addition and rejection approach each other.

17·10 Turbine temperatures. Figure 17·9b compares the performance of simple and regenerative cycles at three turbine-inlet temperatures—1000, 1500, and 2000 F. These cycles are based on 60 F inlet air, 85% machine efficiencies, 70% regenerator effectiveness, 100% combustion efficiency, and no pressure losses.

Each of the simple and regenerative cycles has an optimum pressure ratio for each turbine temperature. The regenerative cycles are more efficient than the simple cycles at the same turbine temperatures, and at markedly lower pressure ratios. As we might expect, rising turbine temperature improves cycle thermal efficiency. Metallurgical limitations keep most practical temperatures below 1500 F in our modern gas turbines, but these curves show why designers make continuing efforts to find ways of using the highest possible temperature by introducing a variety of blade and bucket cooling methods.

Each of the regenerative curves in Fig. 17·9b intersects the simple-cycle curves at some pressure ratio. As pressure ratio rises, the compressor discharge temperature T_2 in Fig. 17·10 approaches the turbine discharge temperature T_5. With a high enough pressure ratio T_2 will exceed T_5. Then the turbine exhaust would cool the compressor exhaust in the regenerator instead of heating it. This, of course, lowers the efficiency below that of the simple cycle.

17·11 A simple-cycle gas turbine. Example: The ideal gas-turbine engine of Fig. 17·1 takes in air at 80 F and 14.7 psia to compress it at constant S through a pressure ratio of 4.0. The combustor raises the air temperature to 1500 F. The heated air expands to 14.7 psia at constant S in the turbine. Assume that $k = 1.3$ and $c_p = 0.28$. Find (a) the compression work; (b) the heat input in the combustor; (c) the expansion work; (d) the thermal efficiency from (a), (b), and (c); (e) the thermal efficiency from Eq. (17·8); (f) the proportion of turbine work absorbed by the compressor; (g) the energy in the exhaust air above atmospheric conditions; (h) the pressure at the turbine inlet.

Solution: From Eq. (17·7)

(a) $T_2 = T_1 P_r^{(k-1)/k} = (460 + 80)4^{(1.3-1)/1.3} = 742°R$
$W_c = c_p(T_2 - T_1) = 0.28(742 - 540) = 56.5$ Btu per lb air

(b) $Q_a = c_p(T_3 - T_2) = 0.28[(1500 + 460) - 742] = 341$ Btu per lb air

(c) $T_4 = \dfrac{T_3}{P_r^{(k-1)/k}} = \dfrac{1960}{1.375} = 1425°R$
$W_t = c_p(T_3 - T_4) = 0.28(1960 - 1425) = 150$ Btu per lb air

(d) Thermal eff $= \dfrac{W}{Q_a} = \dfrac{W_t - W_c}{Q_a} = \dfrac{150 - 56.5}{341} = 0.274$ or 27.4%

(e) Thermal eff $= 1 - \dfrac{1}{P_r^{(k-1)/k}} = 1 - \dfrac{1}{1.375} = 0.274$ or 27.4%

(f) $\dfrac{W_c}{W_t} = \dfrac{56.5}{150} = 0.377$ or 37.7%

(g) Energy in exhaust $= c_p(T_4 - T_1) = 0.28(1425 - 540) = 248$ Btu per lb air

$Q_r = Q_a - W = 341 - 93.5 = 247.5$ Btu per lb air

(h) $P_3 = P_1 P_r = 14.7 \times 4 = 58.8$ psia

17·12 A regenerative-cycle gas turbine. Example: The ideal gas-turbine engine of Fig. 17·3 has same operating factors as in Art. 17·11 above, but in addition an ideal regenerator with 100% effectiveness. Find (a) the compression work, (b) the heat transferred to the pressurized air in the regenerator, (c) the heat input to the air in the combustor, (d) the expansion work in the turbine, (e) the thermal efficiency from W and Q_a, (f) the thermal efficiency from Eq. (17·10), (g) the energy in the exhaust air above atmospheric, (h) the pressure at the turbine inlet, (i) the ratio of compressor work to turbine work.

Solution: (a) By Eq. (17·7) $T_2 = 742°R$ and $W_c = 56.5$ Btu per lb air (see Art. 17·11).

(b) From Art. 17·11, $T_5 = 1425°R$

$Q_t = c_p(T_5 - T_6) = c_p(T_3 - T_2) = 0.28(1425 - 742) = 191.3$ Btu per lb air

(c) $Q_a = c_p(T_4 - T_3) = 0.28(1960 - 1425) = 149.8$ Btu per lb air

(d) $W_t = c_p(T_4 - T_5) = 0.28(1960 - 1425) = 149.8$ Btu per lb air

(e) $e_t = \dfrac{W}{Q_a} = \dfrac{W_t - W_c}{Q_a} = \dfrac{149.8 - 56.5}{149.8} = 0.623$ or 62.3%

(f) $e_t = 1 - \dfrac{T_1}{T_4} P_r^{(k-1)/k} = 1 - \dfrac{540}{1960} \times 4^{(1.3-1)/1.3} = 0.621$ or 62.1%

(g) $Q_r = Q_a - W = 149.8 - 93.5 = 56.3$ Btu per lb air (compare with Q_r of Art. 17·11)

(h) $P_4 = P_1 P_r = 14.7 \times 4 = 58.8$ psia

(i) $\dfrac{W_c}{W_t} = \dfrac{56.5}{149.8} = 0.377$ or 37.7%

17·13 An intercooled regenerative cycle. Example: The ideal regenerative cycle of Art. 17·12 uses two-stage compression intercooled to T_1 with $P_i = \sqrt{P_1 P_2}$. Compute all the performance factors of Art. 17·12.

Solution: (a) $P_i = \sqrt{14.7 \times 4 \times 14.7} = 2 \times 14.7$ psia

P_r per stage $= 2 \times \dfrac{14.7}{14.7} = 2$

$T_2 = T_i = 540 \times 2^{(1.3-1)/1.3} = 634°$R

W_c per stage $= c_p(T_i - T_1) = 0.28(634 - 540) = 26.35$ Btu per lb air

Total $W_c = 2 \times 26.35 = 52.7$ Btu per lb air

(b) From Art. 17·12 at turbine exhaust $T_5 = 1425°$R $= T_3$ at regenerator air outlet, see Fig. 17·3b.

$Q_t = c_p(T_5 - T_6) = c_p(T_3 - T_2) = 0.28(1425 - 634) = 221.6$ Btu per lb air

(c) $Q_a = c_p(T_4 - T_3) = 0.28(1960 - 1425) = 149.8$ Btu per lb air

(d) $W_t = c_p(T_4 - T_5) = 0.28(1960 - 1425) = 149.8$ Btu per lb air

(e, f) $e_t = \dfrac{W}{Q_a} = \dfrac{W_t - W_c}{Q_a} = \dfrac{149.8 - 52.7}{149.8} = 0.648$ or 64.8%

(g) $Q_r = Q_a - W = 149.8 - 97.1 = 52.7$ Btu per lb air

(h) $P_4 = P_1 P_r = 14.7 \times 4 = 58.8$ psia

(i) $\dfrac{W_c}{W_t} = \dfrac{52.7}{149.8} = 0.352$ or 35.2%

17·14 A reheat-regenerative cycle. Example: The ideal regenerative cycle of Art. 17·12 uses two stages of heating both to 1500 F with the first turbine expanding to $P_i = \sqrt{P_1 P_2} = 29.4$ psia. Compute all performance factors of Art. 17·12.

Solution: (a) From Art. 17·12, $T_2 = 742°$R; $W_c = 56.5$ Btu per lb air

$\dfrac{P_4}{P_i} = \dfrac{P_i}{P_5} = \dfrac{P_2}{P_i} = \dfrac{P_i}{P_1} = P_r = 2$ (for each turbine)

$\dfrac{T_4}{T_{i1}} = P_r^{(1-k)/k} = 2^{(1.3-1)/1.3} = 1.1737$

$T_{i1} = \dfrac{T_4}{1.1737} = \dfrac{1960}{1.1737} = 1670°$R

$$\frac{T_{i2}}{T_5} = \frac{1960}{T_5} = 1.1737$$

$$T_5 = 1670°R = T_3$$

(b) $Q_t = c_p(T_5 - T_6) = c_p(T_3 - T_2) = 0.28(1670 - 742) = 259.8$ Btu per lb air

(c) $Q_a = c_p(T_4 - T_3) + c_p(T_{i2} - T_5) = 0.28(1960 - 1670)2 = 162.4$ Btu per lb air

(d) $W_t = c_p(T_4 - T_{i1}) + c_p(T_{i2} - T_5) = 0.28(1960 - 1670)2 = 162.4$ Btu per lb air

(e, f) $e_t = \dfrac{W}{Q_a} = \dfrac{W_t - W_c}{Q_a} = \dfrac{162.4 - 56.5}{162.4} = 0.652$ or 65.2%

(g) $Q_r = Q_a - W = 162.4 - 105.9 = 56.5$ Btu per lb air

(h) $P_4 = P_1 P_r = 14.7 \times 4 = 58.8$ psia

(i) $\dfrac{W_c}{W_t} = \dfrac{56.5}{162.4} = 0.348$ or 34.8%

17·15 An intercooled reheat-regenerative cycle. Example: An ideal cycle works at states defined in Arts. 17·12 to 17·14 and referred to by the numbering of Fig. 17·6. Compute the performance factors.

Solution: From Art. 17·13, $T_2 = T_4 = 634°R$; $W_c = 52.7$ Btu per lb air
From Art. 17·14, $T_5 = T_7 = T_9 = 1670°R$

$$Q_t = c_p(T_5 - T_4) = 0.28(1670 - 634) = 295.0 \text{ Btu per lb air}$$

From Art. 17·14, $W_t = 162.4$ Btu per lb air $= Q_a$

$$e_t = \frac{W_t - W_c}{Q_a} = \frac{162.4 - 52.7}{162.4} = 0.675 \text{ or } 67.5\%$$

$Q_r = Q_a - W = 162.4 - 109.7 = 52.7$ Btu per lb air
$P_4 = 58.8$ psia

$$\frac{W_c}{W_t} = \frac{52.7}{162.4} = 0.325 \text{ or } 32.5\%$$

Table 17·1 Summary of ideal gas-turbine cycles with 14.7-psia 80-F compressor-inlet air and 1500-F turbine inlet air. Cycle pressure ratio = 4.0

Cycle	Btu per lb air				$e_t, \%$	W_c/W_t
	Q_a	Q_r	W	Q_t		
Simple	341.0	247.5	93.5	0	27.4	0.377
Regenerative	149.8	56.3	93.5	191.3	62.3	0.377
Intercooled-regenerative	149.8	52.7	97.1	221.6	64.8	0.352
Reheat-regenerative	162.4	56.5	105.9	259.8	65.2	0.348
Intercooled-reheat-regenerative	162.4	52.7	109.7	295.0	67.5	0.325

GAS-TURBINE CYCLES

The effect of successively improving the simple cycle by the various devices in Arts. 17·13 to 17·15 can be done best by studying Table 17·1.

17·16 An irreversible simple-cycle gas turbine. Example: To find the effects of irreversible compression in the compressor and irreversible expansion in the turbine, let us recompute Art. 17·11 assuming that both the compressor and turbine engine efficiencies are 85%. From Eq. (17·13)

$$e_c = \frac{c_p(T_1 - T_{2s})}{c_p(T_1 - T_2)}$$

Substituting and canceling c_p's, we get

$$0.85 = \frac{540 - 742}{540 - T_2}$$

$$T_2 = 778°R$$

From Eq. (17·14)

$$e_e = \frac{c_p(T_3 - T_4)}{c_p(T_3 - T_{4s})}$$

Substituting and canceling c_p's,

$$0.85 = \frac{1960 - T_4}{1960 - 1425}$$

$T_4 = 1505°R$
$Q_a = c_p(T_3 - T_2) = 0.28(1960 - 778) = 331.0$ Btu per lb air
$Q_r = c_p(T_4 - T_1) = 0.28(1505 - 540) = 270.2$ Btu per lb air
$W_c = c_p(T_2 - T_1) = 0.28(778 - 540) = 66.6$ Btu per lb air
$W_t = c_p(T_3 - T_4) = 0.28(1960 - 1505) = 127.4$ Btu per lb air
$W = W_t - W_c = 127.4 - 66.6 = 60.8$ Btu per lb air
$\quad = Q_a - Q_r = 331.0 - 270.2 = 60.8$ Btu per lb air

$$\frac{W_c}{W_t} = \frac{66.6}{127.4} = 0.523$$

$$e_t = \frac{W}{Q_a} = \frac{60.8}{331.0} = 0.1838 \text{ or } 18.38\%$$

Compared with the ideal cycle of Art. 17·11, the irreversible processes pull thermal efficiency down from 27.4 to 18.38%, showing the importance of equipment efficiency in a gas-turbine cycle. In the actual cycle the compressor takes 52.3% of the gross work developed by the turbine compared with only 37.7% in the ideal cycle.

17·17 An irreversible regenerative gas-turbine cycle. Example: Now let us recompute the example of Art. 17·12 by assuming 85% compressor and turbine engine efficiencies and 85% regenerator effectiveness.

In Art. 17·16 we found that

$$T_2 = 778°R \quad \text{and} \quad T_5 = 1505°R$$

in terms of Fig. 17·10.

From Eq. (17·15) we get

$$e_r = \frac{T_3 - T_2}{T_5 - T_2} \quad \text{(also see Fig. 17·10)}$$

$$0.85 = \frac{T_3 - 778}{1505 - 778}$$

$T_3 = 1396°R$
$Q_a = c_p(T_4 - T_3) = 0.28(1960 - 1396) = 157.9$ Btu per lb air
$Q_t = c_p(T_3 - T_2) = 0.28(1396 - 778) = 173.0$ Btu per lb air
$Q_t = c_p(T_5 - T_6) = 0.28(1505 - T_6) = 173.0$ Btu per lb air
$T_6 = 887°R$
$Q_r = c_p(T_6 - T_1) = 0.28(887 - 540) = 97.1$ Btu per lb air
$W = Q_a - Q_r = 157.9 - 97.1 = 60.8$ Btu per lb air

$$\frac{W_c}{W_t} = \frac{66.6}{127.4} = 0.523$$

$$e_t = \frac{W}{Q_a} = \frac{60.8}{157.9} = 0.385 \text{ or } 38.5\%$$

Compared with the ideal cycle of Art. 17·12 the irreversible processes pull thermal efficiency down from 62.3 to 38.5%. The heat transferred in the regenerator drops from 191.3 to 173.0 Btu per lb of air. Compared with the simple actual cycle, the regenerator pulls thermal efficiency up from 18.38 to 38.5%.

REVIEW TOPICS

1. With the aid of a circuit sketch describe the operation of an open simple-cycle gas-turbine engine.

2. Show the operation of a simple-cycle gas-turbine engine on a P-V graph. Describe each process, and give the steady flow energy equation that applies.

3. Derive the thermal-efficiency relation for a simple-cycle gas-turbine engine. What other cycles does the gas-turbine cycle resemble?

4. Describe the operation of the simple-cycle gas turbine with the aid of a T-S graph. Explain the meaning of all significant areas.

5. Describe the regenerative gas-turbine cycle with the aid of P-V and T-S graphs. How does this cycle improve the thermal efficiency of the gas-turbine engine compared with the simple cycle?

6. Derive the thermal-efficiency equation for the regenerative gas-turbine cycle.

7. Draw and explain the use of intercooling in the simple cycle and regenerative cycle of gas-turbine engines.

8. Draw and explain the use of reheating in the simple and regenerative cycles of the gas-turbine engine.

GAS-TURBINE CYCLES

9. Draw and explain the T-S diagram of intercooled and reheated gas-turbine cycles with and without regeneration.

10. Demonstrate the effect of irreversible compression and irreversible expansion in a simple-cycle gas-turbine engine on a T-S graph.

11. Show by rough trend curves how thermal efficiency varies with pressure ratio for the simple-cycle gas turbine with various degrees of irreversibility of processes.

12. Show by rough trend curves how the thermal efficiency of a regenerative cycle varies with pressure ratio and degrees of irreversibilities of cycle processes.

13. Define the meaning and write the equation for regenerator effectiveness.

PROBLEMS

1. A steady-flow compressor takes in air at 100 F and 14.3 psia and compresses it at constant entropy to six times the intake pressure. With $c_p = 0.24$ Btu per F, find the compressor work input.

2. A combustor receives air at 120 psia and 650 F and discharges it at 1650 F with no loss in pressure. With $c_p = 0.27$ and $c_v = 0.20$, find the heat input to the air.

3. A gas turbine receives air at 95 psia and 1500 F and expands it at constant entropy to 15 psia. Assuming that $k = 1.33$ and $c_p = 0.27$, what shaft output does the turbine develop?

4. A gas-turbine cycle takes in air at 60 F and 14.7 psia and rejects it at 950 F and 14.7 psia. With $c_p = 0.25$, what heat does the cycle reject?

5. In a gas-turbine cycle the turbine develops gross shaft work of 80 Btu per lb of air while the compressor needs 45 Btu per lb of air to pressurize the air. What is the net work output of the cycle?

6. An ideal simple-cycle gas turbine works with a pressure ratio of 4.5. If the average specific heats for the compression and expansion processes are $c_v = 0.20$ and $c_p = 0.27$ Btu per lb per F, what is the thermal efficiency of the cycle?

7. An ideal regenerative-cycle gas turbine works with a pressure ratio of 3.8 when its intake air is at 70 F and the turbine inlet temperature is 1450 F. Assuming that the average specific heats are $c_v = 0.21$ and $c_p = 0.28$, find the ideal thermal efficiency of the cycle.

8. A compressor takes in air at 14.0 psia and 80 F and pressurizes it to 70 psia. The compressor work input is 10% more than the work of ideal compression at constant entropy with $k = 1.35$. Find the compressor work input and actual efficiency.

9. A gas turbine takes in air at 85 psia and 1300 F and expands it to 15.0 psia. The turbine develops 9% less work than an ideal expansion at constant entropy with $k = 1.3$ and $c_p = 0.26$. Find the turbine work and engine efficiency.

10. A regenerator takes in air at 220 F from a compressor and discharges it at a temperature 80 F less than the turbine exhaust temperature. If the turbine exhausts at 880 F, what is the effectiveness of the regenerator?

11. An ideal simple-cycle gas-turbine engine takes in air at 40 F and 14.0 psia to compress it at constant entropy through a pressure ratio of 6.5. The combustor raises the air temperature to 1300 F. The heated air expands at constant entropy to 15.0 psia in the turbine. If $k = 1.35$ and $c_p = 0.27$, find (a) the work input to the compressor, (b) the heat input to the combustor, (c) the work done in the turbine, (d) the thermal efficiency of the cycle, (e) the proportion of work done by the turbine absorbed in the compressor, (f) the energy rejected by the cycle.

12. An ideal regenerative-cycle gas-turbine engine takes in air at 10 F and 13.5 psia to compress it to 60 psia at constant entropy with $k = 1.31$. The combustor raises the air to 1100 F, which expands in the turbine to 14.0 psia at constant entropy with $k = 1.31$. If $c_p = 0.28$ and the regenerator effectiveness is 100%, find (a) the compression work, (b) the heat transferred to the pressurized air in the regenerator, (c) the energy input to the combustor, (d) the work of expansion in the turbine, (e) the thermal efficiency of the cycle, (f) the heat rejected to the atmosphere, (g) the ratio of compressor and turbine works.

13. An ideal intercooled regenerative cycle uses two-stage compression to 130 psia from an atmospheric pressure of 14.9 psia and temperature of 50 F. For $k = 1.3$ and $c_p = 0.28$ and a turbine inlet temperature of 1350 F, find (a) compression work, (b) heat transfer in regenerator, (c) the combustor heat input, (d) the expansion work, (e) the cycle thermal efficiency, (f) the heat rejected from the regenerator exhaust, (g) the heat rejected by the intercooler, (h) the ratio of compressor and turbine works.

14. An ideal reheat-regenerative gas-turbine cycle uses two stages of heating with an intermediate pressure $P_i = \sqrt{P_1 P_2}$, an intake state of 120 F and 14.0 psia, and a turbine inlet and reheat temperature of 1550 F. When turbine inlet pressure is 84 psia, the exhaust pressure is 14.0 psia, and $k = 1.3$ and $c_p = 0.28$, find (a) the compression work, (b) the heat transfer in the regenerator, (c) the combustor heat input, (d) the turbine expansion work, (e) the cycle thermal efficiency, (f) the heat rejected from the cycle, (g) the ratio of compressor and turbine works.

15. An ideal intercooled reheat-regenerative cycle uses two stages of heating and compression with $P_i = \sqrt{P_1 P_2}$. The cycle inlet state is 120 F and 14.0 psia, the outlet pressure is 14.0 psia, and $k = 1.3$ and $c_p = 0.28$. The turbine inlet pressure is 120 psia, and temperatures are 1450 F. Find (a) the compression work, (b) the heat transfer in the regenerator, (c) the total combustor heat input, (d) the turbine shaft work, (e) the cycle thermal efficiency, (f) the cycle heat rejection, (g) the ratio of compressor and turbine works.

16. In Prob. 11 replace the ideal compressor and turbine with an actual compressor and turbine having 90% machine efficiencies. Compute all the performance factors (*a*) to (*f*).

17. In Prob. 12 replace the ideal compressor, turbine, and regenerator with a compressor having 85% compression efficiency, a turbine having 88% engine efficiency, and a regenerator with 75% effectiveness. Compute all the performance factors (*a*) to (*g*).

CHAPTER 18

Nozzle Gas Flow

We have studied the gas-turbine engine as an energy-converting device without knowing how the components work internally. Now let us look at the insides, first the turbine, later the compressor. Figure 18·1 shows a few of the principal elements of a gas turbine that change the internal energy of the gas into mechanical shaft work.

18·1 Gas flow. Hot pressurized gas at P_1 enters the spaces between the arc-shaped blades of the stationary nozzles and flows to the space ahead of the moving buckets at a lower pressure P_2. The nozzles direct the gas flow to impinge on moving buckets mounted on a turbine wheel or disk. Stationary nozzles may cover only part of the rim of the stationary nozzle diaphragm, the remainder being blocked closed. But the moving buckets completely cover the rim of the turbine wheel.

Nozzles reorganize the random motion of the gas molecules so that they largely move in a common direction and form a high-speed jet of gas. The jet is made to pass between the slower moving buckets in a prearranged pattern so that it produces a force on the buckets and does mechanical work on the turbine wheel carrying the buckets. The jet flows through passages between the moving buckets and turns to leave at about right angles to the motion of the buckets at a pressure P_3.

18·2 Energy transfer. In passing through the bucket passages the jets give up much of their kinetic energy by impacting the buckets and increasing their momentum. If you are rusty on impact and momentum see Chap. 5. Depending on bucket design, $P_3 = P_2$ or P_3 may be less than P_2.

Figure 18·1 shows two distinct energy processes in turbines: (1) changing the internal energy of the gas to kinetic energy of the gas mass by persuading the gas molecules to flow in a given direction through the nozzles and (2) changing the kinetic energy of the gas to mechanical shaft energy in the moving buckets by transferring momentum from the gas molecules to the buckets. Nozzles and buckets need to be placed in proper

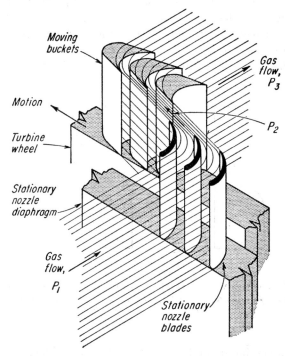

Fig. 18·1 Gas-turbine nozzles direct the gas flow to enter moving buckets tangentially. Slower buckets turn jets to develop work force at the expense of internal energy in the gas.

relation to each other, but otherwise the two processes are largely independent as far as energy conversion is concerned. Let us study each of these two processes in more detail.

18·3 Nozzles. In Art. 16·9 we showed that the steady flow energy equation for flow through nozzles reduces to

$$\frac{w}{2gJ}(v_2{}^2 - v_1{}^2) = w(H_1 - H_2) \qquad (18\cdot 1)$$

where w = rate of flow, lb per sec
v = gas jet speed, fps
H = gas enthalpy, Btu per lb

In Art. 16·12 we learned about the relation between flow and the pressure ratio across the nozzle. Now let us do some mathematical analysis of these relations and speculate some more about molecular behavior in flowing gases.

While Fig. 18·1 shows the practical way of building nozzles for turbines, a more fundamental way of studying nozzle behavior finds how gas flow behaves in passing through a simple hole as in Fig. 18·2. The converging nozzle in Fig. 18·2a has a smooth, curved entrance section followed by a parallel-walled section; this can be used for the higher range of pressure ratios across the nozzle, P_2/P_1. For lower pressure ratios we shall find that the entrance section has to be followed by a flared section to

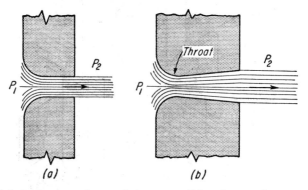

Fig. 18·2 (a) A simple converging nozzle has a parallel-wall exit to form a smooth jet of gas. (b) A converging-diverging nozzle needs a diverging exit section to form a smooth leaving jet.

form the converging-diverging nozzle of Fig. 18·2b. The narrowest cross section is called the throat of the nozzle.

As gas flows through a nozzle, it passes from a higher to a lower pressure region. But what is the pressure of the gas flow? It depends on how and where we measure it. Usually, because the gas molecules diffuse into the nozzle opening from all directions, the gas pressure at the inlet will be the same as the source, P_1.

Tapping the walls of a nozzle we find that pressure drops as the nozzle cross section changes. Pressure measured through an opening at right angles to the flow is called *static pressure*. We assume that this pressure describes the state of the gas at that cross section, Fig. 18·3.

18·4 Stagnation states. Aiming the pressure tap upstream, Fig. 18·3, measures the *total* or *stagnation pressure*. If there are no frictional losses in the nozzle, stagnation pressure stays constant along the length of the nozzle from inlet to outlet.

In defining stagnation pressure we assume that the gas slug entering the pressure tap comes to rest, compressing itself at constant entropy. The following gas flow diverted by the gas stalled in the tap keeps the slug compressed by molecular impingement. Initially this moving gas slug has an enthalpy H_1 and a kinetic energy $v_1^2/2gJ$. When it comes to rest in the pressure gage, it has only an enthalpy H_0, called the *stagnation enthalpy*. Since no energy has been added or subtracted during this compression process,

$$H_1 + \frac{v_1^2}{2gJ} = H_0 \qquad (18\cdot2)$$

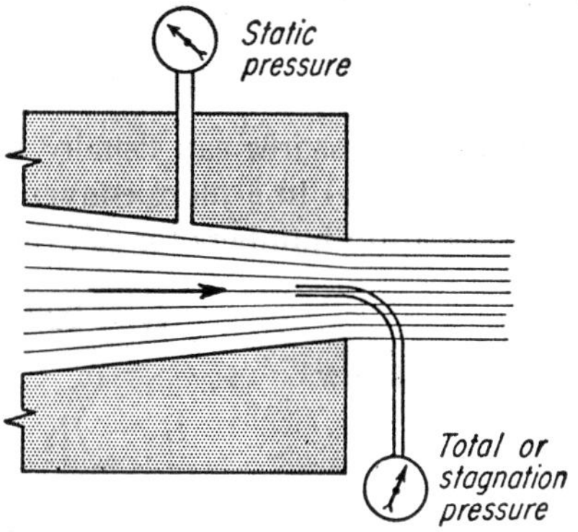

In the usual case the velocity of approach of the gas entering a nozzle is so small that $H_1 = H_0$ and stagnation pressure = $P_0 = P_1$. As the velocity of approach rises, however, H_0 and P_0 become larger than H_1 and P_1.

The same line of reasoning applies to the *stagnation temperature*. Since $H_1 - H_0 = c_p(T_1 - T_0)$ it follows that:

$$T_1 + \frac{v_1^2}{2gJc_p} = T_0 \qquad (18\cdot3)$$

Fig. 18·3 Static pressure of gas flow is measured at right angles to the flow, while the stagnation pressure is measured parallel to the flow.

18·5 Nozzle pressures. A nozzle develops gas velocity by inducing the molecules leaving the source to travel in a more uniform direction as they go through the nozzle. Most gas molecules in the nozzle at any instant travel in the general direction of the gas mass flow. Some of them impact the nozzle walls to produce the static pressure. But since the majority travel in the direction of flow, stagnation pressure will be higher than static pressure, assuming they all travel at the average velocity.

As we reduce gas flow to zero by raising the outlet pressure, more and more of the molecules take on random directions and hit the walls to raise the static pressure with steady stagnation pressure. At zero flow the static and stagnation pressures become equal.

18·6 Nozzle flow. In computing nozzle flows we must recognize that the mass rate of gas flow under steady conditions remains constant at all cross sections of the nozzle; then

$$w = \frac{A_1 v_1}{V_1} = \frac{A_2 v_2}{V_2} = \frac{A_3 v_3}{V_3} = \text{etc.} \qquad (18\cdot4)$$

where A = nozzle cross section, sq ft
v = gas flow velocity, fps
V = gas specific volume, cu ft per lb
w = gas flow, lb per sec

The velocity of the jet leaving the nozzle plays an important part in turbine design; it can be computed from Eq. (18·1) by transposing and assuming a flow rate of 1 lb per sec:

$$v_2 = \sqrt{2gJ(H_1 - H_2) + v_1^2} \tag{18·5}$$

The static pressures can usually be measured most easily, so we transform Eq. (18·5) to a more convenient form. First we shall assume $v_1 = 0$ and substitute numbers for gJ; then

$$v_2 = 223.8\sqrt{H_1 - H_2} \tag{18·6}$$

We also know

$$H_1 - H_2 = c_p(T_1 - T_2) \tag{18·7}$$

$$\frac{T_2}{T_1} = \left(\frac{P_2}{P_1}\right)^{(k-1)/k} \tag{18·8}$$

Equation (18·8) applies to a constant-entropy expansion of the gas while it flows through the nozzle; this usually holds nearly true. Then transforming Eq. (18·7),

$$c_p(T_1 - T_2) = c_pT_1\left(1 - \frac{T_2}{T_1}\right) = c_pT_1\left[1 - \left(\frac{P_2}{P_1}\right)^{(k-1)/k}\right] \tag{18·9}$$

Substituting in Eqs. (18·6), (18·7),

$$v_2 = 223.8\sqrt{c_pT_1\left[1 - \left(\frac{P_2}{P_1}\right)^{(k-1)/k}\right]} \tag{18·10}$$

To complete our tools for nozzle-performance analysis we have to recall that for a constant-entropy expansion

$$\frac{V_2}{V_1} = \left(\frac{P_1}{P_2}\right)^{1/k} \tag{18·11}$$

With Eqs. (18·4), (18·10), and (18·11) we are ready to analyze what happens to a perfect gas as it flows through a nozzle. First we shall assume that we have air as our source with $P_1 = 100$ psia $= 14{,}400$ psfa, $T_1 = 1900°R$, $k = 1.33$, $R = 53.3$, $V_1 = 7.03$ cu ft per lb. The approach velocity $v_1 = 0$, so in effect $P_1 = P_0$; this holds for all the properties.

Next we shall find the properties of the gas, its velocity and the cross-sectional area the nozzle must have for a flow of 1 lb per sec at successively lower static pressures. We conveniently express this as a pressure ratio P_2/P_1.

Table 18·1 lists the calculation results, and Fig. 18·4 shows the variation of specific volume, velocity, and nozzle cross-sectional area with gas-pressure ratio. At unity ratio with equal pressures at inlet and outlet, the velocity is zero, though we can visualize individual molecules drifting erratically from one end to the other in all directions through the nozzle. As the outlet pressure drops, more molecules drift toward the lower pressure than the other way to give a net mass flow through the nozzle.

Table 18·1 Theoretical nozzle performance[a]

P_2/P_1	P_2, psia	V_2, cu ft/lb	v_2, fps	A_2, sq ft
1.0	100	7.03	0	
0.9	90	7.61	832	0.00915
0.8	80	8.31	1,203	0.00691
0.6	60	10.32	1,783	0.00579
0.54	54	11.17	1,945	0.00574
0.4	40	13.44	2,333	0.00576
0.2	20	23.58	2,967	0.00795
0.1	10	39.72	3,415	0.01164

[a] Air flow, 1 lb per sec; P_1, 100 psia or 14,400 psfa; R, 53.2; T_1, 1900°R; k, 1.33; c_p, 0.28.

As Fig. 18·4 shows, the velocity v builds up rapidly at first, with a diminishing rate of increase up to the critical pressure ratio (see Appendix H). At less than the critical the rate of increase again rises. We shall talk about the meaning of critical pressure ratio a little later. The continuing rise in v with drop in pressure ratio raises an interesting question. The

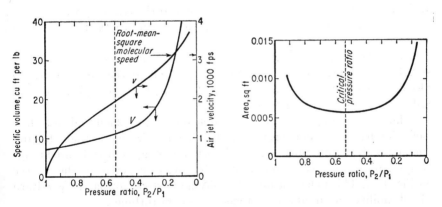

Fig. 18·4 All factors controlling nozzle gas flow vary with the overall pressure ratio.

rms molecular velocity of air molecules at 1900°R is 3,138 fps. Yet Eq. (18·10) for zero pressure ratio and 1900°R gives a maximum value of 5,175 fps. Obviously the equation does not hold for the full range of pressure ratios as plotted in Fig. 18·4. The molecular velocity corresponds to a pressure ratio of 0.156. This is inconsistent: If the entire gas mass travels at molecular velocity, all the molecules must be moving parallel to the axis of the nozzle and none of them can hit the wall to produce a static pressure. But the whole process is complicated by the Maxwell distribution of molecular velocities, which plays an important part in not too clear a fashion. Despite these inconsistencies the equation does hold for the higher pressure ratios, at least down to the critical pressure. As we shall see, shock conditions at lower pressure ratios make the inconsistencies largely an academic question.

The specific volume grows at an increasing rate with the drop in pressure ratio, Fig. 18·4. This follows from the perfect-gas equation and assumes that the static pressure and constant-entropy expansion properly describe the state of the gas at a given cross section of the nozzle.

The most interesting feature of nozzle design is the variation in A as the pressure ratio drops, Fig. 18·4. From infinity at $P_2/P_1 = 1$ it falls to a minimum at the critical pressure ratio and then grows at an increasing rate for the lower pressure ratios.

If we want to build a nozzle with a uniform pressure drop along its length, the shape of the area curve gives us the profile of the nozzle wall section. Fortunately, nozzles need not have uniform pressure drops, so we can make them as in Fig. 18·2. For overall pressure ratios greater than or equal to the critical, we use the converging parallel-wall nozzle with the minimum opening sized to pass the design w for the given P_2/P_1.

For the initial conditions used in Table 18·1 and Fig. 18·4 the nozzle cross-sectional area becomes a minimum at $P_2/P_1 = 0.54$. The pressure ratio at which A becomes a minimum is called the *critical pressure ratio*. As we saw in Fig. 18·2b, the section with minimum area in a nozzle is called the *throat*.

When the overall pressure ratio across a converging-diverging nozzle equals critical pressure ratio or less, the pressure at the throat is always the critical pressure. For the same ratio range, flow through the nozzle will be governed by the area of the throat.

While nozzles appear to be very simple devices, we shall find that their behavior has some unexpected quirks that require close study for full understanding. Equations of flow will be found to apply to a limited range of pressure ratios, even more limited than we discovered for the v curve in Fig. 18·4. This behavior must be fully understood so that we can learn the flexibility and limitations of gas and steam turbines.

In an actual nozzle with given inlet and outlet conditions we found that the cross-sectional area fixes the properties of the gas at that section. In addition the pressure ratio at the section determines the velocity of the gas flowing through that section. Now we shall go one step further and find how pressure ratio and area control *mass rate of flow* of a gas through a nozzle. We use same basic equations:

$$w = \frac{A_2 v_2}{V_2} \qquad (18\cdot4)$$

where w = gas flow, lb per sec
A_2 = nozzle sectional area, sq ft
v_2 = gas velocity, fps
V_2 = gas specific volume at section, cu ft per lb

For constant-entropy gas expansion from stagnation,

$$\frac{V_2}{V_0} = \left(\frac{P_0}{P_2}\right)^{1/k} \qquad (18\cdot12)$$

where V_0 = gas specific volume for stagnation at nozzle inlet, cu ft per lb
P_0 = stagnation pressure at nozzle inlet, psfa
V_2 = gas specific volume at section, cu ft per lb
P_2 = gas pressure at nozzle section, psfa
k = ratio of specific heats

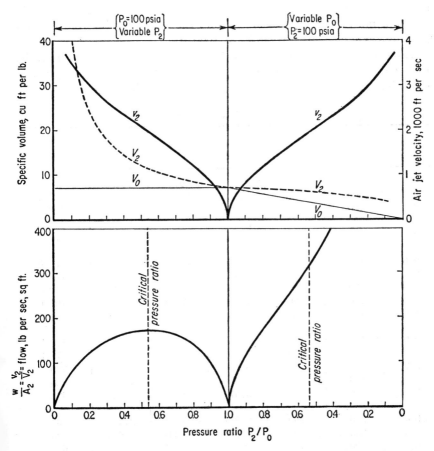

Fig. 18·5 The graphs, left, show the property variations as pressure ratio rises from 0 to 1.0 with variable P_2; the graphs, right, give data for 1.0 to 0 ratios with variable inlet P_0.

We derived the equation for gas velocity:

$$v_2 = 223.8 \sqrt{c_p T_0 \left[1 - \left(\frac{P_2}{P_0}\right)^{(k-1)/k}\right]} \qquad (18\cdot13)$$

With these in hand let us transpose Eq. (18·4) to get

$$\frac{w}{A_2} = \frac{v_2}{V_2} \qquad (18\cdot14)$$

This gives us the ratio of pound per second flow per square foot of nozzle area. We get the right-hand side from Eqs. (18·12) and (18·13).

To get a broader understanding of nozzle behavior we shall study two sets of conditions. In the first we shall place a nozzle to take air at 100 psia and 1900°R as constant *inlet* conditions. We shall vary the exhaust from 0 to 100 psia to vary the pressure ratio P_2/P_0 from 0 to 1.0. When at 1.0 pressure ratio we shall keep the *exhaust* pressure constant at 100 psia and raise the inlet pressure to vary the pressure ratio from 1.0 to 0, keeping the inlet temperature $T_0 = 1900°R$.

Table 18·2 lists the properties calculated for the given pressure ratios with aid of Eqs. (18·4) through (18·14) for the two nozzle operating conditions. Figure 18·5 shows the variations listed in Table 18·2. Graphs

Table 18·2 Theoretical nozzle performance[a]

P_2/P_0	P_0, psia	P_2, psia	V_0, cu ft/lb	V_2, cu ft/lb	v_2, fps	$w/A = v_2/V_2$, lb/sec, sq ft
0.1	100	10	7.15	40.4	3,415	84.6
0.2	100	20	7.15	23.98	2,967	123.7
0.3	100	30	7.15	17.68	2,625	148.5
0.4	100	40	7.15	14.24	2,333	163.8
0.54	100	54	7.15	11.37	1,945	171.2
0.6	100	60	7.15	10.49	1,783	170.0
0.7	100	70	7.15	9.35	1,503	160.8
0.8	100	80	7.15	8.46	1,203	142.3
0.9	100	90	7.15	7.74	832	107.5
1.0	100	100	7.15	7.15	0	0
0.9	111.1	100	6.43	6.96	832	119.7
0.8	125	100	5.71	6.75	1,203	178.3
0.7	143	100	5.00	6.54	1,503	230
0.6	168	100	4.25	6.24	1,783	286
0.5	200	100	3.57	6.01	1,055	342
0.4	250	100	2.857	5.70	2,333	409
0.3	333.3	100	2.142	5.30	2,625	495
0.2	500	100	1.428	4.80	2,967	619
0.1	1000	100	0.715	4.04	3,415	846

[a] Air flow, 1.0 lb per sec; T_0, 1900°R; k, 1.33; c_p, 0.28; R, 54.1.

for the two conditions are really separate units but joined at 1.0 pressure ratio to emphasize the difference in behavior for the two methods of operation.

18·7 Variable back pressure. In Art. 18·6 we discovered the variation in v_2 and V_2 for variable back-pressure operation. The flow per unit nozzle sectional area w/A_2 rises from zero at zero pressure ratio to a maximum at critical pressure ratio and then drops to zero at 1.0 pressure ratio. This curve has to be interpreted carefully—a hasty conclusion might lead us to the absurd condition that nozzle flow is zero at zero pressure ratio.

We also learned that we need a converging-diverging nozzle between 0 and 0.54 (critical) pressure ratio. Between 0.54 and 1.0 pressure ratio we need a converging nozzle. In the converging nozzle the exit section has the smallest area; this area will control the mass flow of gas through the nozzle. If we take a nozzle with an exit area A_2 of 0.01 sq ft, we would get the following mass flows for conditions in Table 18·3.

Table 18·3

P_2/P_0	w/A_2	Lb per sec
1.0	0	0
0.9	107.5	1.075
0.7	160.8	1.608
0.54	171.2	1.712

For pressure ratios less than critical we need a converging-diverging nozzle that has a minimum throat area. This throat area controls the mass flow rate of gas passing through the nozzle. For overall pressure ratios less than critical we shall find the *critical pressure always at the throat*. (If we use a converging nozzle in this range, the critical pressure will establish itself at the exit section.) The w/A_2 curve in the lower pressure range then aids in finding the exit area needed for the nozzle.

For any pressure ratio we first find the flow based on the throat area; knowing the flow, we divide by the w/A_2 for the pressure ratio to find the exit area needed. Values in Table 18·4 assume a 0.01-sq-ft throat.

Table 18·4

P_2/P_0	Throat w/A_2	Flow, lb per sec	Exit w/A_2	Exit A_2
0.54	171.2	1.712	171.2	0.01000
0.4	171.2	1.712	163.8	0.01045
0.2	171.2	1.712	123.7	0.01383
0.1	171.2	1.712	84.6	0.02025

18·8 Throat properties. Since throat properties control the flow through converging-diverging nozzles, we find formulas for calculating them directly very useful. The critical pressure at the throat is

$$P_t = P_0 \left(\frac{2}{k+1}\right)^{k/(k-1)} \quad (18 \cdot 15)$$

(See Appendix H.) The critical temperature at the throat is

$$T_t = \frac{2T_0}{k-1} \quad (18 \cdot 16)$$

since

$$\frac{T_t}{T_0} = \left(\frac{P_t}{P_0}\right)^{(k-1)/k}$$

Theoretically the throat gas properties depend only on the stagnation properties and the ratio of specific heats.

18·9 Sonic velocity. The velocity of the gas flowing through the throat of a nozzle at pressure ratios equal to or less than critical has a special value. To find this we substitute P_t/P_0 from Eq. (18·15) into Eq. (18·13); next substitute $T_0 = (k+1)T_t/2$ from Eq. (18·16) into Eq. (18·13). Then remembering that $223.8 = \sqrt{2gJ}$, simplify and find

$$v_t = \sqrt{gkRT_t} \quad (18 \cdot 17)$$

This is also the velocity of sound in a gas at temperature T_t found from experimental evidence. This shows that, for a nozzle working at pressure ratios of critical or less, sonic velocity of the gas establishes itself at the throat, at the critical pressure ratio.

The curves of Fig. 18·5 show that the gas velocity v_2 is subsonic at pressure ratios more than critical and supersonic at less than critical.

The ratio of actual gas velocity to sonic velocity at given temperature is known as *Mach number M*. So if we know the velocity v of a gas at a temperature T, then

$$M = \frac{v}{\sqrt{gkRT}} \quad (18 \cdot 18)$$

Subsonic velocities have Mach numbers less than 1.0, sonic velocities have a Mach number of 1.0, and supersonic velocities have Mach numbers larger than 1.0. In Chap. 16 we learned that at ambient air temperatures, the ratio of molecular to sonic velocity equals 1.465. When expanding a gas through a nozzle at l-p ratios, we shall usually find that the exit Mach number is less than 2.0. In most power turbines and compressors, Mach numbers run less than 1.0.

18·10 Variable inlet pressure. Now let us study the second part of Table 18·2 and Fig. 18·5 where we keep a constant back pressure of 100 psia on the nozzle and vary the inlet pressure from 100 psia on up to lower the pressure ratio toward zero. Figure 18·5 shows that the stagnation volume V_0 drops proportionately to zero at zero pressure ratio. The exit specific volume V_2 also drops with pressure ratio. The change in gas volume for this type of operation is much less than for variable back pressure at the same ratios.

The interesting feature lies in the behavior of the ratio w/A_2, the flow in pounds per second, per square foot. For this operation the flow steadily increases with drop in pressure ratio, no maximum being reached except near zero pressure ratio. Thus we can raise gas flow through a nozzle practically proportionately by simply raising the inlet pressure. Note that sonic velocity again establishes itself at the throat at critical pressure ratio. This time mass flow keeps rising because we squeeze more molecules into the nozzle inlet, shown by shrinking specific volume V_0, as pressure rises.

18·11 Nozzle static pressures. Up to this point we can see that for best operation a nozzle should be tailored to fit a specific set of conditions as given by overall pressure ratio, inlet properties, type of gas, and mass flow rate needed. In actual operation a nozzle will run off design conditions some of the time, and so its behavior will differ from that calculated. Let us see what happens.

Figure 18·6A shows a converging-diverging nozzle built to run between pressure a at the inlet and pressure n at the outlet (see graph below). Experience shows that the flare of the diverging portion should be kept to about 12° or less. Static pressure along length of the nozzle smoothly drops to critical f at the throat section (where gas flows at sonic velocity, $M = 1$). As it enters the diverging section, the pressure keeps dropping through g and k until it reaches n in the outlet region.

Flow through the nozzle for this design condition appears orderly without any turbulence in the nozzle or at its inlet and outlet. The velocity through the nozzle sections grows as pressure drops toward the outlet. The gas jet leaves at supersonic velocity, M greater than 1.0.

Next, we put the nozzle into an off-design condition by raising the outlet pressure to m. We then find that static pressure in the nozzle still drops smoothly from a to f, g and k. But at k pressure suddenly rises to l a very short distance downstream, and then it smoothly rises to m at the outlet. Figure 18·6B shows that a stationary shock wave has established itself at section k–l. As gas flows through the shock wave, it suddenly rises in pressure and density and its speed drops from supersonic to subsonic. Mass flow through the nozzle is not affected because this depends on the critical pressure at the throat. But exit velocity will be much lower, showing a loss in efficiency. Mass flow remains unchanged because of higher outlet density (smaller V_2).

When we raise the outlet pressure to i, the shock wave grows weaker as it moves further into nozzle at g–h, Fig. 18·6C. Static pressure through the nozzle still drops smoothly from a to f and g. From g to h pressure rises suddenly through the shock wave at that section. Again mass flow has not been altered but exit velocity has dropped some more.

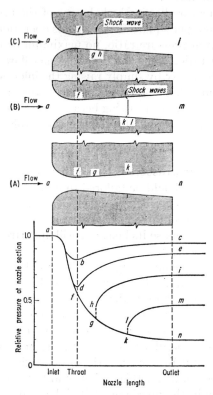

Fig. 18·6 A nozzle designed for specific conditions has smooth static pressure variations along the length a–n; when running at off-design conditions, the pressure drop varies in many different ways.

After the shock wave, the pressure rises smoothly from h to outlet pressure i.

As we raise the outlet pressure to e, the shock wave weakens some more, moves into the throat, and then disappears. Now static pressure along the nozzle drops from a as before, but pressure at the throat d is higher than critical (at f) for the previous conditions. There are no sudden pressure changes, and no shock waves appear. Static pressure rises smoothly from the throat to the outlet pressure e. The nozzle acts as a venturi tube for this pressure condition. Gas velocity increases from the inlet to the throat sections but does not reach sonic speed. Beyond the throat the gas velocity slows as static pressure rises to the outlet at *e*. Decreasing velocity against rising pressure is a *diffusing action*—reverse of a nozzle jet action. Velocity throughout the nozzle is subsonic, and v_2 and mass flow are considerably less than at back pressure *i*.

Raising the outlet pressure to *c* raises the static pressure throughout almost full length of the nozzle with a lower throat pressure at *b*. This further diminishes mass flow and nozzle exit gas velocity.

This generalized rundown shows that gas flowing through a hole does not act in a simple manner. Here, we see one of main reasons why turbines and compressors cannot work at best efficiency over their full output range.

REVIEW TOPICS

1. With the aid of a sketch describe the internal actions of a gas stream flowing through the nozzles and moving buckets of a gas turbine.

2. Write the steady flow energy equation for a nozzle, and describe the terms.

3. What is the static pressure of a gas flowing through a nozzle?

4. What is the stagnation pressure of a gas stream flowing through a nozzle?

5. What is the total pressure of a gas stream flowing through a nozzle? How does it vary along the length of an ideal nozzle?

6. Write and explain the energy equation for stagnation enthalpy.

7. Write and explain the equation for stagnation temperature.

8. Describe the action of a gas stream flowing through a nozzle in terms of general molecular actions.

9. Write and explain the steady-flow mass rate equation for a gas flowing through a nozzle.

10. Derive the equation for nozzle exit gas velocity in terms of inlet temperature and pressure ratio.

11. What is the throat of a nozzle?

12. With the aid of the nozzle exit-velocity equation, compute the flow, velocity, and cross-sectional area of a nozzle for a full range of pressure ratios.

13. What is the critical pressure ratio of a gas? What relation does it have to the throat of a nozzle?

14. How does the ratio of the mass rate of a gas flow to cross-sectional area vary with pressure ratio in a nozzle?

15. Derive the equation for the critical-pressure ratio of a gas in terms of k.

16. Derive the equation for the critical-temperature ratio in terms of k.

17. Write and explain the equation for sonic velocity in a gas. What relation does sonic velocity have to flow of gas through a converging-diverging nozzle?

18. What is the Mach number of a gas flow?

19. How does the flow through a nozzle vary when back pressure is kept constant and the inlet pressure varied?

20. Describe the variations in static pressure of a gas flowing through a nozzle as the back pressure is raised from design conditions when the design pressure is less than the critical pressure. What happens in a shock wave?

PROBLEMS

1. Helium at a pressure of 50 psia flows through a converging nozzle to an exhaust pressure of 30 psia. The inlet temperature is 250 F, and flow can be assumed at constant entropy. Find the exit velocity when helium $c_p = 1.251$ and $c_v = 0.754$ Btu per lb, F; the inlet velocity is zero.

2. Hydrogen at a pressure of 80 psia flows through a converging nozzle to an exhaust pressure of 55 psia. The inlet temperature is 300 F, and flow is at constant entropy. Find the exit velocity when hydrogen $c_p = 3.41$ and $c_v = 2.42$ Btu per lb, F.

3. From the data in Probs. 1 and 2 find the critical-pressure ratio and critical-temperature ratio of helium and hydrogen.

4. From the data in Probs. 1 and 2 find the sonic velocity in hydrogen and helium and compare with the sonic velocity in air for temperatures of -50, 0, 100, 200, 1000, and 2000 F.

5. Recompute Table 18·1 for air flow of 1 lb per sec with $P_1 = 200$ psia, $T_1 = 1200°R$, $c_v = 0.19$, $c_p = 0.26$. Plot the results.

CHAPTER 19

Turbine Nozzles and Buckets

Now that we know how nozzles form high-speed jets from a mass of pressurized high-temperature gas, let us learn how turbine buckets convert the gas-jet kinetic energy to mechanical shaft work. For the most part, steam acts almost the same as the gas we shall talk about.

19·1 Shaft work. In the air engine, Art. 12·11, we converted gas internal energy into shaft work by letting the bouncing gas molecules do work on a moving piston. In our high-speed gas jet we have molecules organized to flow in about the same direction. Their stagnation pressure, ideally, equals the pressure of the gas at its source.

The steady-flowing gas jet makes it unnecessary to use a piston and cylinder with their limiting reciprocating motion. Instead we can use a steady rotating motion—the basic form of mechanical shaft work. To make the jet give up its kinetic energy we must reduce its speed while it works on a moving surface. We do this by changing the momentum of the jet as it exerts a working force on a moving surface.

Figure 19·1 shows the principal working elements of a turbine. The gas jet leaving the nozzle, placed at a sharp angle to the turbine wheel,

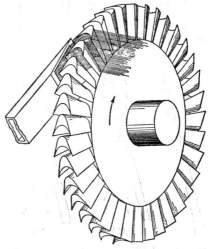

Fig. 19·1 In an impulse turbine, gas jet from the nozzle flows between moving buckets to develop force for turning the wheel.

flows through the passages between the buckets mounted on the wheel perimeter. Keeping the tangential speed of the buckets slower than the jet speed makes the jet turn, so it leaves in a general axial direction and at a lower speed than that with which it entered the buckets. In this way the jet gives up most of its kinetic energy and does work on the buckets. In other words, the change in momentum of the jet develops a force on the buckets between which the jet passes.

19·2 Basic law. Recalling Newton's Second Law of Motion,

$$F = ma = \frac{m(v_2 - v_1)}{t}$$
$$= \frac{w}{g}(v_2 - v_1) \tag{19·1}$$

where F = force, lb
m = mass, slugs
v_2 = velocity of jet leaving buckets, fps
v_1 = velocity of jet leaving nozzle and entering buckets, fps
w = gas flow, lb per sec
g = 32.2 ft per sec^2

Speeds can be either positive or negative depending on their direction. Speeds are vector quantities, and we cannot just add or subtract them arithmetically. Before we get deeper into this, let us see how jet and bucket speeds are generally related in an impulse turbine.

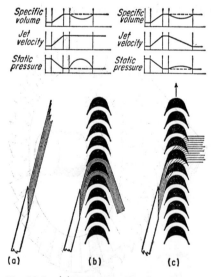

Fig. 19·2 (a) The jet blowing free from the nozzle dissipates itself downstream. (b) With the turbine wheel locked, the buckets turn the jet through a large angle. (c) When the buckets move, the jet turns in a smaller angle and leaves at lower speed.

19·3 Impulse turbine. Figure 19·2 shows what happens to a gas jet leaving a nozzle for three conditions. In a the jet blows free as it leaves the nozzle and ultimately dissipates in turbulence; this is effectively throttling. In b we look at the edge of the turbine wheel and nozzle and redraw the buckets as if they were arranged on a flat surface. By locking the turbine wheel so it cannot move, we make the stationary buckets act as guides to force the gas jet to turn through almost a 180° angle. This develops maximum force on the buckets, but no work because the force does not move. Ideally the jet leaves the buckets with the same velocity that it enters the buckets. This means the jet retains its kinetic energy and transfers none to the turbine wheel and shaft.

The graphs, Fig. 19·2, show what happens to the gas jet as it flows through the nozzle and buckets. The static pressure drops while the gas flows through the nozzle; then it rises while the jet direction changes but falls to the nozzle discharge pressure as it leaves the buckets. The jet velocity rises while flowing through the nozzle but ideally stays constant as it passes through the bucket passages. The specific volume of the gas acts inversely to the static pressure.

Figure 19·2c shows the buckets moving while the jet flows through the passages, the buckets moving at about half the jet speed. This turns the jet through less than a 90° angle in the bucket passages. The graphs, Fig. 19·2, show that the pressure and specific volume behave as about in b with less change but the jet velocity drops markedly when flowing through the buckets. The jet develops a force about half that in b, the locked-wheel condition, but since the force can move, it does work on the moving buckets. The jet does this work by giving up much of its kinetic energy and so slowing down.

In the impulse turbine the gas static pressure is the same at the bucket entrance and exit sections. If the pressure is allowed to drop, the buckets act as moving nozzles and we have a reaction turbine—more about this later.

19·4 Velocity relations. Figures 19·1 and 19·2 suggest that we can choose a wide variety of angles between the nozzle axis and the turbine-wheel plane. We can also have ratios of bucket speed to jet speed varying from zero to unity. Figure 19·3a shows how these variables are related. Vector v_1 shows the direction of the gas jet leaving the nozzle, its length measuring the speed. Vector v_b shows the direction and speed of the buckets. The angle A is the angle between the nozzle axis and the wheel plane.

By subtracting v_b vectorially from v_1, we get v_{1r}, the jet velocity relative to the moving buckets. As we expect, the latter is much less than v_1, but it makes an angle N with the wheel plane appreciably larger than A. To make the jet enter the bucket passages smoothly, the entering edge of the bucket must be made equal to N rather than A. A number of factors determines the best leaving angle for the bucket, N', in actual turbines, but ideally we can make $N' = N$. For the ideal condition there is no friction between the jet and the bucket face and the jet leaving velocity relative to the bucket $v_{2r} = v_{1r}$, the relative jet entering velocity.

We must find the jet absolute velocity leaving the bucket to learn how much kinetic energy was given up by the jet in doing work on the bucket. We do this by adding another v_b vector to the end of v_{2r} and complete the triangle with v_2, the absolute velocity of the leaving jet.

The shaded arc in Fig. 19·3a shows how the vector study fixes the general shape of the turbine bucket. For turbines running at variable speed or with large load variations at constant speed, the entering edge of the bucket would probably have a round nose rather than the sharp edge shown. This reduces the turbulence of the jet entering bucket passages and gives better aerodynamic flow characteristics. The leaving edge would also be rounded to a lesser degree to give better mechanical strength. The dashed line shows the flow of the gas stream relative to the bucket. Figure 19·3b rearranges the velocity vectors about a single v_b to make certain trigonometric relations easier to follow.

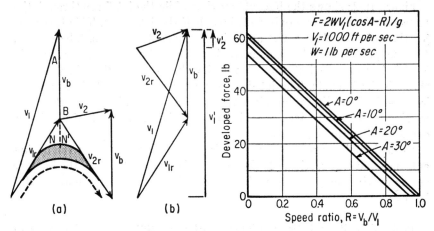

Fig. 19·3 Velocity vectors help calculate the proper entrance and exit angles of the bucket for the best energy transformation.

Fig. 19·4 An impulse turbine develops maximum force with the wheel locked at zero speed but produces no shaft work.

19·5 Force developed. Equation (19·1) gives the force developed by changing the momentum of a flow of fluid. For the turbine wheel, only the component of the velocity parallel to the bucket motion plays a part in developing the moving force. Then for Fig. 19·3b,

$$F = \frac{w}{g}(v_1' - v_2') \tag{19·2}$$

$$= \frac{w}{g}(v_1 \cos A - v_2 \cos B) \tag{19·3}$$

$$= \frac{w}{g}(v_{1r} \cos N + v_{2r} \cos N') \tag{19·4}$$

We shall limit our study to ideal symmetrical buckets where $N = N'$ and $v_{1r} = v_{2r}$, then

$$F = \frac{w}{g} 2v_{1r} \cos N \tag{19·5}$$

$$= \frac{w}{g} 2(v_1 \cos A - v_b) \tag{19·6}$$

Letting $R = v_b/v_1$, the bucket-jet speed ratio, we have

$$F = \frac{w}{g} 2v_1(\cos A - R) \qquad (19\cdot7)$$

This shows that the force acting on the buckets depends on the initial jet velocity, the angle between the jet and the wheel, and the bucket-jet speed ratio for a given jet flow rate.

Figure 19·4 shows how the developed force varies ideally with jet-wheel angle and speed ratio for a jet flow of 1 lb per sec and jet speed of 1,000 fps. Each set of conditions would have its own unique bucket shape. The developed force is maximum at zero speed ratio, that is, locked-wheel condition. But as the wheel speeds up, the developed force drops to zero as the bucket speed approaches the jet speed.

A jet-wheel angle $A = 0$ gives maximum force at any given speed ratio. As the angle grows, the force drops at an increasing rate, showing the need to keep A as small as conditions will allow.

19·6 Bucket efficiency. Now we must find the conditions that will develop maximum shaft energy output. Let us do this in terms of power developed by the buckets:

$$P = Fv_b = \frac{2v_1 w}{g}(\cos A - R)v_b \qquad (19\cdot8)$$

This is the output of the turbine in terms of foot-pounds per second. The input is the kinetic energy of the jet, $K = wv_1^2/2g$, also in foot-pounds per second. When the ratio of output to input reaches a maximum, we have the best arrangement. We call the ratio the bucket efficiency:

$$e = \frac{P}{K} = \frac{(wv_1^2/g)(\cos A - R)v_b}{wv_1^2/2g} \qquad (19\cdot9)$$
$$= 4R(\cos A - R) \qquad (19\cdot10)$$

Figure 19·5 shows the relation of bucket efficiency and speed ratio for various jet-wheel angles. At $A = 0$, the best efficiency of 100% is achieved at a jet-wheel speed ratio of 0.5. The maximum efficiency for each given A is at a speed just one-half the speed at which F becomes zero.

Comparing Figs. 19·5 and 19·4 we find that at the best bucket efficiency for each A, the developed force is just one-half the force produced at locked-wheel condition, $R = 0$. This means that at startup a turbine produces maximum torque, a good feature for starting under load. Keep in mind that the bucket has a unique shape for each combination of A and R.

An interesting feature is the absolute velocity of the gas leaving the buckets, v_2. Figure 19·6 shows how this varies with jet angle and speed ratio. We can have quite high leaving velocities and still get respectable work outputs. For example, at $A = 30°$ and $R = 0.433$, we have $v_2 = 500$ fps for $v_1 = 1,000$ fps. While the ratio of exit to entering velocity is $500/1,000 = 0.5$, the ratio of corresponding kinetic energies is $500^2/1,000^2 = 0.25$; the major part of the energy has been converted even for these relatively unfavorable conditions. The corresponding bucket efficiency is 75% as shown in Fig. 19·5. At only one condition, $R = 0.5$

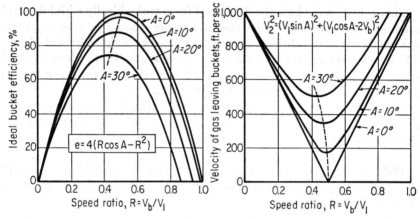

Fig. 19·5 Impulse turbines work at best efficiency with bucket speed about half jet speed.

Fig. 19·6 Gas velocity leaving the buckets varies inversely as the ideal bucket efficiency.

and $A = 0°$, can we have $v_2 = 0$, as shown in Fig. 19·6. This is a theoretical ideal condition not obtainable in practice.

19·7 Compounding. In Chap. 18 we learned that the nozzle jet velocity depended on the pressure drop of the gas flowing through. On a given turbine we usually have a given overall pressure drop available. If we passed the gas through a set of parallel nozzles placed between the total pressure drop, we might find that we get very high gas jet velocities.

The corresponding bucket velocity we might have to use for good efficiency could put a severe centrifugal strain on the turbine wheel, making it impossible to run safely. To avoid this condition, we can place two or more sets of nozzles and buckets in series between the inlet and outlet pressures. All the gas flowing through the first set of nozzles and buckets would expand to an intermediate pressure, producing a moderate jet velocity. From the first set of buckets, the gas would then expand through the next set of nozzles and buckets at a moderate velocity. Any number of sets could be used to achieve the best jet velocity in each set or stage between the turbine inlet and final outlet pressures.

This arrangement, Fig. 19·7, is called pressure compounding. While the pressure drops steadily through the stages, the gas speed alternately rises

TURBINE NOZZLES AND BUCKETS 289

in the nozzles and drops in the buckets. The gas specific volume rises as the pressure drops, so succeeding stages must have larger cross-sectional areas to handle the increasing volume flow. This can be done by increasing the number of nozzles per stage, by raising the height of the nozzles and buckets, or by increasing the diameter of the diaphragms supporting the nozzles and of the wheel holding the buckets.

19·8 Velocity compounding. Another method of getting reasonable bucket speeds with large pressure drop and high jet speed is velocity compounding. Figure 19·8 shows the velocity diagram of this arrangement; Fig. 19·9 shows the physical layout. A set of nozzles produces a high-speed air jet at velocity v_1 that enters a first row of buckets, moving at v_b, with a relative velocity v_{1r}. Ideally the jet leaves the buckets at $v_{2r} = v_{1r}$. The directions of these two vectors fix the entering and leaving angles of the buckets.

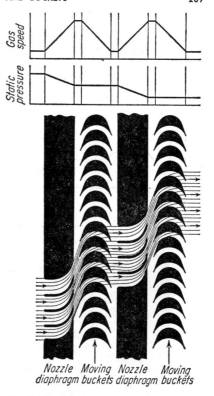

Fig. 19·7 Pressure-compounded impulse turbine has a smaller pressure drop in individual stages.

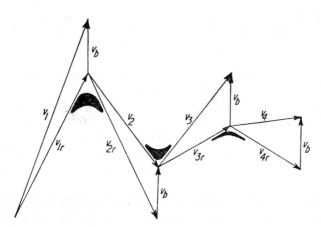

Fig. 19·8 A velocity-compounded turbine stage uses two rows of impulse buckets with an intermediate row of stationary reversing blades.

Adding v_b to v_{2r} gives the absolute velocity v_2 of the leaving gas flow. The gas then enters between a set of stationary reversing blades to be redirected to a second row of moving buckets at ideal velocity $v_3 = v_2$.

Subtracting v_b from v_3 gives the relative entering velocity v_{3r} for the gas in the second row of buckets. Ideally $v_{4r} = v_{3r}$, the leaving relative

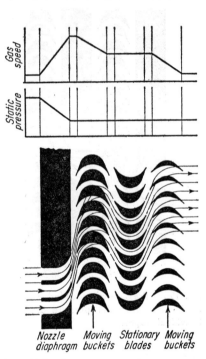

Fig. 19·9 Static pressure stays constant in the three blading rows of the compounded stage.

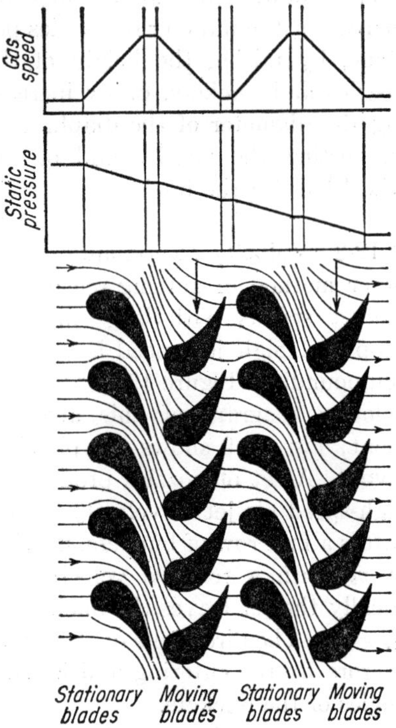

Fig. 19·10 Gas pressure drops in both moving and stationary blading of a reaction turbine.

velocity from the second-row buckets. Finally we add v_b to find v_4, the absolute gas velocity leaving the compound stage.

The graphs at the top of Fig. 19·9 show that gas-flow speed rises in nozzles, drops in first-row moving buckets, stays constant in stationary blades, and then drops again in second-row moving buckets. On the other hand, gas static pressure drops in the nozzles and stays essentially constant as it flows through the buckets and stationary reversing blades.

19·9 Efficiency. The efficiency of velocity compounding is lower than that of pressure compounding, as we can see by the higher leaving velocity v_4. Frictional losses also run higher. But velocity compounding offers an advantage in taking a large pressure drop in one step and reducing the temperature at which the moving buckets run.

19·10 Reaction blading. We found that impulse turbines ideally have gas-pressure drops only in the stationary nozzles. When we direct gas flow through a nozzle as in Fig. 19·9, changing the direction and rate of flow, there is a strong reaction force acting on the nozzle diaphragm tending to turn it oppositely to the moving buckets. For this reason the turbine casing must be securely anchored to prevent overturning.

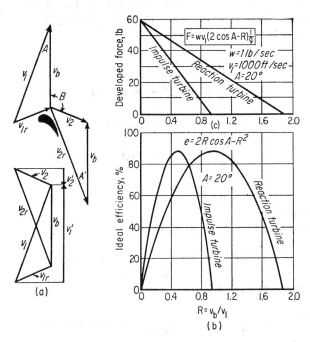

Fig. 19·11 Reaction turbines have performance comparable to impulse turbines at different speed ratios; vectors are for 50% reaction.

Reaction turbines put this force to work by allowing gas static pressure to drop in both moving blading and stationary blading (nozzles), Fig. 19·10. But the gas absolute velocity rises in the stationary blading and drops in the moving blades.

Figure 19·11a shows the velocity diagrams of the gas. Gas leaves the stationary blading with speed v_1. By subtracting moving-blading speed v_b, we find the relative entering velocity v_{1r} of the gas into the moving blading. But the moving blading turns the gas through an angle of almost 90° and because of the pressure drop accelerates the gas flow from v_{1r} to v_{2r}.

Subtracting the blading speed v_b we find the absolute leaving velocity v_2 of the gas. The gas enters the stationary blading of the following stage to be accelerated by a pressure drop to v_1 for that stage.

The blading cross section, Fig. 19·10 and 19·11a, has been developed by aerodynamic principles to give minimum losses. It works at good efficiency over a wide range of flow rates. Figure 19·10 shows the gas-flow path by the stream lines; remember that the gas accelerates while it flows through the moving blades (relative to the blades).

19·11 Performance. We can run an analysis for the reaction turbine that parallels what we did for the impulse turbine. Assuming that the enthalpy drop through stationary blading equals that through moving the blading, we have $v_1 = v_{2r}$. Then from Fig. 19·11a, force acting on the moving blades is

$$\begin{aligned} F &= \frac{w}{g}(v_1' - v_2') \\ &= \frac{w}{g}(v_1 \cos A - v_2 \cos B) \\ &= \frac{w}{g}[v_1 \cos A - (v_{2r} \cos A - v_b)] \\ &= \frac{w}{g}(2v_1 \cos A - v_b) \\ &= \frac{w}{g}v_1(2 \cos A - R) \end{aligned} \qquad (19\cdot11)$$

where $R = v_b/v_1$.

The power developed by the blading is

$$P = Fv_b = \frac{wv_1v_b}{g}(2 \cos A - R) \qquad (19\cdot12)$$

The ideal blading efficiency is

$$e = \frac{P}{(w/2g)(v_1^2 + v_{2r}^2)} = 2R \cos A - R^2 \qquad (19\cdot13)$$

Figure 19·11c shows how F varies with R for $w = 1$ lb per sec, $v_1 = 1{,}000$ fps, and $A = 20°$, for both an impulse and a reaction turbine. Both types develop the same force at locked-rotor conditions. The force drops to zero at twice the blade-jet speed ratio for the reaction turbine.

Figure 19·11b compares blade efficiency for impulse and reaction turbines for the given conditions. Maximum efficiency develops at slightly less than 1.0 speed ratio for the reaction turbine, just twice that for the impulse turbine. Developed forces for both turbines are equal at their maximum blade efficiencies as shown in Fig. 19·11c. The force at maximum efficiency for both units is one-half the locked-rotor force.

19·12 Design factors. Pressure drop across moving blades in a reaction turbine makes some of the air flow leak between blade tips and casing. This clearance must be kept small, usually with the aid of thin sealing strips that wear quickly on accidental contact. Keeping pressure drop per stage small helps minimize the leakage. This means more stages per turbine, about double those needed for a pressure-compounded impulse turbine.

In an impulse turbine, nozzles may occupy only part of the total periphery of the nozzle diaphragm. The gas jet then works on only part of the total number of buckets on a wheel at a time. Since the gas pressure on the inlet and outlet sides of the buckets is constant and equal, the jet has no tendency to break up.

In a reaction turbine, however, unequal pressure on the inlet and outlet sides of the moving blading makes an open path for gas flow through all the blading passages. If nozzles produced a jet of gas over only part of the periphery, the jet would quickly break up in seeking to flow through all the blade passages, at great loss in efficiency. Because of this, reaction turbines must have nozzles (stationary blading) around the full periphery of all stages. That is, all stages must flow full of gas to maintain good blading efficiency.

19·13 Condition line. The steady-flow equation for a gas turbine, Art. 16·8, showed that the work output was

$$W_2 = H_1 - H_2 + \frac{w}{2gJ}(v_1^2 - v_2^2) \qquad (19·14)$$

If the entering and leaving velocities of the gas are relatively small, this boils down to simply

$$W_2 = H_1 - H_2 \qquad (19·15)$$

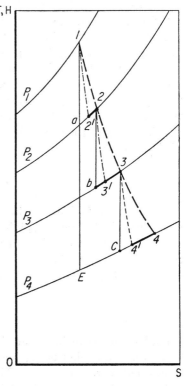

Fig. 19·12 The turbine condition curve for a three-stage unit shows the changing state of gas.

Let us examine the process on a T-S or H-S diagram. The diagrams are exactly proportional for a perfect gas, since $\Delta T = c_p \, \Delta H$. In Fig. 19·12, if an impulse turbine received gas at P_1 and expanded it at constant entropy from state 1, the gas would leave at state E at the exhaust pressure P_4. The ideal work done by the gas would then be

$$W_2 = H_1 - H_E \qquad (19·16)$$

An actual turbine, however, cannot use all the available energy and would exhaust the gas at state 4. The actual work is then

$$W_2 = H_1 - H_4 \tag{19·17}$$

The performance of the turbine would be measured by its engine efficiency:

$$e_e = \frac{H_1 - H_4}{H_1 - H_E} \tag{19·18}$$

Since we have studied the internal processes of a turbine, we can make a more detailed analysis of what happens to the gas when flowing through a turbine. In Fig. 19·12 we have the states plotted for a three-stage pressure-compounded impulse turbine. The line 1–2–3–4 is the *condition curve* or *line* of the turbine and shows the state of the gas at the three sets of nozzle inlets and exhausts. But the condition line is a trend line; a more detailed analysis shows that process 1–2′ is the gas expansion through the first-stage nozzles from P_1 to P_2. An ideal nozzle would expand the gas at constant S from 1 to a. The difference $H_1 - H_2' = KE$, kinetic energy of the gas jet leaving the nozzle, at pressure P_2. The nozzle efficiency is then

$$e_n = \frac{H_1 - H_2'}{H_1 - H_a} \tag{19·19}$$

Process 2′ to 2 shows an increase in entropy and enthalpy as the gas flows through the buckets, giving up some of its kinetic energy as it works on the buckets to develop mechanical work W. The enthalpy increase comes from part of the jet KE being lost in friction and turbulence and staying in the gas as internal energy. Stage efficiency is then

$$e_s = \frac{H_1 - H_2}{H_1 - H_a} \tag{19·20}$$

Each of the following stages has similar processes plotted in Fig. 19·12 for nozzle and bucket flows, friction and turbulence losses being returned as internal energy to the gas. This reconversion of energy is called *reheat*. It has an advantageous effect in raising the overall turbine-engine efficiency above individual stage efficiencies in multistage turbines.

19·14 Pressure lines. These lines on the T-S and H-S diagram, Fig. 19·12, diverge with increasing entropy. This makes the sum of the stage constant-S expansions greater than the single constant-S expansion from the inlet state, H_1, at P_1 to P_4:

$$H_1 - H_a + H_2 - H_b + H_3 - H_c > H_1 - H_E$$

This helps in recovering part of the lost work in each stage but, of course, never recovers all of it. The ratio of these enthalpy differences is called the *reheat factor:*

$$RF = \frac{H_1 - H_a + H_2 - H_b + H_3 - H_c}{H_1 - H_E} \qquad (19\cdot21)$$

This compares the total available drop by stages with the drop available in a single expansion.

Example: A turbine with three stages takes gas at 1760°R and 58.8 psia and expands it to 15.26 psia. Each stage has an efficiency of 80%. Find the overall turbine-engine efficiency and the reheat factor.

For these conditions the state at each point in Fig. 19·12 is as shown in Table 19·1.

Table 19·1

State	T,°R	P, psia	H, Btu/lb
1	1760	58.8	438.83
a	1586	39.1	392.01
2	1621	39.1	401.37
b	1445	25.0	354.69
3	1480	25.0	364.03
c	1300	15.26	316.94
4	1336	15.26	326.33
E	1239	15.26	301.27

Substituting in the RF formula above, we get a reheat factor of

$$RF = \frac{46.82 + 46.68 + 46.95}{438.83 - 301.27} = \frac{140.45}{137.56} = 1.021$$

This means that multistaging increases the available energy drop by 2.1%. This recovers some of the losses to friction and turbulence.

The overall turbine-engine efficiency is

$$e_e = \frac{438.83 - 326.33}{438.83 - 301.27} = 0.818 \text{ or } 81.8\%.$$

This compares with individual stage efficiency of 80.0% and shows the advantage of multistaging. If we expanded the gas in one stage through the pressure range, the engine efficiency would probably be less than 80.0%. Available work lost in the turbine is

$$H_4 - H_E = 326.33 - 301.27 = 25.06 \text{ Btu per lb of gas}$$

REVIEW TOPICS

1. How does a jet of gas transfer some of its kinetic energy to an object doing mechanical work?

2. Draw a sketch of a single-stage turbine wheel and nozzle to demonstrate the mechanics of converting the kinetic energy of the jet to mechanical shaft work.

3. Draw a developed row of buckets of an impulse turbine and show the changes in specific volume, jet velocity, and gas static pressure through the nozzle and buckets for (a) a locked-rotor condition and (b) the best bucket-speed condition.

4. Draw the velocity-vector diagram for an impulse bucket, and explain the meaning of each vector.

5. Write and explain the basic force equation from Newton's Second Law of Motion.

6. Derive the equation for the force acting on a moving bucket exerted by a gas jet flowing through the bucket passages.

7. Write the equation for power developed by a gas jet.

8. Write the equation for the bucket efficiency.

9. What is the best bucket-jet speed ratio for an ideal single-stage impulse turbine wheel?

10. What torque does an ideal impulse turbine develop at standstill compared with its best running speed?

11. Draw the cross section of a pressure-compounded impulse turbine, and show the variation of gas speed and static pressure.

12. Draw the gas-velocity vectors for a pressure-compounded impulse-turbine stage.

13. Draw the cross section of a velocity-compounded impulse-turbine stage, and show the variation in jet speed and pressure.

14. Draw the velocity-vector diagram of the gas flowing through a velocity-compounded impulse-turbine stage.

15. Draw the cross section of a reaction-turbine blading section, and show the variation of gas velocity and pressure.

16. Develop the equations for working force, power, and blading efficiency of a reaction turbine.

17. Why does a reaction turbine have more stages for a given pressure range than an impulse turbine?

18. Why must reaction-turbine stages have full gas flow through all blading on one wheel and not an impulse turbine?

19. Describe the condition line of a multistage gas turbine with the aid of an H-S chart.

20. What is the equation for the engine efficiency of a turbine?

21. What is the equation for nozzle efficiency?

22. What is the equation for the stage efficiency of an impulse turbine.

23. What is meant by reheat in a multistage turbine?

24. Write the equation for the reheat factor of a multistage turbine.

PROBLEMS

1. A bank of nozzles in a gas turbine passes a flow of 20 lb per sec of gas into the buckets on a turbine wheel. The gas enters the bucket passages at 3,200 fps and leaves at 1,200 fps. What force does the gas exert on the buckets under ideal conditions?

2. Check several points on the curves in Fig. 19·4 for the given gas velocity and flow rate. What happens to the force developed as the gas velocity varies? As the flow rate varies?

3. Check several points on the curves in Fig. 19·5 for the given conditions.

4. Check several points on the curves of Fig. 19·6 for the given conditions.

5. Check several points on the curves of Fig. 19·11b.

6. Check several points on the curves of Fig. 19·11c.

7. A gas turbine receives hot combustion gas with an enthalpy of 433 Btu per lb. The gas exhausts from the turbine with an enthalpy of 267 Btu per lb. The gas enters with a velocity of 800 fps and leaves at 250 fps. What work does the gas do in the turbine per pound of flow.

8. A gas turbine admits hot gas (ideal air) at 1450 F and 75 psia and exhausts it at 15 psia. Assuming that entering and leaving gas velocities are equal, find the work output per pound of gas if the turbine runs with an engine efficiency of 86%.

9. For the gas turbine in Prob. 8, find the work output if it has three stages with equal pressure ratios. Each stage has an efficiency of 86%. Find (a) the work output of the gas per stage, (b) the total output of the turbine, (c) the reheat factor, and (d) the temperature of the gas leaving the turbine.

CHAPTER 20

Steady-flow Compressors

We studied gas-turbine engines and the basic theory of turbines in Chaps. 18 and 19. Now we are ready to look at steady-flow air compressors, which some of these engines use in a complete plant cycle.

The basic theory of compressors that we studied in Chap. 12 as applied to reciprocating machines also applies to the wide variety of steady-flow compressors.

20·1 Axial-flow compressors. These compressors, Fig. 20·1, have a rotor carrying blades that pass between stationary blades carried by the casing. Figure 20·2 shows a cross section of two stages of this type of compressor. The blades are airfoil sections, designed on aerodynamic principles to reduce friction and turbulence.

In Fig. 20·2 streamlines with arrows show the direction of air flow through the two stages. The fixed blades form diffusers. These act oppositely to nozzles; they receive a high-speed gas flow through their narrow end, slow it down because of the widening cross section, and so allow it to compress itself to a higher pressure before leaving the larger end. In effect, the gas molecules entering catch up with molecules ahead and so raise the density and pressure.

The moving blades "bite" into the gas stream and "push" it along the general direction of the diffuser axes. The blades strike the individual gas molecules and "bat" them along to increase their speed and energy. That is, the blades do work on the gas. When the molecules "catch up" with one another (as a continuous process) in the diffusers, they produce a higher pressure and have a higher temperature.

The basic molecular action in the axial-flow compressor resembles that in a piston and cylinder, the difference being that flow and compression are continuous in the former and intermittent in the latter. Gas flows in a generally helical path around the rotor for the blade arrangement shown in Fig. 20·2. Other blading angles may be used.

Fig. 20·1 An axial-flow compressor resembles a turbine in the arrangement of the rotor and casing.

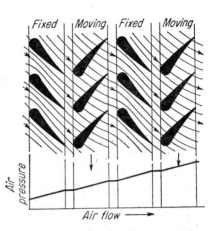

Fig. 20·2 Axial-flow blading uses sections of air foils (propellers) to work at the best practical compression efficiency.

20·2 Centrifugal compressors. These compressors, Fig. 20·3, are another form of steady-flow pressurizer. In the three-stage unit shown, also called a blower, air enters the casing at left, then enters the center annular space of the first-stage impeller. Curved blades (outlined by the

Fig. 20·3 Three-stage centrifugal compressor has three shrouded impellers carrying curved blades to pressurize large air volumes.

welding marks) work on the gas, giving it more energy by exerting a centrifugal force on it.

Gas leaves the impeller periphery to enter surrounding diffusing passages in the casing. Here the gas slows down as its pressure rises, and the passages lead it to the center inlet of the second stage. In this fashion gas pressure and temperature rise in three steps before the gas leaves the outlet at the right.

Figure 20·4 shows another form of centrifugal compressor rotor or impeller. This one runs inside a close-fitting casing that has an opening around the inner portion where the radial fins or blades are curved. The casing holds a diffusing passage surrounding the entire periphery of the impeller. Gas enters between the curved portions of the radial blades (the impeller turns counterclockwise) and is forced out by centrifugal action of the blades to enter the diffuser, in which it is pressurized.

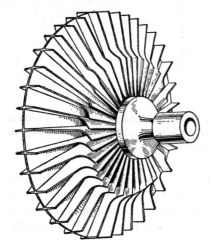

Fig. 20·4 Single-stage centrifugal compressor impeller has radial blades to pressurize the air and rotates inside a close-fitting stationary casing.

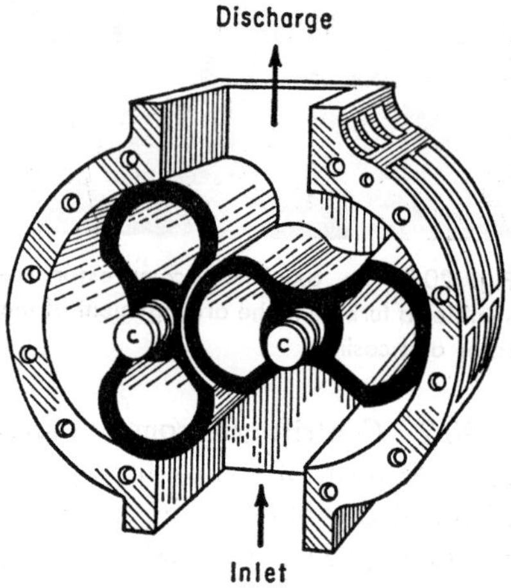

Fig. 20·5 Positive-displacement rotary blower "packs" the air into a higher pressure region in separate volumes.

20·3 Rotary compressors. These compressors, built in several forms, also can be considered steady-flow devices, though they handle gas in batches rather than in a truly steady stream. Figure 20·5 shows a compressor with two rotors, each having two lobes that continuously touch each other as they turn. With clockwise rotation of the left rotor and counterclockwise of the other, air is carried from the bottom to the top of the casing.

Spaces between the rotors and casing trap the gas at its inlet pressure. Forcing the incoming gas into the higher pressure region produces pressurization. Close-contact fits between rotors and casing prevent gas from leaking back into the l-p source. In effect, compressor "packs" the h-p region with gas to produce higher pressure.

Figure 20·6 shows a sliding-vane compressor; it has a rotor turning in a cylinder, the rotor axis being lower than the cylinder axis. The rotor carries spring-loaded vanes that rub along the surface of the cylinder, forming leakproof chambers that vary in volume as the rotor turns.

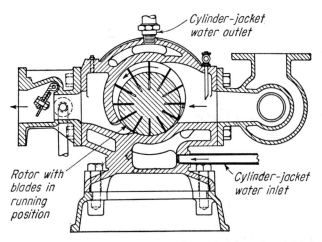

Fig. 20·6 Rotary compressor with sliding vanes forms chambers that shrink in volume as the rotor turns from the inlet to outlet ports.

The entering gas gets trapped in these chambers in the upper part of the cylinder. As the rotor turns, the chamber volume shrinks and compresses the gas. At the lower left port the chambers discharge the compressed gas to the h-p region. The flush fit between the lower part of the rotor and cylinder prevents h-p gas from leaking back to the inlet.

CHAPTER 21

Gas Properties

21·1 Gas tables. In the energy and work equations, Chaps. 4 to 6, specific heats of the gas play an important part. The easiest way to use these equations is to assume that the specific heats (or their ratio k) stay constant for the conditions considered. But Chap. 6 showed that temperature and pressure change the values of c_p, c_v, and k. So where we need to do accurate work, the variation in these factors must be taken into account.

To avoid involved computations using integral calculus Keenan and Kaye compiled the *Gas tables*. An abstract of the tables for air at 1 psia pressure is reproduced in Table 21·1. This table is accurate enough for most thermodynamic design work. When it is used for pressures above 400 psia, the errors in calculating may be on the order of ½% or less.

The complete *Gas tables* include 64 tables for a variety of gases and their mixtures as well as certain mathematical quantities. Let us learn how to use the air table; the original table has 1° intervals to 3000°R and 10° intervals to 6500°R. Using the abstract table here will not give us the accuracy of the original. For important work use the original tables.

21·2 Constant-entropy processes. These are the foundation of the gas table. In the table the pressure and temperature at the beginning of a constant-entropy process are related to the pressure and temperature at the end of the process through the ratio of the relative pressures, p_r.

Example 1: Air at the start of a constant-entropy compression process is at 14.7 psia and 550°R. If the process ends at 850°R, what is the corresponding pressure?

From Table 21·1, at 550°R find $p_{r1} = 1.4779$; at 850°R find $p_{r2} = 6.856$; then

$$\frac{P_2}{P_1} = \frac{p_{r2}}{p_{r1}}$$

$$P_2 = 14.7 \times \frac{6.856}{1.4779}$$

$$= 68.2 \text{ psia}$$

Table 21·1   Thermodynamic properties of air (temperature)[a]

T	h	p_r	u	v_r	ϕ	T	h	p_r	u	v_r	ϕ
400	95.53	0.4858	68.11	305.0	0.52890	890	213.80	8.081	152.80	40.80	0.72163
410	97.93	0.5295	69.82	286.8	0.53481	900	216.26	8.411	154.57	39.64	0.72438
420	100.32	0.5760	71.52	270.1	0.54058	910	218.72	8.752	156.34	38.52	0.72710
430	102.71	0.6253	73.23	254.7	0.54621	920	221.18	9.102	158.12	37.44	0.72979
440	105.11	0.6776	74.93	240.6	0.55172	930	223.64	9.463	159.89	36.41	0.73245
450	107.50	0.7329	76.65	227.45	0.55710	940	226.11	9.834	161.68	35.41	0.73509
460	109.90	0.7913	78.36	215.33	0.56235	950	228.58	10.216	163.46	34.45	0.73771
470	112.30	0.8531	80.07	204.08	0.56751	960	231.06	10.610	165.26	33.52	0.74030
480	114.69	0.9182	81.77	193.65	0.57255	970	233.53	11.014	167.05	32.63	0.74287
490	117.08	0.9868	83.49	183.94	0.57749	980	236.02	11.430	168.83	31.76	0.74540
500	119.48	1.0590	85.20	174.90	0.58233	990	238.50	11.858	170.63	30.92	0.74792
510	121.87	1.1349	86.92	166.46	0.58707	1000	240.98	12.298	172.43	30.12	0.75042
520	124.27	1.2147	88.62	158.58	0.59173	1010	243.48	12.751	174.24	29.34	0.75290
530	126.66	1.2983	90.34	151.22	0.59630	1020	245.97	13.215	176.04	28.59	0.75536
540	129.06	1.3860	92.04	144.32	0.60078	1030	248.45	13.692	177.84	27.87	0.75778
550	131.46	1.4779	93.76	137.85	0.60518	1040	250.95	14.182	179.66	27.17	0.76019
560	133.86	1.5742	95.47	131.78	0.60950	1050	253.45	14.686	181.47	26.48	0.76259
570	136.26	1.6748	97.19	126.08	0.61376	1060	255.96	15.203	183.29	25.82	0.76496
580	138.66	1.7800	98.90	120.70	0.61793	1070	258.47	15.734	185.10	25.19	0.76732
590	141.06	1.8899	100.62	115.65	0.62204	1080	260.97	16.278	186.93	24.58	0.76964
600	143.47	2.005	102.34	110.88	0.62607	1090	263.48	16.838	188.75	23.98	0.77196
610	145.88	2.124	104.06	106.38	0.63005	1100	265.99	17.413	190.58	23.40	0.77426
620	148.28	2.249	105.78	102.12	0.63395	1110	268.52	18.000	192.41	22.84	0.77654
630	150.68	2.379	107.50	98.11	0.63781	1120	271.03	18.604	194.25	22.30	0.77880
640	153.09	2.514	109.21	94.30	0.64159	1130	273.56	19.223	196.09	21.78	0.78104
650	155.50	2.655	110.94	90.69	0.64533	1140	276.08	19.858	197.94	21.27	0.78326
660	157.92	2.801	112.67	87.27	0.64902	1150	278.61	20.51	199.78	20.771	0.78548
670	160.33	2.953	114.40	84.03	0.65263	1160	281.14	21.18	201.63	20.293	0.78767
680	162.73	3.111	116.12	80.96	0.65621	1170	283.68	21.86	203.49	19.828	0.78985
690	165.15	3.276	117.85	78.03	0.65973	1180	286.21	22.56	205.33	19.377	0.79201
700	167.56	3.446	119.58	75.25	0.66321	1190	288.76	23.28	207.19	18.940	0.79415
710	169.98	3.623	121.32	72.60	0.66664	1200	291.30	24.01	209.05	18.514	0.79628
720	172.39	3.806	123.04	70.07	0.67002	1210	293.86	24.76	210.92	18.102	0.79840
730	174.82	3.996	124.78	67.67	0.67335	1220	296.41	25.53	212.78	17.700	0.80050
740	177.23	4.193	126.51	65.38	0.67665	1230	298.96	26.32	214.65	17.311	0.80258
750	179.66	4.396	128.25	63.20	0.67991	1240	301.52	27.13	216.53	16.932	0.80466
760	182.08	4.607	129.99	61.10	0.68312	1250	304.08	27.96	218.40	16.563	0.80672
770	184.51	4.826	131.73	59.11	0.68629	1260	306.65	28.80	220.28	16.205	0.80876
780	186.94	5.051	133.47	57.20	0.68942	1270	309.22	29.67	222.16	15.857	0.81079
790	189.38	5.285	135.22	55.38	0.69251	1280	311.79	30.55	224.05	15.518	0.81280
800	191.81	5.526	136.97	53.63	0.69558	1290	314.36	31.46	225.93	15.189	0.81481
810	194.25	5.775	138.72	51.96	0.69860	1300	316.94	32.39	227.83	14.868	0.81680
820	196.69	6.033	140.47	50.35	0.70160	1310	319.53	33.34	229.73	14.557	0.81878
830	199.12	6.299	142.22	48.81	0.70455	1320	322.11	34.31	231.63	14.253	0.82075
840	201.56	6.573	143.98	47.34	0.70747	1330	324.69	35.30	233.52	13.958	0.82270
850	204.01	6.856	145.74	45.92	0.71037	1340	327.29	36.31	235.43	13.670	0.82464
860	206.46	7.149	147.50	44.57	0.71323	1350	329.88	37.35	237.34	13.391	0.82658
870	208.90	7.450	149.27	43.26	0.71606	1360	332.48	38.41	239.25	13.118	0.82848
880	211.35	7.761	151.02	42.01	0.71886	1370	335.09	39.49	241.17	12.851	0.83039

Table 21·1 Thermodynamic properties of air (temperature)[a] (*Continued*)

T	h	p_r	u	v_r	φ	T	h	p_r	u	v_r	φ
1380	337.68	40.59	243.08	12.593	0.83229	2500	645.78	435.7	474.40	2.125	0.99497
1390	340.29	41.73	245.00	12.340	0.83417	2600	674.49	513.5	496.26	1.8756	1.00623
1400	342.90	42.88	246.93	12.095	0.83604	2700	703.35	601.9	518.26	1.6617	1.01712
1410	345.52	44.06	248.86	11.855	0.83790	2800	732.33	702.0	540.40	1.4775	1.02767
1420	348.14	45.26	250.79	11.622	0.83975	2900	761.45	814.8	562.66	1.3184	1.03788
1430	350.75	46.49	252.72	11.394	0.84158	3000	790.68	941.4	585.04	1.1803	1.04779
1440	353.37	47.75	254.66	11.172	0.84341	3100	820.03	1083.4	607.53	1.0600	1.05741
1450	356.00	49.03	256.60	10.954	0.84523	3200	849.48	1241.7	630.12	0.9546	1.06676
1460	358.63	50.34	258.54	10.743	0.84704	3300	879.02	1418.0	652.81	0.8621	1.07585
1470	361.27	51.68	260.49	10.537	0.84884	3400	908.66	1613.2	675.60	0.7807	1.08470
1480	363.89	53.04	262.44	10.336	0.85062	3500	938.40	1829.3	698.48	0.7087	1.09332
1490	366.53	54.43	264.38	10.140	0.85239	3600	968.21	2067.9	721.44	0.6449	1.10172
1500	369.17	55.86	266.34	9.948	0.85416	3700	998.11	2330.3	744.48	0.5882	1.10991
1550	382.42	63.40	276.17	9.056	0.86285	3800	1028.09	2618.4	767.60	0.5376	1.11791
1600	395.74	71.73	286.06	8.263	0.87130	3900	1058.14	2934.4	790.80	0.4923	1.12571
1650	409.13	80.89	296.03	7.556	0.87954	4000	1088.26	3280	814.05	0.4518	1.13334
1700	422.59	90.95	306.06	6.924	0.88758	4100	1118.46	3656	837.40	0.4154	1.14079
1750	436.12	101.98	316.16	6.357	0.89542	4200	1148.72	4067	860.81	0.3826	1.14809
1800	449.71	114.03	326.32	5.847	0.90308	4300	1179.04	4513	884.28	0.3529	1.15522
1850	463.37	127.18	336.55	5.388	0.91056	4400	1209.42	4997	907.81	0.3262	1.16221
1900	477.09	141.51	346.85	4.974	0.91788	4500	1239.86	5521	931.39	0.3019	1.16905
1950	490.88	157.10	357.20	4.598	0.92504	4600	1270.36	6089	955.04	0.2799	1.17575
2000	504.71	174.00	367.61	4.258	0.93205	4700	1300.92	6701	978.73	0.2598	1.18232
2100	532.55	212.1	388.60	3.667	0.94564	4800	1331.51	7362	1002.48	0.2415	1.18876
2200	560.59	256.6	409.78	3.176	0.95868	4900	1362.17	8073	1026.28	0.2248	1.19508
2300	588.82	308.1	431.16	2.765	0.97123	5000	1392.87	8837	1050.12	0.2096	1.20129
2400	617.22	367.6	452.70	2.419	0.98331						

[a] Reprinted with permission from Joseph H. Keenan and Joseph Kaye, Gas Tables, "Thermodynamic Properties of Air, Products of Combustion and Component Gases," Copyright, 1948, John Wiley & Sons, Inc.

Symbols and units:

T = absolute temperature, °R
h = enthalpy, Btu per lb
p_r = relative pressure
u = internal energy, Btu per lb
v_r = relative volume
ϕ = entropy at constant pressure = $\int c_p (dT/T)$

Temperatures and specific volumes of air at the beginning and end of a process are related through the ratio of relative volumes v_r.

Example 2: Find the specific volume of the air at the end of the process described in Example 1.

$$V_1 = \frac{RT_1}{144 P_1}$$
$$= \frac{53.37 \times 550}{144 \times 14.7}$$
$$= 13.86 \text{ cu ft per lb}$$

From the table, at 550°R find $v_{r1} = 137.85$; at 850°R find $v_{r2} = 45.92$; then

GAS PROPERTIES

$$\frac{V_1}{V_2} = \frac{v_{r2}}{v_{r1}}$$

$$V_2 = 13.86 \times \frac{45.92}{137.85}$$

$$= 4.617 \text{ cu ft per lb}$$

Example 3: What work has been done on the air in the constant-entropy process of Example 1?

Work done on the air

$$W = h_2 - h_1$$

From the table corresponding to 550 and 850°R find $h_1 = 131.46$ and $h_2 = 204.01$, so

$$W = 204.01 - 131.46$$
$$= 72.55 \text{ Btu per lb air}$$

Example 4: What is the change of entropy of process in Example 1?

By definition $\Delta S = 0$. To check this from the table corresponding to 550 and 850°R find $\phi_1 = 0.60518$ and $\phi_2 = 0.71037$; then

$$S_2 - S_1 = (\phi_2 - \phi_1) - \frac{R}{J} \log_e \frac{P_2}{P_1}$$

$$= 0.71037 - 0.60518 - \frac{53.37}{778.16} \log_e \frac{68.2}{14.7}$$

$$= 0.10519 - 0.686 \log_e 4.64$$

$$= 0.10519 - 0.10519 = 0$$

21·3 Actual gases. Much-used gases like air justify the effort made to compile and compute the thermodynamic properties of air as in the Keenan and Kaye tables. But the wide variety of gases used in industry today make table compilation for each one a long and expensive process.

The expedient way to compute the behavior of any gas would be to use some modification of the perfect-gas law. Many efforts have been made in this direction. Let us look at a few to see their form and limitations.

21·4 Van der Waals' equation. The Dutch physicist van der Waals in 1873 proposed a gas-law modification that took account of molecular forces and diameters. The perfect-gas equation assumes that the molecules are dimensionless points having no volume of their own and that they do not exert intermolecular forces of attraction on one another.

Since molecules do occupy a finite volume, the volume through which they travel is smaller by the amount of their own volume. The net result is that gas molecules will strike the walls of the container holding them more often than if they had no volume of their own. This means that the actual pressure they exert will be higher than the perfect-gas equation indicates. To account for this van der Waals proposed that

$$\frac{P_{ideal}}{P_{true}} = \frac{\text{true volume}}{\text{total volume}} = \frac{V - b}{V} \qquad (21 \cdot 1)$$

Then
$$P = \frac{RT}{V - b} \qquad (21 \cdot 2)$$

Gas molecules do exert mutual forces of attraction on one another, so that, when their density increases, their pressure on the container wall will be less than ideal as given by the perfect-gas equation. Van der Waals proposed that this effect varied with the square of the gas volume, so that

$$P_{true} = P_{ideal} - \frac{a}{V^2} \qquad (21 \cdot 3)$$

These include proposed constants a and b that hopefully would apply to all gases, but actually they vary for gases and also vary with the general temperature range for any one gas. Combining these factors we get the general equation

$$P = \frac{RT}{V - b} - \frac{a}{V^2} \qquad (21 \cdot 4)$$

The magnitude of b must be of the general order of the specific volume of the liquefied gas. This means that it will have a strong correcting effect on gases under high pressures, where the density is high. Conversely, at low pressures the correction will be minor.

The value of a/V^2 becomes increasingly larger as the gas density rises. This would be in the right direction according to theoretical considerations.

Van der Waals' equation gives good results on many gases in ranges where they are not near their critical state, (see Art. 24·8). The equation does not give usable results for saturated vapors. For detailed discussion and comparisons of this equation see other advanced texts on thermodynamics.

21·5 Compressibility factor. According to the perfect-gas equation we can write

$$\frac{PV}{RT} = 1 \qquad (21 \cdot 5)$$

To adapt this equation for actual gases we can assume that

$$\frac{PV}{RT} = Z \qquad (21 \cdot 6)$$

where Z is a compressibility factor. The value of Z depends on the pressure and temperature of a gas. To use Eq. (21·6) we must have experimental data on Z for the gas we are working with.

GAS PROPERTIES

To reduce the number of tables or graphs for Z of actual gases, it is found that they can be grouped according to type when the relations are expressed in *reduced properties*. These are the ratios of the actual value of the property to the value of the property at the *critical point* (see Art. 24·8) of the gas. Reduced properties are

$$P_r = \frac{P}{P_c} \qquad T_r = \frac{T}{T_c} \qquad V_r = \frac{V}{V_c}$$

Two or more fluids having the same values for their reduced properties are in *corresponding states*. Figure 21·1 shows the correlation of Z to reduced temperatures and pressures for some 10 gases. The maximum deviation of any one gas from these averaged curves runs about 1%.

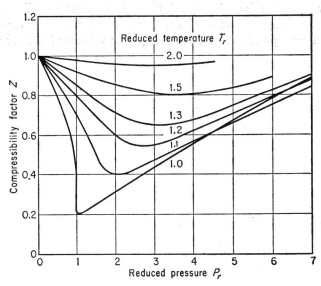

Fig. 21·1 Generalized compressibility chart can be used for methane, ethylene, ethane, propane, n-butane, iso-pentane, n-heptane, nitrogen, carbon dioxide, and steam with maximum error of 1% deviation from true gas behavior.

REVIEW TOPICS

1. What are the advantages of using gas tables instead of gas equations?
2. What accuracy is achieved in calculations using the gas tables for all pressure ranges?
3. Write the van der Waals equation for an actual gas, and explain the function of the added constants.
4. Does the van der Waals equation hold in all ranges of gas states?
5. Describe the utility of the compressibility factor for computing gas states.

6. What is meant by reduced properties?
7. When are gases in corresponding states?

PROBLEMS

1. Air at 1600°R and 90 psia expands at constant entropy to atmospheric pressure in a gas turbine. If the exhaust pressure is 15 psia, what is the exhaust temperature of the air?

2. Find the specific volume of the air exhausting from the gas turbine in Prob. 1. Use $R = 53.37$.

3. What work has been done by the air of Prob. 1 during the constant-entropy expansion?

4. If the gas turbine of Prob. 1 runs at 82% engine efficiency, find (a) the work done per pound of air, (b) the exhaust temperature, (c) the increase in entropy of the gas, (d) the specific volume of the exhaust air.

CHAPTER 22

Properties of Mixed Gases

We used air as the working fluid in studying power cycles. This assumed that air is a gas with specific characteristics relating pressure, temperature, density, internal energy, enthalpy, specific heat, and entropy. Actually, air is a mixture of gases: nitrogen, oxygen, carbon dioxide, argon, and some rare elements.

In many practical problems, especially combustion, we must deal with gas and gas-vapor mixtures. We have a host of data on individual gas elements and we can combine them to figure the characteristics of any mixture. Here we'll deal only with gas mixtures.

22·1 Avogadro's law. In Art. 3·9 we learned that this law said *Equal volumes of perfect gases held under exactly the same temperature and pressure have equal numbers of molecules.*

For this condition the masses or densities of equal volumes of gas are in proportion to their molecular weights; for instance,

$$D_H : D_N : D_C = 2.02 : 28.0 : 44.0$$

where D_H = density of hydrogen, lb per cu ft
D_N = density of nitrogen, lb per cu ft
D_C = density of carbon dioxide, lb per cu ft

We also learned that this leads to a universal gas constant $R_u = 1,545$, and individual gas constants can be figured from $R = R_u/M$, where M is the molecular weight of the gas with the gas constant R. Using the three gases above as an example,

$$R_H = \frac{1,545}{2.02} = 765 \qquad R_N = \frac{1,545}{28.0} = 55.2$$

$$R_C = \frac{1,545}{44.0} = 35.2$$

310   BASIC THERMODYNAMICS

We met the *mole volume*, $V_M = MV$ where V is the gas specific volume in cubic feet per pound. This can be used in the perfect-gas equation: $PV_M = R_u T$. In Chap. 3 we also showed that a mole volume (mole, for short) of any gas at 2,116-psf pressure and 492°R temperature is 359.2 cu ft. Reducing gas quantities to moles simplifies calculations.

22·2 Gibbs-Dalton law. Back in 1802 Dalton, a chemist, postulated the law about gas mixtures that said, in effect: *Gases occupying a common volume each fill that volume and behave as if the others were not present,* Fig. 22·1a.

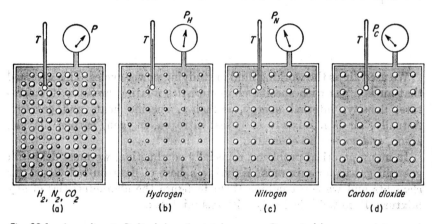

Fig. 22·1 According to Dalton's law the total pressure P exerted by a gas mixture as in (a) is the sum of the partial pressures of the individual gases (b), (c), and (d) when they are all at the same temperature and held in the same total volume of the container.

With this as a basis let us first recognize that the total mass of the mixture is the sum of the individual gas masses:

$$m_t = m_H + m_N + m_C \tag{22·1}$$

using the three gases above. Each gas occupies the same total volume by definition, so

$$V'_t = V'_H = V'_N = V'_C \tag{22·2}$$

Relating this to the gas specific volumes,

$$V'_t = m_H V_H = m_N V_N = m_C V_C \tag{22·3}$$

What about the temperature of the mixture? Since molecules must slam into one another at random, they exchange energy quickly and so average out the total energy in the mixture. This means that the gases all come to a common temperature, so that

$$T = T_H = T_N = T_C \tag{22·4}$$

PROPERTIES OF MIXED GASES 311

According to Dalton's law each gas must exert its own pressure, which can be figured as

$$P_H = \frac{R_H T}{V_H} = \frac{R_H T}{V'_t/m_H}$$

$$P_N = \frac{R_N T}{V_N} = \frac{R_N T}{V'_t/m_N}$$

$$P_C = \frac{R_C T}{V_C} = \frac{R_C T}{V'_t/m_C} \quad (22 \cdot 5)$$

With each gas pounding against the container walls the total pressure must be

$$P = P_H + P_N + P_C \quad (22 \cdot 6)$$

$$= \frac{T}{V'_t}(m_H R_H + m_N R_N + m_C R_C)$$

$$= \frac{R_u T}{V'_t}\left(\frac{m_H}{M_H} + \frac{m_N}{M_N} + \frac{m_C}{M_C}\right)$$

These last equations sum up the first part of the Gibbs-Dalton law: *The pressure of a gas mixture is the sum of the pressures that each gas would exert if it alone occupied the volume.*

The second part of the Gibbs-Dalton law can be taken as a basic definition: *Properties of a gas mixture are equal to the sums of those properties for each component when it occupies the total volume by itself; these properties are internal energy, enthalpy, and entropy.*

Remembering that E = internal energy in Btu per pound, then

$$m_t E = m_H E_H + m_N E_N + m_C E_C \quad (22 \cdot 7)$$

Letting h = enthalpy in Btu per pound, we have

$$m_t h = m_H h_H + m_N h_N + m_C h_C \quad (22 \cdot 8)$$

Remembering that S = entropy in Btu per pound, F,

$$m_t S = m_H S_H + m_N S_N + m_C S_C \quad (22 \cdot 9)$$

Letting c = specific heat of a gas either at constant pressure or at constant volume, then

$$m_t c = m_H c_H + m_N c_N + m_C c_C \quad (22 \cdot 10)$$

The best way to learn how to use these equations and appreciate some of the relationships is in an example.

Example: A 10-cu-ft tank holds 1 lb H_2, 2 lb N_2, and 3 lb CO_2 at 70 F. Find the specific volume, pressure, specific enthalpy, and specific entropy of the individual gases and of the mixture, and the mixture density.

$$V_H = {}^{10}\!/_1 = 10.00 \text{ cu ft per lb}$$
$$V_N = {}^{10}\!/_2 = 5.0 \text{ cu ft per lb}$$
$$V_C = {}^{10}\!/_3 = 3.33 \text{ cu ft per lb}$$
$$V_t = \frac{10}{1+2+3} = 1.667 \text{ cu ft per lb}$$

Using $R = R_u/M$

$$P_H = \frac{1545 \times 530 \times 1}{10 \times 2.02} = 40{,}530 \text{ psfa}$$
$$P_N = \frac{1545 \times 530 \times 2}{10 \times 28} = 5{,}850 \text{ psfa}$$
$$P_C = \frac{1545 \times 530 \times 3}{10 \times 44} = 5{,}583 \text{ psfa}$$
$$P_t = 40{,}530 + 5850 + 5583 = 51{,}963 \text{ psfa}$$

From Keenan and Kaye *Gas tables* find

$h_H = 1796.1$ Btu per lb
$h_N = 131.4$ Btu per lb
$h_C = 90.17$ Btu per lb
$mh = 1 \times 1796.1 + 2 \times 131.4 + 3 \times 90.17 = 2329.4$ Btu
$h = 2329.4/(1 + 2 + 3) = 388.2$ Btu per lb mixture

From *Gas tables* find the internal energy:

$E_H = 1260.0$ Btu per lb
$E_N = 93.8$ Btu per lb
$E_C = 66.3$ Btu per lb
$mE = 1 \times 1260.0 + 2 \times 93.8 + 3 \times 66.3 = 1646.5$ Btu
$$E = \frac{1646.5}{1+2+3} = 274.4 \text{ Btu per lb mixture}$$

From *Gas Tables* find the entropies:

$S_H = 12.52$ Btu per lb, F
$S_N = 1.558$ Btu per lb, F
$S_C = 1.114$ Btu per lb, F
$mS = 1 \times 12.52 + 2 \times 1.558 + 3 \times 1.114 = 18.978$
$$S = \frac{18.978}{1+2+3} = 3.163 \text{ Btu per lb, F of mixture,}$$

The density of the gas mixture can be readily figured from the specific volumes of the components by applying Dalton's law (density is the reciprocal of specific volume):

$$D_t = D_H + D_N + D_C$$

or

$$\frac{1}{V_t} = \frac{1}{V_H} + \frac{1}{V_N} + \frac{1}{V_C} \tag{22·11}$$

$$D_t = \frac{1}{10.0} + \frac{1}{5.0} + \frac{1}{3.33} = 0.6 \text{ lb per cu ft of mixture}$$

This checks with the first calculation, $V_t = 1.667$ cu ft per lb, and is based on the principle that all gases occupy the same volume.

22·3 Volumetric analysis. In analyzing gases we often want to make a volumetric analysis of a mixture. This assumes that each gas is held in a volume at the pressure and temperature of the mixture. Figure 22·2

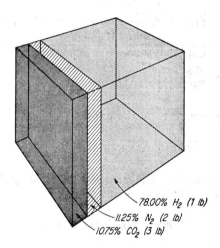

Fig. 22·2 Heavier gases occupy a much smaller volume than lighter gases for each pound when they are at same pressure and temperature.

shows one way of analyzing the problem. The total volume holding the final mixture has its interior divided by partitions so that each component gas has the same pressure and temperature as the final mixture. Then for each gas

$$PV = \frac{R_u T}{M} \quad \text{and} \quad V = \frac{R_u T}{MP}$$

Using the three gases in the above example the total volume for each gas component in the mixture is

$$m_H V_H = m_H \frac{1{,}545T}{M_H P}$$

$$m_N V_N = m_N \frac{1{,}545T}{M_N P}$$

$$m_C V_C = m_C \frac{1{,}545T}{M_C P} \tag{22.12}$$

The total volume of the mixture is

$$V_T = \frac{1{,}545T}{P}\left(\frac{m_H}{M_H} + \frac{m_N}{M_N} + \frac{m_C}{M_C}\right) \tag{22.13}$$

The percentage of each component in the mixture by volume (using H as an example) is

$$\%V_H = \frac{(1{,}545T/P)(m_H/M_H) \times 100}{(1{,}545T/P)(m_H/M_H + m_N/M_N + m_C/M_C)}$$

$$= \frac{m_H/M_H}{m_H/M_H + m_N/M_N + m_C/M_C} \times 100$$

$$= \frac{m_H/M_H}{\Sigma(m/M)} \times 100 \tag{22.14}$$

Then the *volumetric analysis* of the gas is

$$\%V_H + \%V_N + \%V_C = 100 \tag{22.15}$$

22.4 Weight analysis. The gravimetric or weight analysis can be figured from total weight or mass of mixture:

$$m_T = m_H + m_N + m_C$$

Then the weight analysis is

$$\%m_H = \frac{m_H}{m_T} \times 100$$

$$\%m_N = \frac{m_N}{m_T} \times 100$$

$$\%m_C = \frac{m_C}{m_T} \times 100 \tag{22.16}$$

and
$$\%m_H + \%m_N + \%m_C = 100 \tag{22.17}$$

22.5 Molal analysis. We can analyze the make-up of a gas in terms of the number of moles of each gas in the mixture. Referring to Table 3.2, recall that 1 mole of H_2 weighs M lb or 2.02 lb, 1 mole of N_2 weighs 28.0 lb, and 1 mole of CO_2 weighs 44.0 lb.

PROPERTIES OF MIXED GASES

Then to find the number n of moles of a gas we need only to find m/M ratio for it in a mixture. The total moles in the mixture are

$$n_t = \frac{m_H}{M_H} + \frac{m_N}{M_N} + \frac{m_C}{M_C} = \sum \frac{m}{M} \qquad (22 \cdot 18)$$

The percentage of each component in the mixture by moles using H_2 as an example is

$$\%n_H = \frac{m_H/M_H}{\Sigma(m/M)} \times 100 \qquad (22 \cdot 19)$$

This is the same as the percentage analysis of the gas by volume, so the two methods are identical.

Now let us try out some of these formulas for the gas mixture we had in the previous example.

Example: First let us check the total volume of 6 lb of gas:

$$V_t = \frac{1{,}545 \times 530}{51{,}963}\left(\frac{1}{2.02} + \frac{2}{28.0} + \frac{3}{44.0}\right)$$
$$= 10.0 \text{ cu ft for 6 lb of gas mixture}$$

Expanding the fractions in the parentheses we get

$$\sum \frac{m}{M} = 0.495 + 0.0714 + 0.0682 = 0.6346$$

Then
$$\%V_H = \frac{0.495}{0.6346} \times 100 = 78.00$$

$$\%V_N = \frac{0.0714}{0.6346} \times 100 = 11.25$$

$$\%V_C = \frac{0.0682}{0.6346} \times 100 = 10.75$$

Total $\quad\quad\quad\quad\quad\overline{100.00\%}$

As we just learned, this analysis is the same for the molal distribution. Weight analysis can be quickly figured:

$$\%m_H = \tfrac{1}{6} \times 100 = 16.67$$
$$\%m_N = \tfrac{2}{6} \times 100 = 33.33$$
$$\%m_C = \tfrac{3}{6} \times 100 = 50.00$$

Total $\quad\quad\quad\overline{100.00\%}$

The analyses vary considerably percentagewise. Let us summarize the quantities in the mixture:

Gas	m, lb	M, no.	V, ft³/lb	MV	%V	%m
H_2	1	2.02	7.81	15.77	78.00	16.67
N_2	2	28.00	0.563	15.77	11.25	33.33
CO_2	3	44.00	0.3585	15.77	10.75	50.00

Inspect these values, keeping in mind that they hold for a temperature of 70 F or 530°R and a pressure of 51,963 psfa or 360.7 psia, when the gases are compartmented according to their volume as in Fig. 22·2.

When the gases occupy total common volume of 10 cu ft we summarize the data as:

Gas	m, lb	M, no.	V, ft³/lb	D, lb/ft³	P, psia	%m
H_2	1	2.02	10.00	0.10	281.3	16.67
N_2	2	28.00	5.00	0.20	40.6	33.33
CO_2	3	44.00	3.33	0.30	38.8	50.00
Total	6		1.67	0.60	360.7	100.00

In the transparent box of Fig. 22·2 suppose we remove partitions and allow the gases to mix. What happens? Each gas expands into the full 10-cu-ft volume, and since the expansions are adiabatic and no work is done, the temperature remains the same for all gases. This means that their individual internal energies and enthalpies also remain constant. But each of their total pressures of 360.7 psia in the separated state drops to their partial pressure in the mixed state.

The gas expansions imply that a theoretical opportunity for doing work has been lost even though the temperature has stayed constant. Each gas could do work against a piston as it expands.

This loss of availability comes to light when we study the entropy changes of the gases while expanding. In Chap. 8 we learned that the change in entropy for a process was $S_2 - S_1 = Q/T$; for a constant-temperature process this can be shown to reduce to

$$S_2 - S_1 = \frac{mR}{J} \log_e \frac{V_2}{V_1} \tag{22·20}$$

Then for each of the gases in the example,

$$S_H = \frac{1.0 \times 1{,}545}{778 \times 202} \log_e \frac{10.0}{7.81} = 0.243 \text{ Btu per F}$$

$$S_N = \frac{2.0 \times 1{,}545}{778 \times 28.0} \log_e \frac{5.0}{0.563} = 0.3097 \text{ Btu per F}$$

$$S_C = \frac{3.0 \times 1{,}545}{778 \times 44.0} \log_e \frac{3.33}{0.358} = 0.3017 \text{ Btu per F}$$

So when we remove the partitions between the gases in Fig. 22·2 and let them mix, the overall temperature and pressure do not change but the entropy increases by

$$S_H + S_N + S_C = 0.243 + 0.3097 + 0.3017 = 0.8544 \text{ Btu per F}$$

or: $0.8544/6 = 0.1424$ Btu per lb, F.

This demonstrates that mixing gases, even without change in temperature and pressure, is an irreversible process. To separate the molecules of gas in a mixture would require individual work on each to recompartment them in the separate state.

An interesting question arises when the compartments in Fig. 22·2 all hold the same gas at a common temperature and pressure. When we remove the partitions from the closed container, the molecules diffuse among one another. Is this an irreversible process? No—because we recreate an identical condition when we reinsert the partitions; though the molecules are different, they are all of the same kind. This is the *Gibbs paradox*.

REVIEW TOPICS

1. State Avogadro's law.
2. Define the universal gas constant.
3. What is a mole volume?
4. State the Gibbs-Dalton law.
5. Are the temperatures of individual gases in a mixture different?
6. Write the equation for the total pressure of a gas mixture in terms of the universal gas constant, mixture temperature, total volume, mass of the gas components, and their molecular weights.
7. Write the equations for total enthalpy, internal energy, and total entropy of a gas mixture.
8. What is the basic condition for the volumetric analysis of a gas mixture?
9. Write the volumetric analysis equation for one component in a mixture in terms of mass and molecular weight.

PROBLEMS

1. A 6-cu-ft tank holds $\frac{1}{2}$ lb of H_2, 1 lb of N_2, and 6 lb of CO at 100 F. Assume c_p for $H_2 = 3.4$, for $N_2 = 0.24$, for $CO = 0.25$. Find (a) the specific volume, (b) the pressure, (c) the enthalpy of the individual gases and of the mixture. (d) Make both a volumetric and a weight analysis of the mixture.
2. Make a molal analysis for the mixture in Prob. 1.
3. Find the specific heats of the gas mixture in Prob. 1.
4. A mixture of 5 lb of CO and $\frac{1}{4}$ lb of H_2 is in a tank at 30 psia and 80 F. Find (a) the tank volume, (b) the partial pressure of each gas.

CHAPTER 23

Energy Sources and Combustion

To quicken our understanding of engine cycles, we have simply called the energy input a quantity Q. This simple assumption dodged an important question—Just how do we get Q? Where does Q come from?

With today's growing research and technology we find that Q can be made a number of ways. Most of our heat engines develop Q directly from fuels injected in their internals; others receive it as a heat transfer.

Traditionally, Q has been produced as the output of a combustion process. In more technical terms it is the energy released by the oxidation of the fuel elements, carbon C, hydrogen H_2, and sulfur S. The high-temperature chemical reaction of these elements with oxygen O_2 (usually from the air) releases energy that produces high-temperature gases.

These gases give us the high thermal head we need to develop efficient energy-work cycles. Before we take a closer look at combustion processes, let us take an even more basic look at overall energy-production possibilities—sources and methods. Table 23·1 summarizes the energy-production and -conversion processes that have been important in the past and new ones that hold promise for the future.

23·1 Energy sources. In Art. 4·6 we met the First Law of Thermodynamics and also noted that energy and matter are interchangeable. The first law as usually stated in thermodynamics says that energy cannot be created or destroyed.

But more accurately, the first law should say that the sum of energy *and matter* involved in a process cannot be created or destroyed. Our scientists have found that matter and energy are interchangeable; one can be converted to the other. In our sun and the stars these changes go on in either direction. But from a practical-application standpoint on

Table 23·1 Basic energy production and conversion processes

Changing Mass to High-temperature Internal Energy

I. Combustion—electron mass changes to internal energy
  A. Chemical reaction causes oxidation of fuel elements
    1. Carbon
    2. Hydrogen
    3. Sulfur
  B. Produces high-temperature gases that supply Q to cycle
II. Nuclear fission—nuclei mass of heavy atoms changes to internal energy
  A. Chain reaction developing neutron projectiles sustains fission process, the physical reaction of fuel elements
    1. Uranium 233
    2. Uranium 235
    3. Uranium 238
    4. Plutonium 239
  B. Produces high temperature in fuel metals and highly radioactive fission products that supply Q to cycle
III. Nuclear fusion—nuclei mass of light atoms changes to internal energy
  A. Confining plasma in limited volume at superhigh temperatures makes nuclei of light atoms collide and fuse
    1. Hydrogen nuclei (proton)
    2. Deuterium (proton + neutron)
    3. Tritium (proton + 2 neutrons)
  B. Produces superhigh temperatures in plasma and short-lived radiations; may produce electric energy in addition to Q with superhigh thermal head

Energy-conversion Processes

I. Primary and secondary processes (high and low temperature)
  A. Expanding gas and vapor produces mechanical energy
  B. Mechanical energy to electric energy
    1. Electric energy to mechanical energy
    2. Electric energy to heat
    3. Electric energy to chemical energy
  C. Mechanical energy to internal energy
  D. Thermionic—internal energy to electric energy
  E. Thermoelectric—internal energy to electric energy
    1. Reversible process to produce thermal change
      a. Heating
      b. Cooling
  F. Magnetohydrodynamic—internal energy to electric energy
II. Energy conversions at ambient temperatures
  A. Chemical battery to produce electric energy
    1. Primary cell
    2. Secondary cell
  B. Fuel cell—substitutes low- or high-temperature chemical reactions for combustion to produce electric energy
III. Solar radiation conversions
  A. Collection for raising internal energy of matter
  B. Photovoltaic effect to produce electric energy

earth, we change matter to energy in significant amounts and only minutely the other way around.

23·2 Combustion. This is a chemical reaction in which a fuel element combines with oxygen and produces a new molecule. The process develops internal energy in the new molecule which shows up as a high temperature of the gas made up of the molecules.

The chemical reaction involves a redistribution of electrons of the oxygen and fuel elements that causes the electrons to lose a small part of their mass. Since 1 lb of mass equals 11.3 billion kwhr or 38.7 trillion Btu of energy, we can see that only a minute part of mass need disappear for the 62,028 Btu produced by the burning of 1 lb of H_2. The mass decrease is so small that it is impossible to measure in ordinary applications—that is why we did not know about it until recent years, when scientists developed more refined methods of nuclear measurement.

We shall pay detailed attention to combustion later, since it is our most important method of making Q.

23·3 Nuclear fission. This process also changes mass to energy but involves the nuclei of heavy atoms. By arranging uranium and plutonium to form a critical mass, we can make neutrons enter the nuclei of these elements. This unbalances the internal forces that hold the nuclei together and makes them split into smaller nuclei. The fissioning makes a minute part of the mass of the nuclei disappear to create energy.

This energy appears as a high velocity of the split parts of the atom. These slam into neighboring atoms, giving up part of their kinetic energy, and in this way raise the internal energy of the fuel mass. The coolant fluid of the cycle passes over the fuel mass to be heated and carry off the energy to the power cycle. Table 23·1 lists the nuclear fuel elements we can use.

23·4 Nuclear fusion. This process works with light atoms, at the other end of the scale from fission, and goes on steadily in the stars. Scientists of all nations are busily trying to make the process work on earth. By energizing gaseous molecules to temperatures about 100 million F we get a wild melee of separate nuclei and electrons—no organized atoms—called a plasma.

In contrast to fission, when the nuclei in a plasma can be made to collide and stick together, a minute part of their mass disappears to speed up the fused nucleus and so raise the temperature and internal energy of the plasma. A coolant again would carry off the released energy to a power cycle. This process seems a long way off from practical realization.

23·5 Conversion processes. The processes we use today are largely based on expanding a high-temperature gas or vapor from a high to a low pressure. The availability of energy depends on the temperature range through which the working fluid can drop.

Most of our engine cycles produce mechanical shaft power to drive a machine or do other duty. Some of the driven machines are electric generators. These produce an electron flow in a closed circuit that we recognize as an electric current, a form of energy.

The electrical form is primarily a convenient way to transmit energy to the point of use. There it can be reconverted to mechanical shaft energy, to internal energy by heating, or to chemical energy or some other form. Most of these are relatively low-temperature processes.

Mechanical energy can, in turn, be used to produce internal energy, as in a compressed-air system, for transmission to the point of use. Some mechanical energy, of course, dissipates by frictional heating.

23·6 Electric energy. Because of its flexibility, this form of energy offers many advantages in practical applications. Many efforts are being made to generate electron flows directly from internal energy to avoid processing energy through a relatively complex and bulky engine cycle. This offers important advantages in some energy systems, especially some of those used by the military services.

23·7 Thermionic generators. These receive energy as high-temperature heat at one of a pair of electrodes held in an atmosphere of ionized gas, "boiling out" free electrons that flow through the gas to the nearby "cold" electrode. This creates an electric potential between the electrodes, the hot one being positive and the cold one negative.

Connecting the electrodes through an outside circuit allows an electric current to flow and carry energy. This direct production of electric energy has been done at thermal efficiencies of 8% and more in laboratories. Workers in this field predict thermal efficiencies of 30% with units of 50-megawatts capacity in the near future.

23·8 Thermoelectric generators. Generators of modern design use specially compounded semiconductor materials to form two junctions of dissimilar materials. When the unlike materials are connected electrically in a closed circuit and one junction is heated to a relatively high temperature and the other one cooled, by heat transfer, an electric current flows through the circuit. The energy generated varies with the heat energy transferred to the hot junction and the temperature difference. Thermal efficiencies of 7% have been reported with temperature differentials of about 1000 F.

Workers predict that this may soon be as high as 14% when improved materials become available. The arrangement also avoids passing heat through an engine cycle.

This circuit can be used in reverse. When we impress an electric potential across the circuit to produce an electric current, one junction heats up and the other cools, so it can be used for refrigerating or heating an object. Present materials produce cooling effects through about a 100 F range. This is expected to be increased to 180 F in the near future. Here we have refrigeration without a reversed engine cycle.

23·9 Magnetohydrodynamics. This is a basic field of study of astrophysicists and plays an important part in attempts to develop the thermonuclear fusion process of generating heat energy. Attempts are now being made to apply this knowledge of high-temperature plasmas (ionized gases) to generate electric energy directly from the internal energy of such gases.

Making high-temperature ionized gas flow through a strong magnetic field forces free electrons to collect on one of a pair of electrodes. This makes the opposite electrode relatively positive. Connecting electrodes through a closed circuit allows transmission of electric energy originating in the internal energy of the hot ionized gas. Proposed cycles anticipate the gas to be still quite hot after leaving the generator. This energy would be the Q to supply a normal engine cycle, a fraction of the total energy produced by the combination plant.

23·10 Chemoelectric production. This depends on chemical reactions to produce electric energy. Unlike combustion, the reactions take place at room temperature. The ordinary dry cell, technically called a *primary* cell, has been a long-standing method of producing low-voltage d-c electric energy.

Another long-standing method has been the storage battery, *secondary* cell; this unit, though, must be charged with electric energy first to cause certain chemical reactions. Then letting the reactions reverse reproduces electric energy; in the strict sense this is not a primary energy source.

Fuel cells using chemical reactions at room temperature are showing promise. By letting fuel elements such as hydrogen or a hydrocarbon with oxygen diffuse through an electrolyte, electric energy can be produced directly, reducing the fuels to water or CO_2 as in combustion, but at room temperature. Reported efficiencies have been as high as 60%, which compares with theoretical efficiencies of 80 to 90%. Some fuel cells work at temperatures as high as 1800 F.

23·11 Solar radiations. These radiations primarily consisting of heat waves have been an intriguing source of Q for many years. In areas where sunshine predominates most of the year, this has limited possibilities. Solar energy has been and is being used to heat water and generate steam. The principal problem is the large area needed to gather big enough amounts of energy to be useful and to develop economical methods of storing the energy for use when the sun does not shine.

Photovoltaic cells have proved their feasibility in producing low-voltage d-c electric energy directly from solar radiation. The energy can be stored in storage batteries for the hours without sunshine. This method is good for low energy demands, generally not for power.

23·12 Combustion. Now let us look at basic details of combustion reactions that release heat in practical energy-work cycles. This involves understanding the chemical reactions used. As a first step let us recall the atomic and molecular weights of the elements and reaction products involved:

Fuel or gas: H_2 C S O_2 N_2 CO CO_2 H_2O
Molecular weight: 2   12   32   32   28   28   44   18

The various solid, liquid, and gas fuels we use all contain the fuel elements in varying proportion. Each element goes through its own precise reaction with oxygen from the air to release precise amounts of energy per unit mass of fuel element.

In the following sets of reaction equations we give (23·1) the basic chemical reaction and heat release, (23·2) the relative mole volume of elements and reaction product, (23·3) the molecular weights of the elements and products, and (23·4) the relative weights per pound of fuel element:

Solid C & S

$$C + O_2 = CO_2 + 14{,}096 \text{ Btu per lb C} \quad (23\cdot1)$$
$$1 + 1 \rightarrow 1 \quad (23\cdot2)$$
$$12 + 32 = 44 \quad (23\cdot3)$$
$$1 + 2.67 = 3.67 \text{ lb per lb C} \quad (23\cdot4)$$

$$2C + O_2 = 2CO + 3960 \text{ Btu per lb C} \quad (23\cdot1)$$
$$2 + 1 \rightarrow 2 \quad (23\cdot2)$$
$$24 + 32 = 56 \quad (23\cdot3)$$
$$1 + 1.33 = 2.33 \text{ lb per lb C} \quad (23\cdot4)$$

$$2CO + O_2 = 2CO_2 + 4346 \text{ Btu per lb CO} \quad (23\cdot1)$$
$$2 + 1 \rightarrow 2 \quad (23\cdot2)$$
$$56 + 32 = 88 \quad (23\cdot3)$$
$$1 + 0.571 = 1.571 \text{ lb per lb CO} \quad (23\cdot4)$$

$$2H_2 + O_2 = 2H_2O + 61{,}031 \text{ Btu per lb } H_2 \quad (23\cdot1)$$
$$2 + 1 \rightarrow 2 \quad (23\cdot2)$$
$$4 + 32 = 36 \quad (23\cdot3)$$
$$1 + 8 = 9 \text{ lb per lb } H_2 \quad (23\cdot4)$$

$$S + O_2 = SO_2 + 3984 \text{ Btu per lb S} \quad (23\cdot1)$$
$$1 + 1 \rightarrow 1 \quad (23\cdot2)$$
$$32 + 32 = 64 \quad (23\cdot3)$$
$$1 + 1 = 2 \text{ lb per lb of S} \quad (23\cdot4)$$

Each of the fuel elements releases widely different amounts of energy per pound, H_2 being the highest. The volume of product of these reactions is one less than the volume of elements that went into reaction. This does not hold for all possible combustion reactions, so interpret carefully.

We listed carbon and sulfur as one mole volume each in their respective volume equations. In some combustion calculations this acts as a convenient mathematical fiction (we shall not go into this here), but actually they are both solids as we use them. The volume of either, relative to the volume of oxygen used, is so small that it is negligible. So in the physical sense one volume of oxygen entering the reaction results in one volume of combustion product for these two fuel elements only.

In all the reaction equations for mass (23·3) the mass of the elements entering the reaction always equals the mass of the combustion product. Again technically the product mass is actually minutely less than the original elements because the disappearing mass converts to the energy released, but the amount is so tiny that we cannot measure it and for design purposes we can forget it. So, practically, mass in equals mass out.

All the equations (23·4) simply restate the molecular-mass relations in terms of 1 lb of fuel. For example, 1 lb of carbon burns with 2.67 lb of oxygen to form 3.67 lb of CO_2 and release 14,096 Btu.

23·13 Air supply. Excepting special applications, we get all our oxygen for combustion from atmospheric air. Air is a mechanical mixture of oxygen and nitrogen with small amounts of rarer gas elements. Proportions by weight are 23.1% O_2 and 76.9% N_2 assuming that the other elements are negligible. Proportions by volume are 21.0% O_2 and 79.0% N_2. These proportions are accurate enough for most work. This means that by weight we must take in $76.9/23.1 = 3.32$ lb N_2 for every pound of O_2 needed by a combustion reaction.

Figure 23·1a reviews the basic process for burning hydrogen. Each cube represents 1 mole of the basic *reactants* and the resulting combustion *products*. The products usually have fewer moles than the original reactants, but the individual product moles are heavier than original reactant moles. The total mass on each side of reaction equation must be equal.

We can give our heat release in terms of Btu per pound or Btu per mole (specifically pound-mole). To get the latter we multiply the Btu per pound by the molecular weight of the fuel element.

Since we use air to get our oxygen supply, we must know how much is needed to burn up our fuel. Air has 79.0% N_2 and 21.0% O_2 by volume, so for each volume of O_2 needed we also have to take $79.0/21.0 = 3.76$ volumes of N_2, or to get one volume (mole) of O_2 we must take $1 + 3.76 = 4.76$ volumes (moles) of air. Figure 23·1b shows this relationship and how the N_2 is unaffected chemically, appearing unchanged on both sides of the equation. The heat release is the same as in Fig.

23·1a—we merely have to handle more inert mass in our reaction. This becomes an important consideration in design of boiler furnaces and i-c engines.

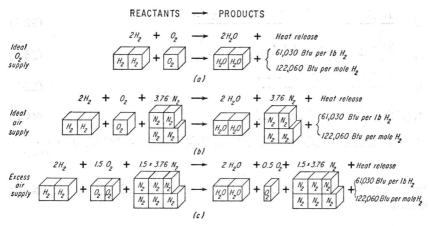

Fig. 23·1 (a) Ideal reaction of hydrogen and oxygen produces water vapor and the highest heat release of known common fuel elements. (b) To get oxygen from the air we have to bring nitrogen into the reaction. Nitrogen passes through unchanged and does not affect the heat release of hydrogen. (c) Most actual reactions need excess air to ensure burning all the fuel so oxygen appears in the combustion products.

23·14 Excess air. Experience shows that we rarely can take exact amounts of fuel and air as in Fig. 23·1b and get *all* the fuel elements to combine with the oxygen, leaving nothing on the product side except products and nitrogen. To ensure burning all the fuels elements, we must supply more oxygen than needed for the ideal equation. This means supplying an excess of air—an excess of both O_2 and N_2—which results in all the fuel combining with O_2 and leaving some uncombined O_2 and, of course, the inert N_2 in the products. Figure 23·1c shows the relation for an air excess of 50%, or a total air supply of 150%. Again we get only the heat release of the ideal reaction.

23·15 Oxygen and air formulas. Most fuels we burn in our present-day engines contain the combustible elements C, H_2, and S. From the set of equations (23·1) to (23·4) we can get a single expression of the oxygen and air needed by a fuel of a given percentage composition. If we let C, H, O, and S represent the weight fractions of the respective elements in 1 lb of fuel, we can write for oxygen needed

$$W_O = 2.67C + 8\left(H - \frac{O}{8}\right) + S \quad \text{lb per lb of fuel} \quad (23·5)$$

Gas fuels will generally be analyzed on a volumetric basis. This must be converted to a weight basis by the following conversion formulas, where CH_4, CO, H_2, H_2S are the volume fractions of the components and M is the molecular weight of the gas mixture:

$$C = \frac{12CH_4 + 12CO + 24C_2H_4 + \cdots}{M} \quad (23\text{-}6)$$

$$H = \frac{2H_2 + 4CH_4 + \cdots}{M} \quad (23\text{-}7)$$

$$O = \frac{16CO + 32CO_2 + \cdots}{M} \quad (23\text{-}8)$$

$$S = \frac{32H_2S + 32SO_2 + \cdots}{M} \quad (23\text{-}9)$$

$$M = 16CH_4 + 28CO + 2H_2 + 28C_2H_4 + 44CO_2 + 34H_2S + \cdots \quad (23\text{-}10)$$

to include all the components. Table 23·2 lists all the constants for combustion equations.

Table 23·2   Properties of common combustibles and products

Gas or fuel	Formula	Mol. wt	Lb/cu ft at 60 F, 14.7 psia	Gas, liquid, or solid	Heating value, Btu Per lb	Heating value, Btu Per cu ft at 60 F, 14.7 psia
Carbon	C	12.00	......	S	14,544	
Hydrogen	H_2	2.02	0.0053	G	62,028	325
Sulfur	S	32.07	......	S	4,050	
Carbon monoxide	CO	28.00	0.0739	G	4,380	323
Methane	CH_4	16.03	0.0423	G	23,670	1012
Acetylene	C_2H_2	26.02	0.0686	G	21,500	1483
Ethylene	C_2H_4	28.03	0.0739	G	21,320	1641
Ethane	C_2H_6	30.05	0.0792	G	22,070	1762
Benzene	C_6H_6	78.05	0.2061	L	18,070	
Hexane	C_6H_{14}	86.11	0.2277	G, L	20,900	4700
Octane	C_8H_{18}	114.14	......	L	20,500	
Oxygen	O_2	32.00	0.0844	G		
Nitrogen	N_2	28.02	0.0739	G		
Air		29	0.0765	G		
Carbon dioxide	CO_2	44.00	0.1145	G		
Sulfur dioxide	SO_2	64.07	0.1692	G		
Water vapor	H_2O	18.02	......	G		

Example 1: Find the theoretical oxygen needed for completely burning 1 lb of coal with the weight analysis: C = 71.7%, O = 8.3%, S = 3.4%, W = 3.5% (water), H = 5.2%, N = 1.3%, and ash = 6.6%.

ENERGY SOURCES AND COMBUSTION

By Eq. (23·5),

$$W_O = 2.67 \times 0.717 + 8\left(0.052 - \frac{0.083}{8}\right) + 1 \times 0.034$$
$$= 2.28 \text{ lb per lb fuel}$$

From Eqs. (23·1) to (23·4) we get Table 23·3.

Table 23·3

Elements	O_2 per lb fuel	O_2 moles per lb fuel
C	$2.67 \times 0.717 = 1.914$	$1 \times 0.717/12 = 0.0598$
H	$8 \times 0.052 = 0.416$	$0.5 \times 0.026 = 0.0130$
S	$1 \times 0.034 = \underline{0.034}$	$1 \times 0.034/32 = \underline{0.0011}$
Total	2.364 lb	0.0739 mole

O_2 needed from weight analysis:

$$O_2 = 2.364 - 0.083 = 2.28 \text{ lb}$$

where there is 0.083 lb of O_2 in the original fuel.
Oxygen needed from mole analysis:

$$O_2 = 0.0739 - \frac{0.083}{32}$$
$$= 0.0713 \text{ mole per lb fuel}$$
$$= 0.0713 \times 32 = 2.28 \text{ lb}$$

We can convert Eq. (23·5) to an air formula by accounting for the following fundamental relations:

Air volumetric analysis: 79% N_2 and 21% O_2
Air weight analysis: 77% N_2 and 23% O_2
Air molecular weight: 29 lb per mole

Standard air conditions: 60% relative humidity, 80 F dry-bulb temperature, 0.013 lb moisture per lb dry air

Mole relations: 3.76 moles N_2 per mole O_2

The dry air theoretically needed for any fuel is

$$W_a = \frac{W_O}{0.232} = 11.5C + 34.5\left(H_2 - \frac{O}{8}\right) + 4.3S \quad \text{lb per lb fuel} \quad (23\cdot11)$$

Depending on the type of engine, the amount of air actually supplied may be more or less than the theoretical needed for complete combustion of the fuel. We can express this as

$$\text{Excess-air percentage} = 100\,\frac{W_A - W_a}{W_a} \quad (23\cdot12)$$

where W_A is the actual weight supplied per lb of fuel;

$$\text{Dilution coefficient} = \frac{W_A}{W_a} \quad (23\cdot13)$$

The volumes of air can be found from the perfect-gas equation.

Example 2: Find the dry air at 14.7 psia 80 F needed to burn completely a fuel gas with an analysis of $CH_4 = 65.0\%$ and $C_2H_6 = 35.0\%$.

By Eq. (23·10): $\quad M = 0.65 \times 16 + 0.35 \times 30 = 20.9$ lb per mole

By Eq. (23·6): $\quad C = \dfrac{12 \times 0.65 + 24 \times 0.35}{20.9} = 77.5\%$ by weight

$H = 100 - 77.5 = 22.5\%$ by weight

$W_a = 11.5 \times 0.775 + 34.5 \left(0.225 - \dfrac{O}{8}\right) + 4.3 \times 0 = 16.67$ lb per lb fuel

$$V = \frac{W_a RT}{P} = 16.67 \times 53.3 \, \frac{80 + 460}{144} \times 14.7 = 226.5 \text{ cu ft of air}$$

Reaction equations for the components are

$$CH_4 + 2O_2 = CO_2 + 2H_2O \quad \text{and} \quad 2C_2H_6 + 7O_2 = 4CO_2 + 6H_2O$$

Component	Moles per 100 moles gas	Moles O_2 needed
CH_4	65.0	$2 \times 65.0 = 130.0$
C_2H_6	35.0	$7/2 \times 35.0 = 122.5$
Total		252.5

Moles of air $= 252.5 + 3.76 N_2 \times 252.5 = 1{,}203$ moles air per 100 moles fuel

$$W_a = \frac{1{,}203 \times 29}{100 \times 20.9} = 16.70 \text{ lb per lb fuel}$$

$V = 1{,}203/100 = 12.03$ cu ft air per cu ft gas (both at the same temperature and pressure)

Example 3: Find the fuel-air ratio for a liquid fuel $C_{12}H_{26}$ when using 30% excess air.

$$2C_{12}H_{26} + 37O_2 = 24CO_2 + 26H_2O$$

$M = 12 \times 12 + 1 \times 26 = 170$ lb per mole of fuel
Moles O_2 theoretically needed $= 37/2$ moles per mole of fuel
Moles air $= 1.3(37/2 + 3.67 \times 37/2) = 114.5$ per mole of fuel

$$\text{Air-fuel ratio} = \frac{114.5 \times 29}{1 \times 170} = 19.6 \text{ lb per lb fuel}$$

ENERGY SOURCES AND COMBUSTION

Example 4: Find the theoretical air needed in pounds and the volume of actual air for 30% excess at 60 F and 14.7 psia for a coal with the weight analysis: C = 70.5, O = 6.0, S = 3.0, H = 4.5, N = 1.0, ash = 11.0, and water = 4.0.

$$W_a = 11.5 \times 0.705 + 34.5 \left(0.045 - \frac{0.068}{8}\right) + 4.3 \times 0.030$$

$$= 9.53 \text{ lb air per lb coal}$$

$$W_A = W_a \times 1.3 = 9.53 \times 1.3 = 12.39 \text{ lb air per lb coal}$$

$$V = \frac{12.39 \times 53.3 \times 520}{144 \times 14.7} = 162.4 \text{ cu ft per lb coal}$$

For most practical cases the water vapor in the air is neglected in these calculations because of its small effect on the results.

23·16 Combustion products. Equations (23·1) to (23·4) show the type of gases produced by the burning of the fuel elements. In addition the combustion products will have N_2, O_2, CO, and unburned fuel elements, as well as water vapor. Thermodynamic equipment that burns fuel must handle combustion products as well as elements supporting combustion.

Weights are found from the following relations:

$$CO_2 = {}^{44}\!/_{12} Y C \qquad (23\text{·}14)$$
$$CO = {}^{56}\!/_{24}(1 - Y)C \qquad (23\text{·}15)$$
$$O_2 = 0.232 X W_a + 1.33(1 - Y)C \qquad (23\text{·}16)$$
$$N_2 = 0.768(1 + X)W_a + N \qquad (23\text{·}17)$$
$$H_2O = {}^{36}\!/_4 H + W + W_{AV} \qquad (23\text{·}18)$$
$$SO_2 = {}^{64}\!/_{32} S \qquad (23\text{·}19)$$

where CO_2, CO, O_2, N_2, SO_2 and H_2O = lb product per lb fuel
  C, H, S, and N = fuel elements, lb per lb fuel
  X = excess air ratio
  Y = ratio of C in fuel oxidized to CO_2
  W_a = theoretical dry air, lb per lb fuel
  W = mechanical moisture in fuel, lb per lb
  W_{AV} = water vapor in air supply = $(1 + X)W_a G/7{,}000$
  G = grains of moisture per lb of dry air from psychrometric chart

Total moles of gas per pound of fuel are found from

$$M = \frac{CO_2}{44} + \frac{CO}{28} + \frac{O_2}{32} + \frac{N_2}{28} + \frac{SO_2}{64} + \frac{H_2O}{18} \qquad (23\text{·}20)$$

The gas volume is then (see Art. 3·9)

$$V_G = 359M \frac{t_g + 460}{492} \times \frac{14.7}{P_g} \qquad (23\cdot21)$$

where V_G = total gas volume, cu ft per lb fuel
M = total moles of gas per lb fuel
t_g = gas temperature, F
P_g = gas pressure, psia

The volumetric analysis is

$$CO_2' = \frac{CO_2}{44M} \qquad CO' = \frac{CO}{28M}$$
$$O_2' = \frac{O_2}{28M} \qquad N_2' = \frac{N_2}{28M} \qquad (23\cdot22)$$
$$SO_2' = \frac{SO_2}{64M} \qquad H_2O' = \frac{H_2O}{18M}$$

Combustion gas consists of dry gases and water vapor. To analyze gases on a dry-gas base, use the same equations but eliminate the water-vapor items in Eqs. (23·20) and (23·22).

Example 5: A fuel oil with an analysis C = 83.7%, H = 12.7%, S = 0.7%, N = 1.7%, and O_2 = 1.2%, is burned with air having a dry-bulb temperature of 80 F and a wet-bulb of 70 F. When using 30% excess air for complete combustion, find (a) the total combustion-gas volume at 400 F and 14.7 psia, (b) the dry-gas analysis based on CO_2, O_2, and N_2.

Using Eqs. (23·11), (23·14) to (23·19), and (23·22),

$$W_a = 11.5 \times 0.837 + 34.5\left(0.127 - \frac{0.012}{8}\right) + 4.3 \times 0.007$$
$$= 13.98 \text{ lb air per lb oil}$$
$$W_{AV} = (1 + X)\frac{W_a G}{7,000}$$

From a psychrometric chart G = 95 grains per lb dry air

$$W_{AV} = (1 + 0.30)13.98 \times {}^{95}\!/_{7,000} = 0.3 \text{ lb vapor per lb dry air}$$

Component gas weights in pounds per pound of fuel are

$CO_2 = {}^{44}\!/_{12}\, 0.837 = 3.07$
$CO = 0$
$O_2 = 0.232 \times 0.30 \times 13.98 = 0.97$
$N_2 = 0.768(1 + 0.30)13.98 + 0.017 = 13.99$
$SO_2 = {}^{64}\!/_{32}\, 0.007 = 0.02$
$H_2O = {}^{36}\!/_{4}\, 0.127 + 0.3 = 1.44$

Gas moles per pound of fuel are

$$CO_2' = 3.07/44 = 0.0698$$
$$O_2' = 0.97/32 = 0.0303$$
$$N_2' = 13.99/28 = 0.4992$$
$$SO_2' = 0.02/64 = 0.0003$$
$$H_2O' = 1.44/18 = 0.0800$$
$$M = 0.6796$$

$$V_G = 0.6796 \times 359 \times \frac{400 + 460}{492} \frac{14.7}{14.7} = 427 \text{ cu ft per lb oil}$$

Gas analysis by volume is

$$CO_2' = \frac{0.0698}{0.0698 + 0.0303 + 0.4992} = 0.1164 = 11.64\%$$

$$O_2' = \frac{0.0303}{0.5993} = 0.0505 = 5.05\%$$

$$N_2' = \frac{0.4992}{0.5593} = 0.8331 = 83.31\%$$

Example 6: Write the complete combustion equation for a coal burned with 35% excess air when its analysis is C = 71.7%, O = 8.3%, H = 5.2%, S = 3.4%, N = 1.3%, ash = 6.6%, and water = 3.5%. First write the mole relations for 100 lb of coal (Table 23·4).

Table 23·4

Element	Weight	Mol. wt	Moles
C	71.7	12	5.98
H	5.2	2	2.60
S	3.4	32	0.11
O	8.3	32	0.26
N	1.3	28	0.05
Totals	89.9		9.00

To burn 100 lb of coal we must satisfy the reaction equation:

$$9.00(C, H, S, O, N) + \left(\frac{3.5}{18}\right)^a H_2O + M_o(O_2 + 3.76^b N_2)1.35$$
$$= 5.98CO_2 + M_o \times 0.35O_2 + 0.11SO_2 + 2.79H_2O^c$$
$$+ [(M_o \times 3.76 \times 1.35) + 0.05]N_2$$

[a] From water = 3.5% [b] From $N_2/O_2 = 0.79/0.21 = 3.76$
[c] From $3.5/18 + 5.2/2 = 2.79$

From the mass balance:
Weight of oxygen in fuel and air = weight of oxygen in combustion gases

Oxygen leaving process:

$$0.35 M_o O_2 = 0.35 \times M_o \times 32 = 11.2 M_o$$
$$5.98 CO_2 = 5.98 \times 32 = 191.5$$
$$0.11 SO_2 = 0.11 \times 32 = 3.5$$
$$2.79 H_2 O = 2.79 \times 16 = 44.7$$
$$\text{Total} \quad 239.7 + 11.2 M_o$$

Oxygen entering process:

$$0.26 O_2 = 0.26 \times 32 = 8.3$$
$$(3.5/18) H_2 O = 0.19 \times 16 = 3.1$$
$$1.35 M_o O_2 = 1.35 M_o \times 32 = 43.2 M_o$$
$$\text{Total} \quad 11.4 + 43.2 M_o$$

Equating, $\quad 11.4 + 43.2 M_o = 239.7 + 11.2 M_o$
$$M_o = 7.14$$

The complete combustion equation for 100 lb of coal is

$$9.00(C, H, S, O, N) + 0.20 H_2O + 6.6 \text{ ash} + 9.64(O_2 + 3.76 N_2)$$
$$= 5.98 CO_2 + 2.5 O_2 + 0.11 SO_2 + 2.79 H_2O + 36.3 N_2 + 6.6 \text{ ash}$$

23·17 Water vapor. Whenever accounting for water vapor in combustion calculations, add it to the left side of the reaction equation and carry it over to the right-hand product side. Water vapor in combustion products (usually at low partial pressure) is treated as a gas. Partial pressure of the vapor is proportional to the molal percentage or volumetric percentage of the vapor in the total gas products. The saturation temperature of steam at this partial pressure is the water dew point or temperature at which the water vapor condenses. Sulfur dioxide and oxygen in combustion gases react with water vapor to form acids. When the gas temperature drops below the dew point, these gases corrode metal surfaces on which they precipitate.

23·18 Combustion conditions. These must meet several needs before the reaction proceeds:

1. Elements entering the reaction must first be heated to ignition temperature. Physically this means that the individual reactant molecules must be speeded up until they have enough energy to collide with one another and combine their electron orbits to form the product molecules. This rearrangement of elements to form new molecules releases the internal energy (by converting a minute amount of electron mass to energy). If the molecules are not fast enough individually, short-range repulsion forces among molecules prevent them from coming into physical contact, and no reaction takes place.

2. Once a reaction starts between a mass of fuel and oxygen, enough

time must be allowed for the reaction to run to completion. It takes time for the fuel molecules to find free oxygen molecules with which they can combine to form product molecules. Remember that there are enormous relative distances between gas molecules. They may pass one another hundreds of times in near misses before they contact. If all fuel molecules met mating oxygen molecules at the same instant, we would have an explosion instead of smooth progressive burning.

3. To insure that fuel molecules meet available oxygen molecules, the two gases must be thoroughly mixed. Combustion chambers and furnaces must be designed to promote mixing of the gases as they flow through; that is, the flow must be turbulent. Figure 23·1b and c shows that there is a large proportion of inert nitrogen molecules in the gases that will get in the way of fuel and oxygen molecules. They increase the need for turbulent flow.

4. We already have talked about the need for excess air in completely burning the available fuel. This is a matter of thermal efficiency—fuel costs money while air is free, at most costing a small amount of fan energy to push it into the combustion space. To control thermal efficiency we measure the fuel-air ratio (pounds of fuel per pound of air). This usually has to be lower than the ideal to burn the fuel completely. Excess-air needs depend on the furnace or engine. The range runs from about 5 to as high as 50% (high excess air means low fuel-air ratio); in gas turbines it may run 200% and more to limit temperature rise. Some i-c engines occasionally use a fuel-air ratio higher than ideal (air deficiency) for short periods to develop maximum power output for peaking—at the expense of good thermal efficiency.

This by no means completely discusses the conditions controlling combustion reactions, but we have covered the primary factors. Combustion involves many complex intermediate reactions with intermediate products that we have not the space to cover here.

23·19 Dissociation. Our theory of internal energy of molecules associates the largest portion with the kinetic energy of the molecule as a whole traveling through space (see Chap. 5). Other parts of internal energy take the form of internal vibrations of the molecules and their nuclei as well as their rotation on an axis and the spin and orbits of the electrons (see Chap. 1).

To start a combustion process we know we must raise the temperature of the reactants to ignition level. As the reaction proceeds, we convert mass to energy—as the temperature rise shows. If we try to push the temperature up by forcing in more and more reactants, we soon come to a point where the temperature stays at a maximum but we get a variety of incomplete combustion products and unburned fuels in the products.

This phenomenon involves *dissociation*. It seems that the combustion reaction partly acts in reverse. At high temperatures some of the product molecules absorb energy and break up into constituent atoms, for example,

$$14{,}096 \text{ Btu per lb } C + CO_2 \rightarrow C + O_2 \qquad (23 \cdot 23)$$
$$61{,}031 \text{ Btu per lb } H_2 + H_2O \rightarrow H_2 + O \qquad (23 \cdot 24)$$

When energy is withdrawn from the burning gases by heat transfer to bring the temperature below the dissociation temperature, the fuel elements and oxygen again combine to release energy and form the products of a completed reaction.

Experiments in an i-c reaction showed that the actual temperature and pressure reached were 4800 F and 600 psia. Without dissociation the maximum temperature should have been 5470 F and the pressure 658 psia.

It seems that above dissociation temperature the product molecules have so much energy in their internal spins and vibrations that the mutual-attraction forces are not strong enough to hold the molecule together as a unit. So the molecule flies apart into its constituent atoms. Breaking the bonds makes these parts absorb energy in the potential form (reconverting to a minute mass increase) and so reduces the internal energy of the system. This seems to establish a fairly uniform dissociation temperature. Removing energy from the reacting gases by heat transfer may keep the energy level low enough to avoid the limiting dissociation temperature.

23·20 Heating value. The heating value of a fuel tells us how many units of energy we may get from it when it is completely oxidized and the products cooled back to the initial temperature of the reactants. We may allow the combustion process to undergo any variety of changes in pressure, temperature, and volume, but convenience limits testing to either constant volume or constant pressure.

23·21 Constant-volume-process calorimeter. This uses a "bomb" immersed in a measured amount of water, Fig. 23·2a. The test procedure involves (1) placing a measured amount of fuel—a few grams—in the bomb, (2) forcing oxygen at about 600 psi into the bomb, (3) igniting the fuel by an electric spark, (4) transferring the heat developed by the completely burned fuel to the surrounding water, (5) measuring the temperature rise—a few degrees—of the water bath. Knowing the amount of water and temperature rise, we can figure the heat transferred:

$$Q = Mc(t_2 - t_1) \quad \text{Btu} \qquad (23 \cdot 25)$$

where M = mass of water receiving Q, lb
c = specific heat of water, Btu per lb
t_2 = temperature of water after combustion, F
t_1 = temperature of water before combustion, F

Then the higher heating value of the fuel is

$$HHV = \frac{Q}{m} \quad \text{Btu per lb} \tag{23·26}$$

where m is the mass of the fuel sample in pounds. This brief explanation gives only the basics of the bomb calorimeter. Actually a variety of corrections for radiation and other factors must be made.

Combustion products within the bomb are essentially cooled to within about 4 F of the reactant initial temperature (this is also the water temperature). This means that any water vapor generated by burning hydrogen in the fuel has been condensed and so contributed its latent heat of evaporation to the fuel heating value.

23·22 Constant-pressure-process calorimeter. This uses a water-cooled coil to absorb heat Q released by the burning fuel, Fig. 23·2b.

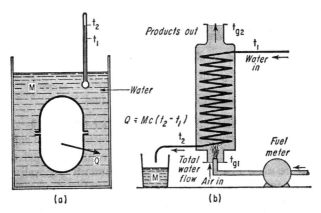

Fig. 23·2. (a) Constant-volume bomb calorimeter holds burning fuels under high oxygen pressure and cools them back to starting temperature by transferring heat to water. (b) Constant-pressure steady-flow calorimeter burns fuels at atmospheric pressure to heat a steady stream of water flowing through coils surrounding furnace.

This unit is used to find the heating values of liquid and gaseous fuels. The test procedure involves (1) adjusting the water-flow rate through the coil and the fuel-feed rate to the burner to get a constant temperature of combustion gases leaving the calorimeter at above 120 F, (2) measuring the total water flow and fuel input for a given period of time, (3) finding the average inlet and outlet temperatures of the water for this period and the average inlet temperature of the reactants and average outlet temperature of the products during this period of test time. These data then are substituted in the above formulas. M is the total water collected during the test period, t_2 is the average water temperature leaving the calorimeter during the test period, t_1 is the average water temperature entering the calorimeter during the test period, and m is the total mass of fuel burned during the test period.

This calorimeter gives the lower heating value of the fuel if it contains hydrogen. If we keep the product temperature above 120 F, the water vapor from burning hydrogen does not condense; so it carries its heat of vaporization out of the calorimeter.

23·23 Constant-volume bomb calorimeter. This gives the higher heating value for a fuel containing hydrogen. The standard test temperature for the reactants is 77 F, the base above which the heating value is measured. Theoretically, the difference between higher and lower heating values for the same fuel is the heat of vaporization of saturated steam at this standard temperature. This is 1050 Btu per lb of vapor. One can be calculated from the other if we know the weight of water vapor produced by the burning hydrogen:

$$HHV_h = HHV_l + 1{,}050w \qquad \text{Btu per lb} \qquad (23\cdot27)$$

where w is the weight of water vapor per pound of fuel. If we know the weight of H_2 per pound of fuel, then

$$w = 9H_2 \qquad \text{lb per lb fuel} \qquad (23\cdot28)$$

When a fuel has no hydrogen, it has no higher or lower heating value, but just one that will be measured directly by either type of calorimeter—if the fuel is liquid or gaseous. The heating value of a fuel depends on the temperature used as a base for its measurement. Generally, the lower the base temperature, the higher the heating value. Corrections can be made to different base temperatures if we know the specific heats of the reactants and products.

23·24 Cycle heat input. Total heat input Q into an actual engine cycle seldom, if ever, equals the heating value of the fuel supplied. Most practical cycles do not have a large enough heat-transfer surface or cannot hold the combustion products long enough to reduce leaving temperature of the products (flue gas or exhaust) to temperature of the incoming reactants.

Practical engines and cycles have small amounts of unburned or incompletely burned fuels in their products. Limited combustion time, incomplete turbulence, too-rapid cooling, and other factors cause this condition.

ENERGY SOURCES AND COMBUSTION

As a general rule we can assume that heat released in a furnace or combustion chamber is less than total heating value of the fuel; in turn the heat input Q to the cycle from combustion gases will be less than total heat released by burned fuel.

23·25 Calculated heating value. The practical way to determine fuel heating value is by using a calorimeter. But knowing the heating value of each fuel element we can also calculate it from an analysis of a fuel, as shown in Table 23·5.

Table 23·5

Fuel element	Weight, %	HHV, Btu per lb	$w \times HHV$
C	70.5	14,096	9938
H_2	4.5	61,031	2746
S	3.0	3984	120
O_2	6.0	0	0
N_2	1.0	0	0
Ash	11.0	0	0
Water	4.0	0	0
Total	100.0		12,804

REVIEW TOPICS

1. What is the basic method of producing Q input to a heat engine?
2. What is the basic process of combustion of ordinary fuels?
3. What essentially happens in a nuclear-fission process?
4. What are the essentials of a fusion process?
5. What are the elements in a thermionic generator?
6. What are the features of a thermoelectric generator?
7. How would a magnetohydrodynamic generator work?
8. What methods are available for producing electric energy from chemical processes.
9. How is solar energy being used to produce usable energy?
10. Write the combustion-reaction equations for carbon, carbon monoxide, hydrogen, and sulfur with oxygen. Also give the mole and weight relations for reactants and combustion products.
11. What are the proportions of oxygen and nitrogen in atmospheric air?
12. Derive the equation for the weight of oxygen needed to burn 1 lb of hydrogen.
13. Write the equations for converting per cent volumes into per cent weights of fuel elements in a conventional fuel.

14. Derive the equation for the weight of air needed to burn 1 lb of fuel having carbon, hydrogen, and sulfur.

15. Write the equation for excess-air percentage.

16. Write the relations for calculating the weights of combustion products for a fuel with a given weight analysis of fuel elements.

17. Write the formula for figuring the total moles of gas in combustion products with a given weight analysis.

18. How is the volumetric analysis of combustion products calculated from the weight analysis?

19. Describe the four major conditions that must be observed for successful combustion of a fuel.

20. How does dissociation limit the combustion process?

21. How can the heating value of fuels be found?

22. Describe the arrangement and method of calculating the heating value of fuels measured in a constant-volume calorimeter.

23. Describe the arrangement and method of calculating the heating value of a fuel burned in a constant-pressure calorimeter.

24. What is meant by higher and lower heating value of fuels? When does this apply?

PROBLEMS

1. Fifty pounds of carbon burns in two-thirds of the air needed for complete combustion. Find (a) relative weights of CO_2 and CO produced, (b) the percentage of the total heating value developed.

2. Find the relative weights of air needed to burn equal weights of carbon, hydrogen, and sulfur.

3. A producer gas has a percentage volumetric composition of $CO_2 = 10.8$, $CO = 18.3$, $CH_4 = 3.1$, $O_2 = 0.1$, $H_2 = 12.9$, $N_2 = 54.6$, $C_2H_4 = 0.2$. Find (a) the weight analysis by elements, (b) the weight analysis by compounds.

4. For anthracite coal with the weight analysis $C = 84.7$, $H = 2.9$, $O = 1.6$, $N = 1.5$, $S = 0.8$, ash $= 5.8$, and water $= 2.7$, find the theoretical weight of oxygen needed for complete combustion and the actual air needed for 35% excess air.

5. For a bituminous coal with the weight analysis $C = 71.6$, $H = 4.8$, $O = 6.3$, $N = 1.3$, $S = 3.4$, ash $= 9.1$, and water $= 3.5$, find (a) the theoretical weight of air needed for complete combustion, (b) the actual weight of air needed for 1.6 dilution, (c) the actual volume of air at 14.7 psia and 80 F.

6. Find the volume of air at 14.6 psia and 70 F needed to burn completely fuel oil of weight analysis $C = 83.65$, $H = 12.70$, $S = 0.75$, $N = 1.70$, and $O = 1.20$ when using 20% excess air.

7. For a natural gas with a volumetric composition of $CH_4 = 59.8$,

$C_2H_6 = 37.6$, $N_2 = 2.2$, $CO_2 = 0.4$, find for complete combustion (a) pounds of air theoretically needed per pound of gas and per cubic foot of gas at 16.8 psia and 80 F, (b) the ratio of air to gas volume at 14.7 psia and 60 F.

8. For a fuel oil with the weight analysis $C = 85.4$, $H = 11.3$, $O = 2.8$, $N = 0.2$, $S = 0.3$, find (a) the hypothetical mole formula for 100 lb of fuel, (b) the complete reaction equation per 100 lb of fuel with 40% excess air, (c) the dry-gas analysis based on CO_2, O_2, N_2.

9. For a fuel gas with the volumetric analysis $CH_4 = 68.0$ and $C_2H_6 = 32.0$, find the volume of actual air required per 1,000 cu ft of gas for complete combustion when using 15% excess air at 14.7 psia and 80 F.

10. A bituminous coal has the following analysis: $C = 79.6$, $H = 4.5$, $S = 0.5$, $O = 3.8$, $N = 1.8$, ash $= 8.6$, water $= 1.2$. Figure the dry-gas analysis for a dilution coefficient of 1.35 assuming that 5% of C burns to CO and air vapor is negligible.

CHAPTER 24

Vapor and Liquid Properties

Now that we understand the behavior of ideal gases and their use in thermodynamic cycles, we are ready to study vapors. To start, let us review the molecular theory of solids, liquids, and vapors. This will help to understand the peculiar behavior of vapors.

24·1 Water. Water is the working fluid most used in power cycles, so let us find out how it behaves in its three phases. In Fig. 24·1a we start with a block of *subcooled* ice at temperature T_1 (below 32 F) completely filling the cylinder and held under a pressure P_1 by a piston pressing on it.

The sketch shows schematically that molecules of water in solid phase hang in space, held there by mutual forces of attraction and repulsion. The kinetic energy of the molecules keeps them bouncing around, each in its own general location. But their energy is not enough to make them break away from the balanced restraining forces acting on them.

24·2 Melting. In Fig. 24·1b we add heat Q_1 to the ice to raise its temperature to T_2. Some of the molecules absorb more of this energy than the others (Maxwellian distribution); so their increased kinetic energy lets them break away successively from the mutual forces and they start wandering erratically past the other molecules. The mutual forces they meet everywhere pull them through haphazard paths as they wander through the ice. These molecules are in the liquid phase; we have a mixture of *saturated* ice and water in Fig. 24·1b. Average temperatures of the ice and the water are equal, even though the phases are different.

This may seem paradoxical, but let us study what happened. In changing phase from solid to liquid, the liquid molecules absorbed more energy then their slower neighbors. But in breaking away from their restraining forces they did mechanical work. This slowed them down to the *average* speed of the ice molecules, but having overcome the initial "inertia," they now can move around.

While a mixture of saturated ice and water is heating, the temperature stays constant, usually at about 32 F. Energy used to make this change of phase is the *enthalpy of fusion* and represents work done by molecules in breaking away from their fixed locations in the solid.

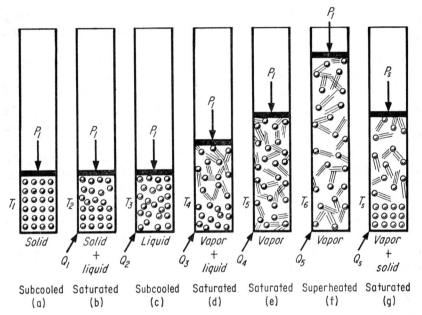

Fig. 24·1 Confining a piece of ice in a cylinder and piston under constant pressure while heating it lets us study how it changes phase from solid (a) to a mixture of solid and liquid (b), then to all liquid (c), and on to a mixture of steam and liquid (d). Continued heating forms saturated steam (e) and finally superheated steam (f). If we make the pressure P low enough, we sublimate ice directly to steam at a low temperature (g).

In Fig. 24·1c we add more heat Q_2 and melt all the ice to water; when it has all changed to the liquid phase, heat addition makes the temperature rise to some temperature T_3. The water is now a *subcooled* liquid. The molecules all wander among one another like a swarm of angry bees. Their mutual forces of attraction are strong enough to hold them in one mass at the bottom of a container. The molecules at the surface of the water have attractive forces acting on them, tending to pull them into the body of water. This compacts them and produces *surface tension:* the effect that allows a needle to float on the surface of a liquid.

24·3 Vaporizing. In Fig. 24·1d we add heat Q_3 to the water and start boiling off some of the water to vapor or steam. During this heating, at some temperature T_4 (depending on value of P_1), some of the molecules absorb more energy than others. This lets them break away completely from the liquid molecules to soar above them and form steam.

Again we find that the *average* temperature of the steam molecules and the liquid molecules is the same. This mixture of *saturated* steam and water does not change its *saturation temperature* while the water and steam are being heated and are in contact with each other. During this *evaporation* process the faster steam molecules do work in breaking away from the attractive forces of the water molecules. This slows them down to the same average temperature (speed).

Work done by the escaping steam molecules is the *enthalpy of evaporation*. In most cases, steam forms as bubbles within the water being heated, at the heat-transfer surface. When the bubble becomes large enough, it breaks away from the heating surface and rises through the water. At the top surface it breaks through and joins the steam accumulating above the water.

In Fig. 24·1e heat addition Q_4 evaporates all the water to steam, the water just having disappeared. It is now all *saturated steam* at temperature $T_5 = T_4$. Adding some more heat Q_5 in Fig. 24·1f raises the temperature to T_6, and we have a *superheated* vapor. As a vapor the molecules are close enough together still to exert slight attractive forces on one another. But if we continue superheating, we eventually expand the volume of the steam and raise its temperature enough so that it acts very much like a perfect gas.

24·4 Condensing. Changing the phase of a substance is a reversible process; by adding or taking away energy as heat or by changing pressure we can make a substance change to any phase wanted. For instance, in Fig. 24·1 we can start with the superheated vapor in f and by taking away heat Q_5 we can get the saturated vapor in e, at temperature T_4.

Further cooling by removing heat Q_3 in d makes some of the saturated steam condense to saturated water. That is, some of the steam particles slow down enough so that the mutual attractive forces bind them together loosely to form water. Removing heat Q_2 in c condenses all the steam and cools the liquid to a *subcooled* water. In b removing heat Q_1 makes the water freeze to ice. Again removing the kinetic energy of the water molecules slows them down to the point where balanced attraction and repulsion forces hold them in one general location to form a solid. Further heat removal subcools the ice and lowers its temperature to T_1, which simply reduces the vibrations of the ice molecules.

24·5 Sublimation. We can make a solid evaporate directly to its vapor if we establish the proper temperature and pressure. In Fig. 24·1g, by making both pressure P_s and temperature T_s low enough, we can make ice sublimate to steam. This would be a mixture of saturated ice and steam. Here when the solid molecules break away from their bonds, they have enough energy to soar into space without going through an intermediate liquid phase.

24·6 P-T relations. If we change the pressure on the piston in Fig. 24·1 we shall find that the temperatures at which the ice melts and the water boils will be different. Volume changes will also be altered. These relations cannot be expressed simply by formulas, so we most conveniently show them by graphs, Figs. 24·2 and 24·3.

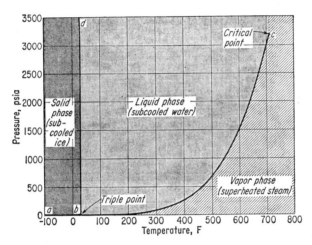

Fig. 24·2 Pressure-temperature diagram shows the relation for changing the phase of water from solid to liquid to vapor at will.

Both figures have the same data; Fig. 24·2 is on arithmetic coordinates, and Fig. 24·3 on logarithmic. Suppose we had our ice initially at 500 psia and −100 F. This would be subcooled in the solid phase, but as we heated it to 32 F, we would find it melting to water, indicated by line bd. This curve is the locus of melting while heating, or fusion while freezing. All mixtures of saturated ice and water lie on this vertical curve. When we work with pressures above 3000 psia, the freezing or melting temperature drops lower than 32 F.

Returning to our 500-psia heating, when all the ice has melted, the water rises in temperature into the subcooled-liquid region. When it reaches 467 F, it changes phase by starting to boil. This lies on the curve bc, which is the locus of all saturated water-steam mixtures for pressures up to 3206.2 psia.

When all the water has boiled to steam, the temperature rises and the steam enters the superheated-vapor phase. As it continues heating, steam enters the gas phase. This has no definite boundary, simply meaning that the steam behaves nearly like a perfect gas, its state being predictable by the perfect-gas equation.

24·7 Low-pressure heating. If we drop the pressure on subcooled ice to 0.04 psia and start heating the ice, it vaporizes at about 27.5 F to steam. After complete vaporization further heating produces an l-p superheated steam. This can be most clearly seen in Fig. 24·3.

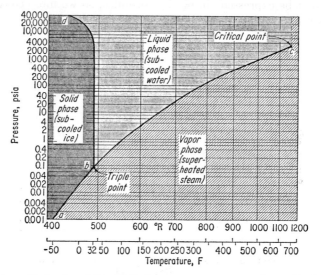

Fig. 24·3 Logarithmic plot of Fig. 24·2 gives more accurate readings of both temperature and pressure at the extremes of their values.

Point b is the *triple point* for water. Saturated ice, saturated water, and saturated steam can all exist simultaneously in a mixture only at 32.02 F and 0.0888 psia. Along curve ab ice sublimes to steam or in reverse steam freezes to ice and, as you see, only at very low pressures and low temperatures.

24·8 Critical point. Now let us look at an h-p process. Under a pressure of 3206.2 psia, heating subcooled ice will make it start melting at about 29 F. When it has all melted, further heating raises the temperature of the liquid. At this pressure (and higher) we never reach the boiling point. On a purely *arbitrary* basis, when we reach the temperature 705.4 F at point c, we call anything hotter a superheated steam, anything cooler a liquid or water.

Point c is called the *critical point* for steam, at the *critical temperature* of 705.4 F and the *critical pressure* of 3206.2 psia. At pressures and temperatures higher than these, water does not experience a change in phase,

VAPOR AND LIQUID PROPERTIES

liquid being indistinguishable from vapor. The fluid acts somewhat like a very dense gas.

Crowding of the molecules at high pressures prevents their ability to get away from one another, and the mutual forces of attraction play a big part in their thermal behavior. These graphs do not show the accompanying changes in specific volume of water and steam; so let us study the behavior of water and steam on the P-V graph. The area measures mechanical work transferred.

24·9 Volume changes. Figure 24·4 shows the P-V-T relations for water and steam. Points a and b (corresponding to Figs. 24·2 and 24·3) are essentially a single point on the zero pressure line in this graph, b being the triple point for water. Point c is the critical point.

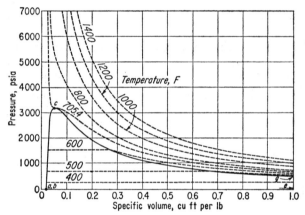

Fig. 24·4 P-V diagram for steam shows that it acts much like a gas at the higher temperatures but not at the lower range of values.

Curve bc is the locus of all saturated-water states. Curve cg is the locus of all saturated-steam states. The dashed lines are constant-temperature lines. Temperatures less than the critical are horizontal lines for saturated-water–steam mixtures. There is only one *saturation temperature* for each pressure; this was also shown by curves bc in Figs. 24·2 and 24·3. In these relations the pressure is called the *saturation pressure*.

Let us study the 500 F curve in Fig. 24·4. Where the curve intersects the saturated-water locus bc, we have saturated water with a specific volume of 0.020432 cu ft per lb under a pressure of 680.8 psia.

If we heat this saturated water, holding the pressure constant, the temperature remains constant at 500 F and the specific volume of the mixture of *saturated steam* and *water* grows. Continued heating eventually boils off all the water to steam when we arrive at the *saturated vapor* locus cg at a specific volume of 0.6749 cu ft per lb.

Still holding the pressure constant and continuing heating, the saturated steam now raises the temperature to form *superheated steam*. The constant-temperature curves slope downward with increasing specific volume.

Returning for a moment to saturated water at 500 F, let us assume that we cool it but hold the pressure constant. Its temperature might drop to, say, 32 F, and its specific volume would shrink slightly from 0.020432 to 0.020403 cu ft per lb. For most purposes we can assume that the subcooled-water region lies on curve bc for saturated water. For accurate work at pressures above about 500 psia we may sometimes need to account for changes in volume, where the shrinkage becomes larger on cooling. More on this later.

At temperatures above the critical in Fig. 24·4 the constant-temperature curves begin to approach the appearance of ideal-gas curves at the higher levels. Notice the distortions in critical-temperature and 800 F curves.

The region between the saturated-water and saturated-steam locus is called the *wet region*, area a–c–g–e, extending off the right end of the chart. As state points for steam move from left to right, the saturated mixture becomes drier and drier and *quality* increases or inversely *moisture* decreases. These are figured in the percentages of weights of steam to weight of mixture and the percentage of weight of water in the mixture.

24·10 Heat transferred. The counterpart of the P-V graph for measuring work is the T-S graph for measuring heat produced or absorbed, Fig. 24·5. Here we have lines of constant pressure, a locus of saturated water cd and a locus for saturated steam cef, and lines of constant moisture.

Let us trace the state changes for 1 lb of ice held in a cylinder and piston under 10 psia constant pressure. If the ice is subcooled at -100 F, it will be at point a in Fig. 24·5. As we add heat to the ice, its temperature and entropy both rise along line ab till the temperature reaches 32 F.

Further heating then starts melting the ice along line bd. The temperature of both ice and water stays constant at 32 F, but the entropy of the mixture of saturated ice and subcooled water rises. We looked at the molecular reasons earlier. At point d all ice has melted to a subcooled water.

Line bd corresponds to the triple point on the P-T plot in Figs. 24·2 and 24·3. Along this line a mixture of saturated ice, saturated steam, and saturated water can exist in one mixture.

Now an important point: All enthalpies for the fluid water in all its states are reckoned from 32 F as a base. This makes both enthalpy and entropy *zero* at this temperature and corresponding saturation pressure. As a result, enthalpy and entropy for ice are both negative values, being less than zero relatively.

Returning to our example, when we add heat to the subcooled water at 32 F and 10 psia, both temperature and entropy rise along the saturated-water locus dc (this is not true at higher pressures; more about this later). When the temperature reaches 193.2 F, the water is saturated and starts to vaporize; that is, it boils to form saturated steam at the same temperature.

Boiling takes place while the temperature of both saturated water and saturated steam stays constant but the entropy of the mixture rises.

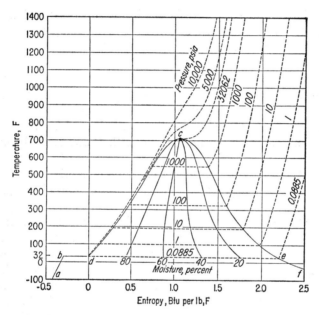

Fig. 24·5 Temperature-entropy diagram for steam features a nearly symmetrical shape of saturated liquid and vapor lines meeting at critical point c.

Moisture percentage of the saturated mixture decreases with rising entropy, being 100% at the saturated-liquid locus dc and 0% at the saturated-vapor locus cef. When heating has completely evaporated the water, we have everything as saturated steam, and its state lies on the saturated-vapor locus.

Further heating of the 10-psia steam makes both its temperature and entropy rise as it becomes superheated steam. At the higher temperatures, superheated steam begins to act like a gas because of its large specific volume. Its behavior can be figured closely from the ideal-gas equation, $PV = RT$.

24·11 T-S graph areas. Remembering the properties of the T-S graph, we see that the total area under the line bd down to absolute zero temperature at -460 F measures the latent enthalpy of fusion for ice melting to water; inversely it measures heat given up by water freezing to ice.

The total area, down to absolute zero, under the saturated-water locus from d to the pressure line for 10 psia measures the heat for warming the 10-psia water from 32 F to its saturation temperature at 193.2 F. The total area under the horizontal 10-psia pressure line in the saturation or wet region measures the latent enthalpy of vaporization at that pressure.

The total area under the 10-psia line in the superheat region measures the heat needed to raise the steam to a given temperature from the saturation temperature.

24·12 Latent enthalpies. Most of the constant-pressure lines in Fig. 24·5 converge into the saturated-water locus as both entropy and temperature drop. At pressures above about 400 psia the lines lie slightly to the left of the saturated-water line, as indicated by the 5,000- and 10,000-psia lines, which do not merge with the saturated-water locus line cd.

For accurate calculations at higher pressures corrections should be made for this *compressed-liquid* state, another term for subcooled liquid. More about this later.

The shape of the wet region between the saturated-water and -steam lines shows that the latent enthalpy of steam is very large at low pressures. As pressure rises, the latent enthalpy shrinks until at the critical point c it reduces to zero. Above the critical pressure there is no boiling action as we heat water to steam. The water gradually expands like a very dense gas, forming no bubbles, and we arbitrarily call the fluid *water* below the critical temperature of 705.4 F and *steam* above this temperature at pressures above the critical of 3206.2 psia. We now have power plants working in this area.

At pressures below the triple point at 32 F and 0.0885 psia we have the latent enthalpy of sublimation where ice sublimes to steam or, inversely, steam condenses to ice (usually as hoar frost). No intermediate liquid phase exists in this range of very low pressures. Notice the large sudden increase in latent enthalpy in Fig. 24·5 below the triple point, line bd in the T-S graph.

We seldom, if ever, work in this region below 32 F when using thermodynamic power cycles. We are considering it just to complete the picture of physical behavior of matter over a wide range of states.

24·13 T-H graph. This graph, Fig. 24·6, reflects the complex variation in specific heats of water and steam. While the specific heats of actual gases also vary, the variations are so gradual that a plot like Fig. 24·6 for gases would be a set of nearly straight lines sloping upward to the right.

Compared with Fig. 24·5, Fig. 24·6 looks like a warped T-S diagram. Here again we can identify the subcooled-ice locus ab, the melting locus bd (corresponding to the triple point), the saturated-liquid locus dc, and the saturated-vapor locus cef.

Again the constant-pressure lines tend to converge at the lower pressures into the liquid locus; at the superheated states they soar off to the

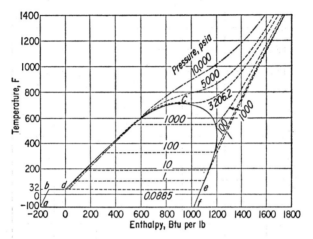

Fig. 24·6 Highly variable specific heats of water and steam and latent enthalpies make a complicated relation between temperature and enthalpy.

upper right. In this area, note that the pressure lines are crowded together much more than on the T-S graph in Fig. 24·5. We seldom use this chart in practical thermodynamic work, but it gives us a good introduction to the range of enthalpy values that we must work with when dealing with steam and water.

24·14 Mollier diagram. This diagram, Fig. 24·7, is the most-used chart in steam work. It plots steam and water enthalpy against entropy with curves for constant pressure, constant temperature, constant moisture, the saturated-liquid and the saturated-vapor locus.

Here again we can identify the subcooled-ice locus ab, the enthalpy of fusion bd, the saturated-liquid locus dc, and the saturated-vapor locus cef. The critical point stands at c. This chart has great utility because we can quickly relate enthalpies for constant-entropy processes. Here again the lower pressure lines converge into the liquid-locus line.

The wet region appears as a very warped version of the T-S diagram. The superheat region in the upper right shows the temperature lines as being practically horizontal for pressures less than 100 psia. This shows that a simple $H = cT$ relation holds for the steam in the upper superheat region. In other words, steam acts very much like an ideal gas in this area. See Appendix K for a larger version of the Mollier chart.

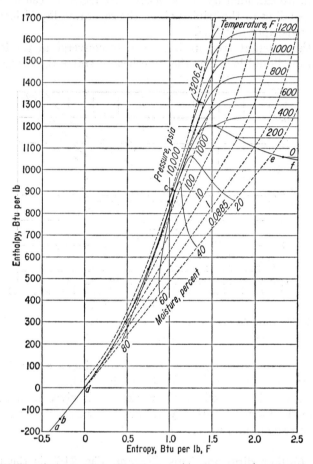

Fig. 24·7 Mollier diagram gives the relation of enthalpy and entropy for steam for constant pressures, constant temperatures, and saturated water and saturated steam. Plotted to large scale, this diagram can be used instead of steam tables.

24·15 Steam tables. By drawing Mollier diagrams to very large scales we can read enthalpies to within 0.1 Btu per lb, but the chart is very large and awkward to handle. The *Steam table* has been the standard way of finding the properties when calculating processes. There are three main tables: (1) saturation—temperature (2) saturation—pressure and (3) superheated vapor. In addition there are tables for (4) compressed liquid and (5) saturation—solid-vapor.

The saturation—temperature table is shown in Table 24·1. This is

Table 24·1 Thermodynamic properties of Steam
Dry saturated steam: temperature table[a]

Temp., F	Abs. press., psia	Specific volume			Enthalpy			Entropy			Temp., F
		Sat. liquid	Evap.	Sat. vapor	Sat. liquid	Evap.	Sat. vapor	Sat. liquid	Evap.	Sat. vapor	
t	p	v_f	v_{fg}	v_g	h_f	h_{fg}	h_g	s_f	s_{fg}	s_g	t
32	0.08854	0.01602	3306	3306	0.00	1075.8	1075.8	0.0000	2.1877	2.1877	32
35	0.09995	0.01602	2947	2947	3.02	1074.1	1077.1	0.0061	2.1709	2.1770	35
40	0.12170	0.01602	2444	2444	8.05	1071.3	1079.3	0.0162	2.1435	2.1597	40
45	0.14752	0.01602	2036.4	2036.4	13.06	1068.4	1081.5	0.0262	2.1167	2.1429	45
50	0.17811	0.01603	1703.2	1703.2	18.07	1065.6	1083.7	0.0361	2.0903	2.1264	50
60	0.2563	0.01604	1206.6	1206.7	28.06	1059.9	1088.0	0.0555	2.0393	2.0948	60
70	0.3631	0.01606	867.8	867.9	38.04	1054.3	1092.3	0.0745	1.9902	2.0647	70
80	0.5069	0.01608	633.1	633.1	48.02	1048.6	1096.6	0.0932	1.9428	2.0360	80
90	0.6982	0.01610	468.0	468.0	57.99	1042.9	1100.9	0.1115	1.8972	2.0087	90
100	0.9492	0.01613	350.3	350.4	67.97	1037.2	1105.2	0.1295	1.8531	1.9826	100
110	1.2748	0.01617	265.3	265.4	77.94	1031.6	1109.5	0.1471	1.8106	1.9577	110
120	1.6924	0.01620	203.25	203.27	87.92	1025.8	1113.7	0.1645	1.7694	1.9339	120
130	2.2225	0.01625	157.32	157.34	97.90	1020.0	1117.9	0.1816	1.7296	1.9112	130
140	2.8886	0.01629	122.99	123.01	107.89	1014.1	1122.0	0.1984	1.6910	1.8894	140
150	3.718	0.01634	97.06	97.07	117.89	1008.2	1126.1	0.2149	1.6537	1.8685	150
160	4.741	0.01639	77.27	77.29	127.89	1002.3	1130.2	0.2311	1.6174	1.8485	160
170	5.992	0.01645	62.04	62.06	137.90	996.3	1134.2	0.2472	1.5822	1.8293	170
180	7.510	0.01651	50.21	50.23	147.92	990.2	1138.1	0.2630	1.5480	1.8109	180
190	9.339	0.01657	40.94	40.96	157.95	984.1	1142.0	0.2785	1.5147	1.7932	190
200	11.526	0.01663	33.62	33.64	167.99	977.9	1145.9	0.2938	1.4824	1.7762	200
210	14.123	0.01670	27.80	27.82	178.05	971.6	1149.7	0.3090	1.4508	1.7598	210
212	14.696	0.01672	26.78	26.80	180.07	970.3	1150.4	0.3120	1.4446	1.7566	212
220	17.186	0.01677	23.13	23.15	188.13	965.2	1153.4	0.3239	1.4201	1.7440	220
230	20.780	0.01684	19.365	19.382	198.23	958.8	1157.0	0.3387	1.3901	1.7288	230
240	24.969	0.01692	16.306	16.323	208.34	952.2	1160.5	0.3531	1.3609	1.7140	240
250	29.825	0.01700	13.804	13.821	218.48	945.5	1164.0	0.3675	1.3323	1.6998	250
260	35.429	0.01709	11.746	11.763	228.64	938.7	1167.3	0.3817	1.3043	1.6860	260
270	41.858	0.01717	10.044	10.061	238.84	931.8	1170.6	0.3958	1.2769	1.6727	270
280	49.203	0.01726	8.628	8.645	249.06	924.7	1173.8	0.4096	1.2501	1.6597	280
290	57.556	0.01735	7.444	7.461	259.31	917.5	1176.8	0.4234	1.2238	1.6472	290
300	67.013	0.01745	6.449	6.466	269.59	910.1	1179.7	0.4369	1.1980	1.6350	300
310	77.68	0.01755	5.609	5.626	279.92	902.6	1182.5	0.4504	1.1727	1.6231	310
320	89.66	0.01765	4.896	4.914	290.28	894.9	1185.2	0.4637	1.1478	1.6115	320
330	103.06	0.01776	4.289	4.307	300.68	887.0	1187.7	0.4769	1.1233	1.6002	330
340	118.01	0.01787	3.770	3.788	311.13	879.0	1190.1	0.4900	1.0992	1.5891	340
350	134.63	0.01799	3.324	3.342	321.63	870.7	1192.3	0.5029	1.0754	1.5783	350
360	153.04	0.01811	2.939	2.957	332.18	862.2	1194.4	0.5158	1.0519	1.5677	360
370	173.37	0.01823	2.606	2.625	342.79	853.5	1196.3	0.5286	1.0287	1.5573	370
380	195.77	0.01836	2.317	2.335	353.45	844.6	1198.1	0.5413	1.0059	1.5471	380
390	220.37	0.01850	2.0651	2.0836	364.17	835.4	1199.6	0.5539	0.9832	1.5371	390
400	247.31	0.01864	1.8447	1.8633	374.97	826.0	1201.0	0.5664	0.9608	1.5272	400
410	276.75	0.01878	1.6512	1.6700	385.83	816.3	1202.1	0.5788	0.9386	1.5174	410
420	308.83	0.01894	1.4811	1.5000	396.77	806.3	1203.1	0.5912	0.9166	1.5078	420
430	343.72	0.01910	1.3308	1.3499	407.79	796.0	1203.8	0.6035	0.8947	1.4982	430
440	381.59	0.01926	1.1979	1.2171	418.90	785.4	1204.3	0.6158	0.8730	1.4887	440
450	422.6	0.0194	1.0799	1.0993	430.1	774.5	1204.6	0.6280	0.8513	1.4793	450
460	466.9	0.0196	0.9748	0.9944	441.4	763.2	1204.6	0.6402	0.8298	1.4700	460
470	514.7	0.0198	0.8811	0.9009	452.8	751.5	1204.3	0.6523	0.8083	1.4606	470
480	566.1	0.0200	0.7972	0.8172	464.4	739.4	1203.7	0.6645	0.7868	1.4513	480
490	621.4	0.0202	0.7221	0.7423	476.0	726.8	1202.8	0.6766	0.7653	1.4419	490
500	680.8	0.0204	0.6545	0.6749	487.8	713.9	1201.7	0.6887	0.7438	1.4325	500
520	812.4	0.0209	0.5385	0.5594	511.9	686.4	1198.2	0.7130	0.7006	1.4136	520
540	962.5	0.0215	0.4434	0.4649	536.6	656.6	1193.2	0.7374	0.6568	1.3942	540
560	1133.1	0.0221	0.3647	0.3868	562.2	624.2	1186.4	0.7621	0.6121	1.3742	560
580	1325.8	0.0228	0.2989	0.3217	588.9	588.4	1177.3	0.7872	0.5659	1.3532	580
600	1542.9	0.0236	0.2432	0.2668	617.0	548.5	1165.5	0.8131	0.5176	1.3307	600
620	1786.6	0.0247	0.1955	0.2201	646.7	503.6	1150.3	0.8398	0.4664	1.3062	620
640	2059.7	0.0260	0.1538	0.1798	678.6	452.0	1130.5	0.8679	0.4110	1.2789	640
660	2365.4	0.0278	0.1165	0.1442	714.2	390.2	1104.4	0.8987	0.3485	1.2472	660
680	2708.1	0.0305	0.0810	0.1115	757.3	309.9	1067.2	0.9351	0.2719	1.2071	680
700	3093.7	0.0369	0.0392	0.0761	823.3	172.1	995.4	0.9905	0.1484	1.1389	700
705.4	3206.2	0.0503	0	0.0503	902.7	0	902.7	1.0580	0	1.0580	705.4

[a] Abridged from "Thermodynamic Properties of Steam," by Joseph H. Keenan and Frederick G. Keyes, John Wiley & Sons, Inc., New York, 1937.

only an abstract of the complete table covering six pages in *Thermodynamic properties of steam* by Keenan and Keyes, published by John Wiley & Sons, Inc., New York, 1936. In this table all the properties of saturated steam and saturated water are listed against even values of temperature in degrees Fahrenheit in the first column under the symbol t.

The second column has the corresponding saturation pressure p. Columns 3, 4, and 5 have the specific volumes of water and steam and their difference. Columns 6, 7, and 8 have the enthalpies for water and steam and their difference; columns 9, 10, and 11 have the entropies for water and steam and their difference.

The complete *Steam tables* have a second dry-saturated-steam table arranged against even values of pressure. The two sets of tables make it easy to interpolate and find properties for uneven values of either pressure or temperature. For lack of space we shall not reproduce an abstract of the pressure table.

The third set of tables, Table 24·2 lists properties of superheated steam against even values of both pressure and temperature. Three properties are listed for each state given by the pressure and the temperature: specific volume v, cu ft per lb; enthalpy h, Btu per lb; specific entropy s, Btu per lb, F.

The complete *Steam tables* for superheated steam cover a total of 36 pages. Table 24·2 is a very brief abstract of these. The figure in parentheses under each pressure is the saturation temperature for that pressure in degrees Fahrenheit.

24·16 High-pressure table. Recent work in superpressure cycles raised the need for extending the *Steam tables* to 10,000 psia pressure. Until further experimental work is completed, interim tables were compiled by extrapolation and scattered data to this higher pressure. Table 24·3 is a brief abstract of the h-p table. This one has been extended over a range of temperatures from 32 to 1600 F. Some workers consider the fluid a compressed liquid below the critical temperature of 705.4 F and superheated steam above. This is purely arbitrary, since superheat has no real meaning when no saturation temperature applies.

Table 24·3, a new effort, was made to include properties way down to 32 F, usually called the compressed-liquid region. Table 24·2 for superheated steam does not list properties below 200 F for the lower pressures and below 700 F for the higher pressures. To get these, Keenan and Keyes' *Steam tables* include a table for compressed liquid. This is a correction table that can be applied to the listed properties for the saturated liquid to figure properties in the subcooled or compressed state.

24·17 Saturation table. Tables 24·2 and 24·3 are relatively easy to use. But the saturation tables need a little study and practice.

For example, at 200 F there is only one saturation pressure p, 11.526

Table 24·2 Thermodynamic properties of steam
Superheated-steam table[a]

Abs. press., psia (sat. temp.)		\multicolumn{13}{c}{Temp., F}
		200	300	400	500	600	700	800	900	1000	1100	1200	1400	1600
1 (101.74)	v	392.6	452.3	512.0	571.6	631.2	690.8	750.4	809.9	869.5	929.1	988.7	1107.8	1227.0
	h	1150.4	1195.8	1241.7	1288.3	1335.7	1383.8	1432.8	1482.7	1533.5	1585.2	1637.7	1745.7	1857.5
	s	2.0512	2.1153	2.1720	2.2233	2.2702	2.3137	2.3542	2.3923	2.4283	2.4625	2.4952	2.5566	2.6137
5 (162.24)	v		90.25	102.26	114.22	126.16	138.10	150.03	161.95	173.87	185.79	197.71	221.6	245.4
	h		1195.0	1241.2	1288.0	1335.4	1383.6	1432.7	1482.6	1533.4	1585.1	1637.7	1745.7	1857.4
	s		1.9370	1.9942	2.0456	2.0927	2.1361	2.1767	2.2148	2.2509	2.2851	2.3178	2.3792	2.4363
10 (193.21)	v		45.00	51.04	57.05	63.03	69.01	74.98	80.95	86.92	92.88	98.84	110.77	122.69
	h		1193.9	1240.6	1287.5	1335.1	1383.4	1432.5	1482.4	1533.2	1585.0	1637.6	1745.6	1857.3
	s		1.7927	1.8595	1.9172	1.9689	2.0160	2.0596	2.1002	2.1383	2.1744	2.2086	2.2413	2.3598
14.696 (212.00)	v			30.53	34.68	38.78	42.86	46.94	51.00	55.07	59.13	63.19	67.25	75.37	83.48
	h			1192.8	1239.9	1287.1	1334.8	1383.2	1432.3	1482.3	1533.1	1584.8	1637.5	1745.5	1857.3
	s			1.8160	1.8743	1.9261	1.9734	2.0170	2.0576	2.0958	2.1319	2.1662	2.1989	2.2603	2.3174
20 (227.96)	v			22.36	25.43	28.46	31.47	34.47	37.46	40.45	43.44	46.42	49.41	55.37	61.34
	h			1191.6	1239.2	1286.6	1334.4	1382.9	1432.1	1482.1	1533.0	1584.7	1637.4	1745.4	1857.2
	s			1.7808	1.8396	1.8918	1.9392	1.9829	2.0235	2.0618	2.0978	2.1321	2.1648	2.2263	2.2834
40 (267.25)	v				11.040	12.628	14.168	15.688	17.198	18.702	20.20	21.70	23.20	24.69	27.68	30.66
	h				1186.8	1236.5	1284.8	1333.1	1381.9	1431.3	1481.4	1532.4	1584.3	1637.0	1745.1	1857.0
	s				1.6994	1.7608	1.8140	1.8619	1.9058	1.9467	1.9850	2.0212	2.0555	2.0883	2.1498	2.2069
60 (292.71)	v				7.259	8.357	9.403	10.427	11.441	12.449	13.452	14.454	15.453	16.451	18.446	20.44
	h				1181.6	1233.6	1283.0	1331.8	1380.9	1430.5	1480.8	1531.9	1583.8	1636.6	1744.8	1856.7
	s				1.6492	1.7135	1.7678	1.8162	1.8605	1.9015	1.9400	1.9762	2.0106	2.0434	2.1049	2.1621
80 (312.03)	v					6.220	7.020	7.797	8.562	9.322	10.077	10.830	11.582	12.332	13.830	15.325
	h					1230.7	1281.1	1330.5	1379.9	1429.7	1480.1	1531.3	1583.4	1636.2	1744.5	1856.5
	s					1.6791	1.7346	1.7836	1.8281	1.8694	1.9079	1.9442	1.9787	2.0115	2.0731	2.1303
100 (327.81)	v					4.937	5.589	6.218	6.835	7.446	8.052	8.656	9.259	9.860	11.060	12.258
	h					1227.6	1279.1	1329.1	1378.9	1428.9	1479.5	1530.8	1582.9	1635.7	1744.2	1856.2
	s					1.6518	1.7085	1.7581	1.8029	1.8443	1.8829	1.9193	1.9538	1.9867	2.0484	2.1056
120 (341.25)	v					4.081	4.636	5.165	5.683	6.195	6.702	7.207	7.710	8.212	9.214	10.213
	h					1224.4	1277.2	1327.7	1377.8	1428.1	1478.8	1530.2	1582.4	1635.3	1743.9	1856.0
	s					1.6287	1.6869	1.7370	1.7822	1.8237	1.8625	1.8990	1.9335	1.9664	2.0281	2.0854
140 (353.02)	v					3.468	3.954	4.413	4.861	5.301	5.738	6.172	6.604	7.035	7.895	8.752
	h					1221.1	1275.2	1326.4	1376.8	1427.3	1478.2	1529.7	1581.9	1634.9	1743.5	1855.7
	s					1.6087	1.6683	1.7190	1.7645	1.8063	1.8451	1.8817	1.9163	1.9493	2.0110	2.0683
160 (363.53)	v						3.008	3.443	3.849	4.244	4.631	5.015	5.396	5.775	6.152	6.906	7.656
	h						1217.6	1273.1	1325.0	1375.7	1426.4	1477.5	1529.1	1581.4	1634.5	1743.2	1855.5
	s						1.5908	1.6519	1.7033	1.7491	1.7911	1.8301	1.8667	1.9014	1.9344	1.9962	2.0535
180 (373.06)	v						2.649	3.044	3.411	3.764	4.110	4.452	4.792	5.129	5.466	6.136	6.804
	h						1214.0	1271.0	1323.5	1374.7	1425.6	1476.8	1528.6	1581.0	1634.1	1742.9	1855.2
	s						1.5745	1.6373	1.6894	1.7355	1.7776	1.8167	1.8534	1.8882	1.9212	1.9831	2.0404
200 (381.79)	v						2.361	2.726	3.060	3.380	3.693	4.002	4.309	4.613	4.917	5.521	6.123
	h						1210.3	1268.9	1322.1	1373.6	1424.8	1476.2	1528.0	1580.5	1633.7	1742.6	1855.0
	s						1.5594	1.6240	1.6767	1.7232	1.7655	1.8048	1.8415	1.8763	1.9094	1.9713	2.0287
220 (389.86)	v						2.125	2.465	2.772	3.066	3.352	3.634	3.913	4.191	4.467	5.017	5.565
	h						1206.5	1266.7	1320.7	1372.0	1424.0	1475.5	1527.5	1580.0	1633.3	1742.3	1854.7
	s						1.5453	1.6117	1.6652	1.7120	1.7545	1.7939	1.8308	1.8656	1.8987	1.9607	2.0181
240 (397.37)	v						1.9276	2.247	2.533	2.804	3.068	3.327	3.584	3.839	4.093	4.597	5.100
	h						1202.5	1264.5	1319.2	1371.5	1423.2	1474.8	1526.9	1579.6	1632.9	1742.0	1854.5
	s						1.5219	1.6003	1.6546	1.7017	1.7444	1.7839	1.8209	1.8558	1.8889	1.9510	2.0084
260 (404.42)	v							2.063	2.330	2.582	2.827	3.067	3.305	3.541	3.776	4.242	4.707
	h							1262.3	1317.7	1370.4	1422.3	1474.2	1526.3	1579.1	1632.5	1741.7	1854.2
	s							1.5897	1.6447	1.6922	1.7352	1.7748	1.8118	1.8467	1.8799	1.9420	1.9995
280 (411.05)	v							1.9047	2.156	2.392	2.621	2.845	3.066	3.286	3.504	3.938	4.370
	h							1260.0	1316.2	1369.4	1421.5	1473.5	1525.8	1578.6	1632.1	1741.4	1854.0
	s							1.5796	1.6354	1.6834	1.7265	1.7662	1.8033	1.8383	1.8716	1.9337	1.9912
300 (417.33)	v							1.7675	2.005	2.227	2.442	2.652	2.859	3.065	3.269	3.674	4.078
	h							1257.6	1314.7	1368.3	1420.6	1472.8	1525.2	1578.1	1631.7	1741.0	1853.7
	s							1.5701	1.6268	1.6751	1.7184	1.7582	1.7954	1.8305	1.8638	1.9260	1.9835
350 (431.72)	v							1.4923	1.7036	1.8980	2.084	2.266	2.445	2.622	2.798	3.147	3.493
	h							1251.5	1310.9	1365.5	1418.5	1471.1	1523.8	1577.0	1630.7	1740.3	1853.1
	s							1.5481	1.6070	1.6563	1.7002	1.7403	1.7777	1.8130	1.8463	1.9086	1.9663
400 (444.59)	v							1.2851	1.4770	1.6508	1.8161	1.9767	2.134	2.290	2.445	2.751	3.055
	h							1245.1	1306.9	1362.7	1416.4	1469.4	1522.4	1575.8	1629.6	1739.5	1852.5
	s							1.5281	1.5894	1.6398	1.6842	1.7247	1.7623	1.7977	1.8311	1.8936	1.9513

[a] Abridged from "Thermodynamic Properties of Steam," by Joseph H. Keenan and Frederick G. Keyes, John Wiley & Sons, Inc., New York, 1937.

Table 24·2 Thermodynamic properties of steam (*Continued*)
Superheated-steam table[a]

Abs. press., psia (sat. temp.)		500	550	600	620	640	660	680	700	800	900	1000	1200	1400	1600
450 (456.28)	v	1.1231	1.2155	1.3005	1.3332	1.3652	1.3967	1.4278	1.4584	1.6074	1.7516	1.8928	2.170	2.443	2.714
	h	1238.4	1272.0	1302.8	1314.6	1326.2	1337.5	1348.8	1359.9	1414.3	1467.7	1521.0	1628.6	1738.7	1851.9
	s	1.5095	1.5437	1.5735	1.5845	1.5951	1.6054	1.6153	1.6250	1.6699	1.7108	1.7486	1.8177	1.8803	1.9381
500 (467.01)	v	0.9927	1.0800	1.1591	1.1893	1.2188	1.2478	1.2763	1.3044	1.4405	1.5715	1.6996	1.9504	2.197	2.442
	h	1231.3	1266.8	1298.6	1310.7	1322.6	1334.2	1345.7	1357.0	1412.1	1466.0	1519.6	1627.6	1737.9	1851.3
	s	1.4919	1.5280	1.5588	1.5701	1.5810	1.5915	1.6016	1.6115	1.6571	1.6982	1.7363	1.8056	1.8683	1.9262
550 (476.94)	v	0.8852	0.9686	1.0431	1.0714	1.0989	1.1259	1.1523	1.1783	1.3038	1.4241	1.5414	1.7706	1.9957	2.219
	h	1223.7	1261.2	1294.3	1306.8	1318.9	1330.8	1342.5	1354.0	1409.9	1464.3	1518.2	1626.6	1737.1	1850.6
	s	1.4751	1.5131	1.5451	1.5568	1.5680	1.5787	1.5890	1.5991	1.6452	1.6868	1.7250	1.7946	1.8575	1.9155
600 (486.21)	v	0.7947	0.8753	0.9463	0.9729	0.9988	1.0241	1.0489	1.0732	1.1899	1.3013	1.4096	1.6208	1.8279	2.033
	h	1215.7	1255.5	1289.9	1302.7	1315.2	1327.4	1339.3	1351.1	1407.7	1462.5	1516.7	1625.5	1736.3	1850.0
	s	1.4586	1.4990	1.5323	1.5443	1.5558	1.5667	1.5773	1.5875	1.6343	1.6762	1.7147	1.7846	1.8476	1.9056
700 (503.10)	v		0.7277	0.7934	0.8177	0.8411	0.8639	0.8860	0.9077	1.0108	1.1082	1.2024	1.3853	1.5641	1.7405
	h		1243.2	1280.6	1294.3	1307.5	1320.3	1332.8	1345.0	1403.2	1459.0	1515.9	1623.5	1734.8	1848.8
	s		1.4722	1.5084	1.5212	1.5333	1.5449	1.5559	1.5665	1.6147	1.6573	1.6963	1.7666	1.8299	1.8881
800 (518.23)	v		0.6154	0.6779	0.7006	0.7223	0.7433	0.7635	0.7833	0.8763	0.9633	1.0470	1.2088	1.3662	1.5214
	h		1229.8	1270.7	1285.4	1299.4	1312.9	1325.9	1338.6	1398.6	1455.4	1511.0	1621.4	1733.2	1847.5
	s		1.4467	1.4863	1.5000	1.5129	1.5250	1.5366	1.5476	1.5972	1.6407	1.6801	1.7510	1.8146	1.8729
900 (531.98)	v		0.5264	0.5873	0.6089	0.6294	0.6491	0.6680	0.6863	0.7716	0.8506	0.9262	1.0714	1.2124	1.3509
	h		1215.0	1260.1	1275.9	1290.9	1305.1	1318.8	1332.1	1393.9	1451.8	1508.1	1619.3	1731.6	1846.3
	s		1.4216	1.4653	1.4800	1.4938	1.5066	1.5187	1.5303	1.5814	1.6257	1.6656	1.7371	1.8009	1.8595
1000 (544.61)	v		0.4533	0.5140	0.5350	0.5546	0.5733	0.5912	0.6084	0.6878	0.7604	0.8294	0.9615	1.0893	1.2146
	h		1198.3	1248.8	1265.9	1281.9	1297.0	1311.4	1325.3	1389.2	1448.2	1505.1	1617.3	1730.0	1845.0
	s		1.3961	1.4450	1.4610	1.4757	1.4893	1.5021	1.5141	1.5670	1.6121	1.6525	1.7245	1.7886	1.8474
1100 (556.31)	v			0.4532	0.4738	0.4929	0.5110	0.5281	0.5445	0.6191	0.6866	0.7503	0.8716	0.9885	1.1031
	h			1236.7	1255.3	1272.4	1288.5	1303.7	1318.3	1384.3	1444.5	1502.2	1615.2	1728.4	1843.8
	s			1.4251	1.4425	1.4583	1.4728	1.4862	1.4989	1.5535	1.5995	1.6405	1.7130	1.7775	1.8363
1200 (567.22)	v			0.4016	0.4222	0.4410	0.4586	0.4752	0.4909	0.5617	0.6250	0.6843	0.7967	0.9046	1.0101
	h			1223.5	1243.9	1262.4	1279.6	1295.7	1311.0	1379.3	1440.7	1499.2	1613.1	1726.9	1842.5
	s			1.4052	1.4243	1.4413	1.4568	1.4710	1.4843	1.5409	1.5879	1.6293	1.7025	1.7672	1.8263
1400 (587.10)	v			0.3174	0.3390	0.3580	0.3753	0.3912	0.4062	0.4714	0.5281	0.5805	0.6789	0.7727	0.8640
	h			1193.0	1218.4	1240.4	1260.3	1278.5	1295.5	1369.1	1433.1	1493.2	1608.9	1723.7	1840.0
	s			1.3639	1.3877	1.4079	1.4258	1.4419	1.4567	1.5177	1.5666	1.6093	1.6836	1.7489	1.8083
1600 (604.90)	v				0.2733	0.2936	0.3112	0.3271	0.3417	0.4034	0.4553	0.5027	0.5906	0.6738	0.7545
	h				1187.8	1215.2	1238.7	1259.6	1278.7	1358.4	1425.3	1487.0	1604.6	1720.5	1837.5
	s				1.3489	1.3741	1.3952	1.4137	1.4303	1.4964	1.5476	1.5914	1.6669	1.7328	1.7926
1800 (621.03)	v				0.2407	0.2597	0.2760	0.2907	0.3502	0.3986	0.4421	0.5218	0.5968	0.6693
	h					1185.1	1214.0	1238.5	1260.3	1347.2	1417.4	1480.8	1600.4	1717.3	1835.0
	s					1.3377	1.3638	1.3855	1.4044	1.4765	1.5301	1.5752	1.6520	1.7185	1.7786
2000 (635.82)	v					0.1936	0.2161	0.2337	0.2489	0.3074	0.3532	0.3935	0.4668	0.5352	0.6011
	h					1145.6	1184.9	1214.9	1240.0	1335.5	1409.2	1474.5	1596.1	1714.1	1832.5
	s					1.2945	1.3300	1.3564	1.3783	1.4576	1.5139	1.5603	1.6384	1.7055	1.7660
2500 (668.13)	v							0.1484	0.1686	0.2294	0.2710	0.3061	0.3678	0.4244	0.4784
	h							1132.3	1176.8	1303.6	1387.8	1458.4	1585.3	1706.1	1826.2
	s							1.2687	1.3073	1.4127	1.4772	1.5273	1.6088	1.6775	1.7389
3000 (695.36)	v								0.0984	0.1760	0.2159	0.2476	0.3018	0.3505	0.3966
	h								1060.7	1267.2	1365.0	1441.8	1574.3	1698.0	1819.9
	s								1.1966	1.3690	1.4439	1.4984	1.5837	1.6540	1.7163
3206.2 (705.40)	v									0.1583	0.1981	0.2288	0.2806	0.3267	0.3703
	h									1250.5	1355.2	1434.7	1569.8	1694.6	1817.2
	s									1.3508	1.4309	1.4874	1.5742	1.6452	1.7080
3500	v									0.0306	0.1364	0.1762	0.2058	0.2546	0.2977	0.3381
	h									780.5	1224.9	1340.7	1424.5	1563.3	1689.8	1813.6
	s									0.9515	1.3241	1.4127	1.4723	1.5615	1.6336	1.6968
4000	v									0.0287	0.1052	0.1462	0.1743	0.2192	0.2581	0.2943
	h									763.8	1174.8	1314.4	1406.8	1552.1	1681.7	1807.2
	s									0.9347	1.2757	1.3827	1.4482	1.5417	1.6154	1.6795
4500	v									0.0276	0.0798	0.1226	0.1500	0.1917	0.2273	0.2602
	h									753.5	1113.9	1286.5	1388.4	1540.8	1673.5	1800.9
	s									0.9235	1.2204	1.3529	1.4253	1.5235	1.5990	1.6640
5000	v									0.0268	0.0593	0.1036	0.1303	0.1696	0.2027	0.2329
	h									746.4	1047.1	1256.5	1369.5	1529.5	1665.3	1794.5
	s									0.9152	1.1622	1.3231	1.4034	1.5066	1.5839	1.6499
5500	v									0.0262	0.0463	0.0880	0.1143	0.1516	0.1825	0.2106
	h									741.3	985.0	1224.1	1349.3	1518.2	1657.0	1788.1
	s									0.9090	1.1093	1.2930	1.3821	1.4908	1.5699	1.6369

[a] Abridged from "Thermodynamic Properties of Steam," by Joseph H. Keenan and Frederick G. Keyes, John Wiley & Sons, Inc., New York, 1937.

Table 24.3 Properties of steam at high pressures
Abridged interim steam table[a]

Pressure, psia		32	100	200	300	400	500	600	700	Temperature, F 800	900	1000	1100	1200	1300	1400	1500	1600
5,500	v	0.01573	0.01588	0.01635	0.01708	0.01810	0.01955	0.02177	0.0262	0.0463	0.0880	0.1143	0.1343	0.1516	0.1674	0.1825	0.1968	0.2106
	h	16.18	82.38	180.5	280.0	381.8	488.3	603.9	741.3	985.0	1224.1	1349.3	1440.8	1518.2	1588.0	1657.0	1722.7	1788.1
	s	0.00016	0.12609	0.28735	0.42774	0.55352	0.67061	0.78510	0.9090	1.1093	1.2930	1.3821	1.4427	1.4908	1.5306	1.5699	1.6035	1.6369
6,000	v	0.01570	0.01585	0.01633	0.01705	0.01805	0.01947	0.02160	0.0257	0.0402	0.0755	0.1013	0.1207	0.1370	0.1518	0.1660	0.1794	0.1924
	h	17.61	83.66	181.6	280.9	382.5	488.6	603.1	737.2	944.9	1192.5	1330.5	1427.2	1507.7	1578.4	1648.1	1715.2	1781.8
	s	0.00010	0.12576	0.28681	0.42695	0.55241	0.66904	0.78249	0.9032	1.0744	1.2642	1.3622	1.4263	1.4765	1.5168	1.5564	1.5908	1.6247
6,500	v	0.01568	0.01583	0.01630	0.01702	0.01801	0.01941	0.02146	0.0252	0.0366	0.0656	0.0906	0.1093	0.1249	0.1390	0.1524	0.1650	0.1771
	h	19.03	84.94	182.8	281.9	383.2	489.0	602.5	733.6	918.2	1162.8	1310.5	1413.0	1496.8	1569.0	1639.8	1708.0	1775.6
	s	0.00004	0.12543	0.28627	0.42617	0.55132	0.66752	0.78004	0.8978	1.0502	1.2375	1.3425	1.4104	1.4626	1.5038	1.5441	1.5790	1.6134
7,000	v	0.01566	0.01581	0.01628	0.01699	0.01798	0.01935	0.02133	0.0248	0.0339	0.0577	0.0815	0.0997	0.1147	0.1282	0.1408	0.1528	0.1643
	h	20.45	86.22	183.9	282.9	383.9	489.4	602.0	730.4	899.4	1132.0	1290.4	1398.7	1485.8	1559.8	1631.7	1700.8	1769.4
	s	-0.00002	0.12510	0.28574	0.42540	0.55024	0.66604	0.77771	0.8930	1.0328	1.2105	1.3233	1.3951	1.4494	1.4916	1.5324	1.5679	1.6027
7,500	v	0.01563	0.01579	0.01626	0.01696	0.01794	0.01930	0.02122	0.0245	0.0319	0.0513	0.0738	0.0915	0.1060	0.1189	0.1310	0.1422	0.1532
	h	21.87	87.50	185.1	283.9	384.7	489.8	601.6	727.1	885.3	1101.1	1270.4	1384.3	1474.9	1550.7	1623.8	1693.6	1763.2
	s	-0.00008	0.12478	0.28520	0.42464	0.54918	0.66460	0.77545	0.8885	1.0194	1.1842	1.3046	1.3802	1.4366	1.4799	1.5214	1.5573	1.5926
8,000	v	0.01561	0.01577	0.01623	0.01693	0.01790	0.01924	0.02112	0.0242	0.0304	0.0458	0.0673	0.0844	0.0985	0.1108	0.1224	0.1331	0.1435
	h	23.27	88.78	186.2	284.9	385.4	490.2	601.3	724.6	875.5	1074.4	1250.8	1369.8	1463.9	1541.5	1616.6	1686.6	1757.1
	s	-0.00017	0.12446	0.28467	0.42388	0.54812	0.66317	0.77328	0.8845	1.0099	1.1614	1.2867	1.3657	1.4243	1.4686	1.5109	1.5472	1.5830
8,500	v	0.01558	0.01574	0.01620	0.01689	0.01787	0.01919	0.02103	0.0240	0.0293	0.0420	0.0617	0.0781	0.0918	0.1037	0.1148	0.1252	0.1350
	h	24.67	90.06	187.4	285.9	386.2	490.6	601.0	722.7	867.8	1052.6	1232.1	1355.2	1453.0	1532.4	1608.2	1679.9	1751.2
	s	-0.00024	0.12414	0.28414	0.42313	0.54706	0.66176	0.77117	0.8806	1.0006	1.1424	1.2698	1.3515	1.4124	1.4579	1.5009	1.5377	1.5739
9,000	v	0.01556	0.01572	0.01618	0.01686	0.01783	0.01914	0.02095	0.0238	0.0284	0.0393	0.0568	0.0726	0.0860	0.0974	0.1081	0.1181	0.1276
	h	26.07	91.33	188.5	286.9	386.9	491.0	600.8	721.0	861.8	1036.3	1213.2	1340.8	1442.0	1523.7	1600.5	1673.0	1745.6
	s	-0.00034	0.12381	0.28362	0.42238	0.54602	0.66036	0.76909	0.8775	0.9939	1.1271	1.2531	1.3378	1.4008	1.4476	1.4911	1.5284	1.5652
9,500	v	0.01554	0.01570	0.01616	0.01684	0.01780	0.01909	0.02088	0.0236	0.0279	0.0371	0.0527	0.0677	0.0807	0.0920	0.1022	0.1118	0.1209
	h	27.47	92.61	189.7	287.9	387.7	491.5	600.6	719.7	857.3	1023.8	1196.0	1326.7	1430.9	1515.5	1592.8	1666.1	1740.1
	s	-0.00042	0.12349	0.28309	0.42164	0.54499	0.65897	0.76709	0.8745	0.9886	1.1156	1.2379	1.3246	1.3895	1.4379	1.4818	1.5195	1.5570
10,000	v	0.01552	0.01568	0.01614	0.01682	0.01776	0.01904	0.02080	0.0234	0.0275	0.0353	0.0495	0.0635	0.0760	0.0872	0.0969	0.1061	0.1149
	h	28.84	93.88	190.8	288.9	388.5	491.9	600.4	718.4	853.3	1012.9	1181.0	1313.3	1419.7	1507.4	1585.5	1659.3	1734.7
	s	-0.00054	0.12317	0.28257	0.42089	0.54396	0.65759	0.76517	0.8714	0.9832	1.1050	1.2243	1.3122	1.3784	1.4286	1.4729	1.5108	1.5491

[a] The complete Interim Steam Table prepared by the Subcommittee on Interim Steam Tables of the ASME Power Division is available from the Order Department of the ASME.

psia. The specific volume of the saturated liquid v_f is 0.01663 cu ft per lb. The specific volume added by vaporization v_{fg} is 33.62 cu ft per lb. The specific volume of the saturated vapor v_g is 33.64 cu ft per lb. Note that $v_g = v_f + v_{fg}$.

Reading across to the enthalpy group of columns we find that

$$h_g = h_f + h_{fg}$$

Substituting for 200 F, this works out to

$$1145.9 \text{ Btu per lb} = 167.99 + 977.9$$

In words, the enthalpy of the saturated vapor equals the sum of the enthalpy of the saturated liquid and the enthalpy of vaporization.

We have a like relation for entropy. The entropy of the vapor equals the sum of the entropy of the saturated liquid and the entropy of vaporization: $s_g = s_f + s_{fg}$. At 200 F we see that $1.7762 = 0.2938 + 1.4824$.

Saturation table for pressure in the Keenan and Keyes tables also contains a set of columns for internal energy defined by $u_g = u_f + u_{fg}$. From the basic definition we can also calculate $u_g = h_g - pv_g/J$. Pressure would be the saturation pressure corresponding to the particular saturation temperature.

Since the steam table is based on zero enthalpy and zero entropy at 32 F and 0.8854 psia, it follows that internal energy at 32 F must be a negative quantity. This difference, however, is so small because of the low pressure that for practical engineering purposes we can usually figure internal energy at 32 F and nearby range as equal to the enthalpy.

24·18 Wet mixtures. Many steam processes involve a mixture of saturated steam and saturated liquid. The liquid usually appears as a mist or fog in these processes for relatively dry mixtures or as a liquid containing steam bubbles for relatively wet mixtures.

These mixtures are measured either by *per cent moisture y* or by *per cent quality x*. These factors are defined by

$$x = \frac{w_g}{w_g + w_f} \times 100 \qquad (24·1)$$

and

$$y = \frac{w_f}{w_g + w_f} \times 100 \qquad (24·2)$$

where w_g = weight of steam in mixture, %
w_f = weight of liquid in mixture, %

To find the properties of a mixture we must know the moisture or quality. Then any of the properties can be found by

$$v = v_g - \frac{yv_{fg}}{100} \qquad \text{cu ft per lb} \qquad (24·3)$$

VAPOR AND LIQUID PROPERTIES

$$v = v_f + \frac{xv_{fg}}{100} \quad \text{cu ft per lb} \qquad (24 \cdot 4)$$

$$h = h_g - \frac{yh_{fg}}{100} \quad \text{Btu per lb} \qquad (24 \cdot 5)$$

$$h = h_f + \frac{xh_{fg}}{100} \quad \text{Btu per lb} \qquad (24 \cdot 6)$$

$$s = s_g - \frac{ys_{fg}}{100} \quad \text{Btu per lb, F} \qquad (24 \cdot 7)$$

$$s = s_f + \frac{xs_{fg}}{100} \quad \text{Btu per lb, F} \qquad (24 \cdot 8)$$

These formulas show why the volume, enthalpy, and entropy of vaporization are given in the saturation tables. They permit rapid calculation of wet mixtures.

24·19 Compressed liquid. For modern steam-station pressures running around 2,000 psia and up, the compressed-liquid region must be treated with care in figuring cycles if great accuracy is needed. A complete steam table for high pressures gives us this directly, Table 24·3. But the accepted steam tables for pressures below 5,500 psia do not. To make this correction we use Keenan and Keyes' Table 4 for compressed liquids (not shown).

Suppose we wanted to get accurate values for water at 2,000 psia and 400 F. From Keenan and Keyes' Table 4 we would find that the saturated liquid at this temperature had $v_f = 0.018639$ cu ft per lb, $h_f = 374.97$ Btu per lb, and $s_f = 0.56638$ Btu per lb per F. For rough work we would use the saturated values. For accuracy we must apply the corrections given in the table.

At 400 F and 2,000 psia we find these items in the table:

$$(v - v_f)10^5 \qquad -19.5$$
$$(h - h_f) \qquad +2.03$$
$$(s - s_f)10^3 \qquad -4.57$$

Then the accurate values are

$$v = \frac{-19.5}{10^5} + v_f$$
$$= -0.000195 + 0.018639$$
$$= 0.018444 \text{ cu ft per lb}$$
$$h = 2.03 + h_f$$
$$= 2.03 + 374.97$$
$$= 377.00 \text{ Btu per lb}$$
$$s = \frac{-4.57}{10^3} + s_f$$
$$= -0.00457 + 0.56638$$
$$= 0.56181 \text{ Btu per lb, F}$$

For this state we see that compressed-liquid values differ in the third place from saturated-liquid state at the given temperature.

24·20 Organic vapors. Many vapors are used in modern heat cycles. Table 24·4 lists some of the properties of Biphenyl, a compound receiving attention in the nuclear-plant field. Note how the properties differ radically from those of water. Biphenyl has both higher fusion and higher critical pressures and temperatures. For some situations this may offer important physical as well as thermodynamic advantages.

Table 24·4 Thermodynamic properties of Biphenyl (Monsanto)

Temp, F t	Press., psia, p	Density, lb per cu ft		Enthalpy, Btu per lb		
		Liquid $1/v_f$	Vapor $1/v_g$	Liquid h_f	Latent h_{fg}	Vapor h_g
156.6	0.015	61.93		0.0	190.9	190.9
200	0.060	60.85	0.00108	17.7	175.0	192.7
300	0.701	58.08	0.0117	61.9	154.0	215.9
400	4.20	55.24	0.0616	112.6	148.0	260.6
500	16.3	52.32	0.236	171.0	135.0	306.0
600	47.0	49.04	0.750	234.8	117.0	351.8
700	110.1	45.38	1.730	301.5	103.0	404.5
800	221.8	40.89	3.355	369.6	91.5	461.1
900	399.7	34.58	6.850	438.4	67.5	505.9
980	607.0	19.60	19.600	493.7	0.0	493.7

24·21 Metallic vapors. See Appendix I for a discussion of sodium liquid and vapor thermodynamic properties.

REVIEW TOPICS

1. Describe the behavior of water as it changes from ice to liquid to vapor while adding heat and holding it under constant pressure.
2. Interpret the events of phase changes of water while being heated and while being cooled.
3. What is meant by enthalpy of fusion and by latent heat?
4. What are subcooled ice and subcooled liquid?
5. What is meant by saturated ice, saturated water, and saturated vapor?
6. What is the triple point of water?
7. How do vaporizing and condensing compare in meaning?
8. What is meant by sublimation of ice?
9. What is described by the critical point of a substance?

VAPOR AND LIQUID PROPERTIES

10. Rough out a *P-V* graph for steam, showing the triple point, the critical point, the saturated-liquid locus, the saturated-vapor locus, and the constant-temperature lines. Name the regions on the graph.

11. Rough out the *T-S* graph for steam, showing the triple point, the critical point, the saturated-liquid locus, the saturated-vapor locus, and the constant-pressure lines. Name the regions on the graph.

12. Rough out the *T-H* chart for steam showing the same features as for the *T-S* chart.

13. Rough out a Mollier diagram for steam showing the triple point, the critical point, the saturated-liquid locus and the saturated-vapor locus, and several constant-pressure lines.

14. What information can be found in the *Steam tables?*

15. Write the equation for per cent moisture of saturated mixtures.

16. Write the equation for per cent quality of saturated mixtures.

17. How much of a correction must be applied to saturated properties of water at different pressure levels?

PROBLEMS

1. Find the temperature, specific volume, enthalpy, and entropy of 2 lb of saturated water at 49.2 psia.

2. Find the pressure, specific volume, enthalpy, and entropy of 3 lb of saturated water at a temperature of 580 F.

3. For ½ lb saturated steam at 680 F find the pressure, specific volume, enthalpy, internal energy, and entropy.

4. For 3 lb of superheated steam at 140 psia and 1200 F find the specific volume, total volume, enthalpy, entropy, and the corresponding saturation temperature.

5. One pound of superheated steam has a specific volume of 0.4752 cu ft per lb, an enthalpy of 1295.7 Btu per lb, and an entropy of 1.4710 Btu per lb, F. What are its pressure and temperature?

6. One pound of superheated steam has a pressure of 8,500 psia and temperature of 1100 F, find the volume, enthalpy, and entropy.

7. Two pounds of Biphenyl vapor has a pressure of 47.0 psia at saturation. Find its volume and enthalpy. What is its latent enthalpy at that pressure? What is its temperature?

8. Water has a temperature of 250 F and pressure of 400 psia. What is its phase?

9. What is the phase of water at a temperature of −10 F and 0.001 psia?

10. What is the base temperature and pressure for the steam tables?

11. What is the base temperature and pressure for Biphenyl?

12. What is the maximum enthalpy for saturated steam? At what pressure and temperature?

360   BASIC THERMODYNAMICS

13. A saturated mixture of water and its vapor is at 422.6 psia pressure with a quality of 23%. Find its specific volume, temperature, enthalpy, internal energy, and entropy.

14. A saturated mixture of water and its vapor is at 680 F temperature with a moisture content of 8%. Find its specific volume, enthalpy, internal energy, and entropy.

15. A saturated mixture of water and its vapor is at 45 F temperature with a moisture content of 10%. Find its specific volume, enthalpy, internal energy, and entropy.

16. A saturated mixture of water and its vapor at 300 F has an enthalpy of 850 Btu per lb. What are its moisture and quality?

17. With the aid of Table 4 in Keenan and Keyes, *Thermodynamic properties of steam*, find the specific volume, enthalpy, and entropy of water at 3000 psia and 620 F.

CHAPTER 25

Steam Processes

Having acquainted ourselves with *Steam tables* and charts, we are ready to learn how to apply them. We use steam like a gas, as a working fluid to convert heat to work, to produce high-speed fluid jets, to convert work to heat, or to produce refrigeration, etc. We make water and its vapor pass through processes just the same as gases, but the water phase often changes, so that the process behaves differently.

25·1 Processes. Figures 25·1 to 25·6 catalog the six basic processes most often used to make up a steam cycle. They are all ideal in that they assume zero friction and radiation losses. They are reversible processes with the exception of the last two.

For each process we have the P-V graph at top, then the energy-balance diagram, next the T-S graph, and a Mollier or H-S chart. With the charts we have the basic equations for each process. None have exponential quantities because all the complex mathematics have been done in the *Steam tables* and charts.

Subscript 1 for W and Q means energy entering the process, subscript 2 means energy leaving. Subscript 1 on all properties means the initial state, and 2 the final state. The reversible processes can be carried on in the opposite order; then all we need to do is interchange the 1 and 2 subscripts to get the proper equations.

25·2 Constant pressure. In Fig. 25·1 we assume the initial state 1 as wet steam (a mixture of saturated vapor and saturated water) in a cylinder and piston. Addition of Q raises the internal energy E and expands the steam, making it do mechanical work W on the piston.

The process proceeds till steam becomes superheated at state 2. This makes the process line cross the saturated-vapor locus in all the charts.

In all three the process, of course, follows a constant-pressure locus, (see Figs. 24·4, 24·5, and 24·7).

Since the process is reversible, the area under the process on the P-V graph measures the work W, and the area under the process on the T-S graph measures heat transferred Q. End points of the process on the

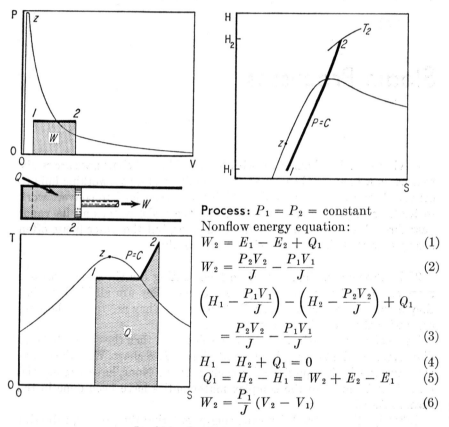

Process: $P_1 = P_2 = $ constant
Nonflow energy equation:
$$W_2 = E_1 - E_2 + Q_1 \tag{1}$$
$$W_2 = \frac{P_2V_2}{J} - \frac{P_1V_1}{J} \tag{2}$$
$$\left(H_1 - \frac{P_1V_1}{J}\right) - \left(H_2 - \frac{P_2V_2}{J}\right) + Q_1$$
$$= \frac{P_2V_2}{J} - \frac{P_1V_1}{J} \tag{3}$$
$$H_1 - H_2 + Q_1 = 0 \tag{4}$$
$$Q_1 = H_2 - H_1 = W_2 + E_2 - E_1 \tag{5}$$
$$W_2 = \frac{P_1}{J}(V_2 - V_1) \tag{6}$$

Fig. 25·1 Constant-pressure process.

Mollier chart give the corresponding enthalpies; areas have no meaning on this chart.

Equation (1) gives the energy balance for the process; (2) and (6) give the work W as measured on the P-V graph. Equation (3) relates the right-hand sides of (1) and (2) and substitutes $H = E + PV/J$. Clearing, we get (5). Once we have these properties in terms of enthalpy, we can easily solve them by finding H's in the steam tables for the particular states. The Mollier chart gives H's directly by finding the intersection of P and T or P and x properties for given states.

25·3 Constant volume. In Fig. 25·2, Q simply raises the internal energy of the steam. The vertical process line on the P-V graph shows that no work can be done because of zero area under a vertical line.

Equation (3) relates the right-hand sides of (1) and (2) and substitutes from $H = E + PV/J$. Equation (5) evaluates Q in terms of H, but (6) shows the basic simple relation in terms of E.

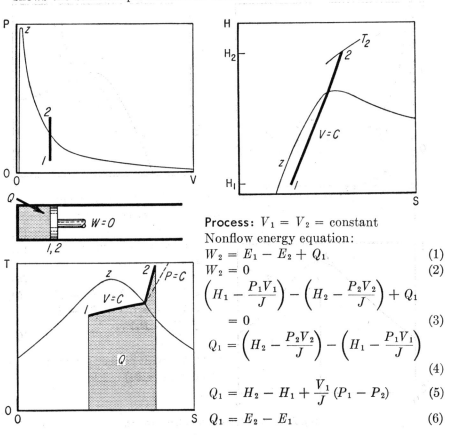

Process: $V_1 = V_2 = $ constant
Nonflow energy equation:

$$W_2 = E_1 - E_2 + Q_1 \quad (1)$$
$$W_2 = 0 \quad (2)$$
$$\left(H_1 - \frac{P_1 V_1}{J}\right) - \left(H_2 - \frac{P_2 V_2}{J}\right) + Q_1 = 0 \quad (3)$$
$$Q_1 = \left(H_2 - \frac{P_2 V_2}{J}\right) - \left(H_1 - \frac{P_1 V_1}{J}\right) \quad (4)$$
$$Q_1 = H_2 - H_1 + \frac{V_1}{J}(P_1 - P_2) \quad (5)$$
$$Q_1 = E_2 - E_1 \quad (6)$$

Fig. 25·2 Constant-volume process.

Here the steam changes from an initial wet mixture at a lower pressure to a superheated steam at a higher pressure. The process line again crosses the saturated-vapor locus. Note that process lines on the T-S and H-S graphs are more steeply inclined than for a constant-pressure process.

25·4 Constant temperature. Here again we start with wet steam and end with superheated steam, Fig. 25·3. Once all the steam becomes saturated, the charts show that its pressure drops as it becomes superheated.

Because property relations are complicated, we seldom evaluate the work W from the P-V relations. The rectangular area for Q on the T-S chart makes it easy to calculate Q. The entropy S can be easily found in the *Steam tables*, knowing P, T, and the moisture for the end states of the process.

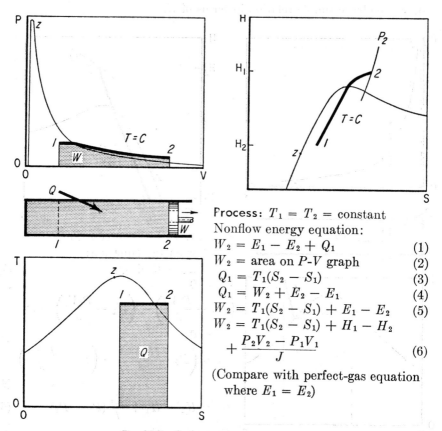

Process: $T_1 = T_2 =$ constant
Nonflow energy equation:

$$W_2 = E_1 - E_2 + Q_1 \qquad (1)$$
$$W_2 = \text{area on } P\text{-}V \text{ graph} \qquad (2)$$
$$Q_1 = T_1(S_2 - S_1) \qquad (3)$$
$$Q_1 = W_2 + E_2 - E_1 \qquad (4)$$
$$W_2 = T_1(S_2 - S_1) + E_1 - E_2 \qquad (5)$$
$$W_2 = T_1(S_2 - S_1) + H_1 - H_2$$
$$+ \frac{P_2 V_2 - P_1 V_1}{J} \qquad (6)$$

(Compare with perfect-gas equation where $E_1 = E_2$)

Fig. 25·3 Constant-temperature process.

Equation (3) gives the direct calculation of Q. Equation (4) comes from (1). Equation (5) relates the right-hand sides of (3) and (4). Expressing E's in terms of H's we get (6) to figure W.

25·5 Constant entropy. In this reversible adiabatic process, Fig. 25·4, we have no heat transferred. We assume that we start out with a relatively higher pressure superheated steam and let it expand to a lower pressure wet-steam state. Process lines again cross the saturated-vapor locus.

The vertical process line on the T-S graph ties in with $Q = 0$, since

STEAM PROCESSES 365

it has no area underneath. The vertical line on the H-S graph shows its particular utility for quickly figuring constant-entropy processes.

Equation (3) shows the basically simple relation for nonflow processes in terms of E; (4) gives it in terms of H. For steady-flow processes we get a simple evaluation for W in terms of H.

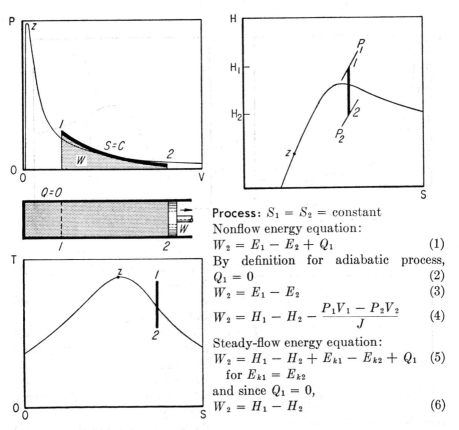

Process: $S_1 = S_2 =$ constant
Nonflow energy equation:
$$W_2 = E_1 - E_2 + Q_1 \qquad (1)$$
By definition for adiabatic process,
$$Q_1 = 0 \qquad (2)$$
$$W_2 = E_1 - E_2 \qquad (3)$$
$$W_2 = H_1 - H_2 - \frac{P_1 V_1 - P_2 V_2}{J} \qquad (4)$$

Steady-flow energy equation:
$$W_2 = H_1 - H_2 + E_{k1} - E_{k2} + Q_1 \qquad (5)$$
for $E_{k1} = E_{k2}$
and since $Q_1 = 0$,
$$W_2 = H_1 - H_2 \qquad (6)$$

Fig. 25·4 Constant-entropy process.

25·6 Irreversible adiabatic. Here we look at a steady-flow process, Fig. 25·5. It is irreversible so we draw it as a dotted line to remind us that areas underneath have no meaning. Again we start with a higher pressure superheated steam and expand it to a wet mixture at lower pressure.

On the T-S graph we define the magnitude of enthalpies H_1 and H_2. As said before, the enthalpies are measured above water at 32 F. In other words, $H = 0$ at 32 F for water. Then the enthalpy for steam at state 1 is made up of three parts: (1) The area under curve a–b–c measures the enthalpy h_f to heat the water at constant pressure P_1 to saturated water at c. (2) The area under the P_1 constant-pressure line $c-g$ measures the enthalpy of vaporization h_{fg} to vaporize the saturated water at c to saturated steam at g. (3) The area under the P_1 constant-pressure line $g-1$ measures enthalpy $H_1 - h_g$ to superheat the saturated steam to state 1. Then the total gray area under the constant-pressure line $a-c-g-1$ measures the enthalpy H_1 for state 1.

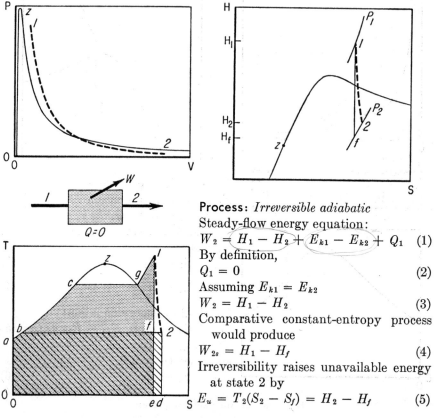

Process: *Irreversible adiabatic*
Steady-flow energy equation:
$$W_2 = H_1 - H_2 + E_{k1} - E_{k2} + Q_1 \quad (1)$$
By definition,
$$Q_1 = 0 \quad (2)$$
Assuming $E_{k1} = E_{k2}$
$$W_2 = H_1 - H_2 \quad (3)$$
Comparative constant-entropy process would produce
$$W_{2s} = H_1 - H_f \quad (4)$$
Irreversibility raises unavailable energy at state 2 by
$$E_u = T_2(S_2 - S_f) = H_2 - H_f \quad (5)$$

Fig. 25·5 Irreversible-adiabatic process.

By similar reasoning, the hatched area under the P_2 constant-pressure curve $a-b-2$ measures the enthalpy H_2 of the final state of the process. Line $1-f$ shows the constant-entropy (reversible adiabatic) process with the same initial state 1 working between the same two pressures P_1 and P_2. Assuming no change in inlet and exit kinetic energies E_k, the net area $b-c-g-1-f$ measures the net work of the reversible process. H_f is unavailable energy at pressure P_2.

For the irreversible process 1–2 the unavailable energy at P_2 increases to H_2; the hatched area $2-d-e-f$ shows the amount of increase. The net work W_1 is then the difference of the two areas 0–a–c–g–1–e and 0–a–b–2–d. This leaves no net area on the chart that can be measured directly as for the reversible constant-entropy process, so we use a dotted line to remind us of this.

In the Mollier chart we compare the irreversible adiabatic 1–2 with the constant-entropy process 1–f between the same two pressures P_1 and P_2. The increase in unavailable energy is $E_u = H_2 - H_f$.

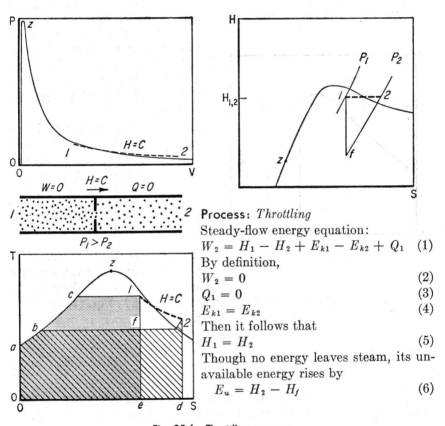

Process: *Throttling*
Steady-flow energy equation:
$$W_2 = H_1 - H_2 + E_{k1} - E_{k2} + Q_1 \quad (1)$$
By definition,
$$W_2 = 0 \quad (2)$$
$$Q_1 = 0 \quad (3)$$
$$E_{k1} = E_{k2} \quad (4)$$
Then it follows that
$$H_1 = H_2 \quad (5)$$
Though no energy leaves steam, its unavailable energy rises by
$$E_u = H_2 - H_f \quad (6)$$

Fig. 25·6 Throttling process.

25·7 Throttling. In the throttling process, Fig. 25·6, we pass a steady flow of steam through an orifice or hole from a higher to a lower pressure without any energy transfer by work or heat. This is the ultimate of the irreversible adiabatic; as for gases it is defined by $H_1 = H_2$.

The P-V graph shows that moderately wet steam can be superheated slightly by throttling; in general, wet steam can be dried a small amount by throttling.

In the T-S diagram, the gray area for H_1 equals the hatched area for

H_2. The process line 1–2 becomes nearly horizontal. Increase in unavailable energy at P_2 becomes a maximum and is measured by the hatched area $2-d-e-f$.

In the Mollier chart the process line runs horizontal, since $H_1 = H_2$. Interestingly, from complete reversibility to complete irreversibility the process line swings 90° from $1-f$ to $1-2$. Net work produced varies with the degree of irreversibility.

Now let us figure steam processes: constant pressure ($P = C$), constant volume ($V = C$), constant temperature ($T = C$), constant entropy ($S = C$), irreversible adiabatic expansion, irreversible adiabatic compression, throttling, and reversible heating. Get out your copy of Keenan and Keyes' *Thermodynamic properties of steam*, and check through the following examples.

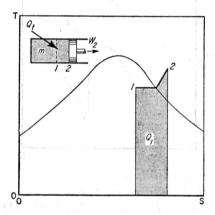

Fig. 25·7 Figuring a constant-pressure process.

The four T-S charts are all drawn to scale. Remember, temperature is *absolute* temperature in degrees Rankine, but we refer only to Fahrenheit temperature in the *Steam tables*. Zero entropy corresponds to 32 F water at 0.08854 psia.

25·8 Constant pressure. (See Fig. 25·7.) For the given example below we find Eq. (1) from Steam Table (ST) 2. The vapor and vaporization factors in Eqs. (2) to (5) we find all at the given pressure of 400 psia.

Example

Given: $P = C$ expansion of 3 lb of wet saturated steam in a cylinder and piston, initially at 400 psia with 15% moisture, finally at 600 F.

Find: T_1, H_1, E_1, V_1, S_1, H_2, E_2, V_2, S_2, Q_1, W_2, ΔE, ΔV, ΔS.

Solution: *State 1.* $P_1 = 400$ psia; $y = 0.15$.

$$T_1 = t_{sat} = 444.59 \text{ F} \tag{1}$$
$$H_1 = h_g - yh_{fg} = 1204.5 - 0.15 \times 780.5 = 1087.4 \text{ Btu per lb} \tag{2}$$
$$E_1 = u_g - yu_{fg} = 1118.5 - 0.15 \times 695.9 = 1014.1 \text{ Btu per lb} \tag{3}$$
$$V_1 = v_g - yv_{fg} = 1.1613 - 0.15 \times 1.1420 = 0.990 \text{ cu ft per lb} \tag{4}$$
$$S_1 = s_g - ys_{fg} = 1.4844 - 0.15 \times 0.8630 = 1.2945 \text{ Btu per lb, F} \tag{5}$$

STEAM PROCESSES

State 2. $T_2 = 600$ F

$H_2 = 1306.9$ Btu per lb (6)
$V_2 = 1.4770$ cu ft per lb (7)
$E_2 = h - \dfrac{P_2 V_2}{J} = 1306.9 - 400 \times 144 \times \dfrac{1.4770}{778} = 1197.5$ Btu per lb (8)
$S_2 = 1.5894$ Btu per lb, F (9)
$W_2 = \dfrac{P_1}{J}(V_2 - V_1)m = \dfrac{400 \times 144}{778}(1.4770 - 0.9900)3 = 108.1$ Btu (10)
$Q_1 = (H_2 - H_1)m = (1306.9 - 1087.4)3 = 658.5$ Btu (11)
$\Delta E = (E_2 - E_1)m = (1197.5 - 1014.1)3 = 550.2$ Btu (12)
$\Delta V = (V_2 - V_1)m = (1.4770 - 0.9900)3 = 1.461$ cu ft (13)
$\Delta S = (S_2 - S_1)m = (1.5894 - 1.2945)3 = 0.8847$ Btu per F (14)

Check: $W_2 = E_1 - E_2 + Q_1 = -550.2 + 658.5 = 108.3$ Btu (15)

The process ends at 400 psia pressure and 600 F temperature. Since saturated *t* for pressure is 444.59 F, the process ends with superheated steam. So we refer to ST 3 to find Eqs. (6) and (7). Since the superheat steam table does not have internal energy *u* (or *E* as we symbolize it) we must figure it from *h*, *P*, and *V* as in Eq. (8). Equation (9) comes directly from the tables. Equations (10) and (11) are standard equations for the process. Pressures must be stated in terms of psfa so we multiply psia by 144.

To get the total changes for 3 lb of steam we must multiply by $m = 3$ in Eqs. (12) to (14). Equation (15) is a convenient check on Eq. (10) and disagrees only by 0.2 because of slide-rule reading errors.

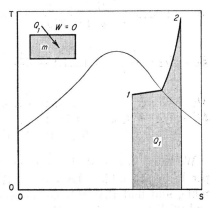

Fig. 25·8 Figuring a constant-volume process.

25·9 Constant-volume process. (See Fig. 25·8.) In the example below, Eq. (1) comes from ST 2 corresponding to 120 psia, as do the factors for Eqs. (2) to (5). Since $T_2 = 1000$ F is higher than any saturation temperature, the steam must be superheated.

Example

Given: $V = C$ heating process of 5 lb of wet steam initially at 120 psia with 30% moisture ending with temperature of 1000 F.

Find: $T_1, H_1, E_1, V_1, P_2, H_2, E_2, V_2, Q_1, W, \Delta E, \Delta V, \Delta S$.

Solution: *State 1.* $P_1 = 120$ psia; $y_1 = 0.30$.

$T_1 = 341.25$ F (1)
$H_1 = h_g - yh_{fg} = 1190.4 - 0.3 \times 877.9 = 927.0$ Btu per lb (2)
$E_1 = u_g - yu_{fg} = 1107.6 - 0.3 \times 795.6 = 868.9$ Btu per lb (3)
$V_1 = v_g - yv_{fg} = 3.7280 - 0.3 \times 3.7101 = 2.6150$ cu ft per lb (4)
$S_1 = s_g - ys_{fg} = 1.5878 - 0.3 \times 1.0962 = 1.2589$ Btu per lb, F

(5)

State 2. $T_2 = 1000$ F.

$V_2 = V_1 = 2.6150$ cu ft per lb (6)
Total $V_2 = 2.6150 \times 5 = 13.075$ cu ft (7)

At $p = 325$
$$v = 2.636 \quad h = 1524.5 \quad s = 1.7863 \tag{8}$$
At $p = 330$
$$v = 2.596 \quad h = 1524.4 \quad s = 1.7845 \tag{9}$$

$$P_2 = 330 - \frac{2.615 - 2.596}{2.636 - 2.596}(330 - 325) = 327.62 \text{ psia} \tag{10}$$

$H_2 = 1524.4$ Btu per lb (11)

$$S_2 = 1.7863 - \frac{327.62 - 325}{330 - 325}(1.7863 - 1.7845) = 1.7854 \text{ Btu per lb, F} \tag{12}$$

$$E_2 = H_2 - \frac{P_2 V_2}{J} = 1524.4 - \frac{327.62 \times 144 \times 2.615}{778} = 1365.9 \text{ Btu per lb}$$

(13)

$Q_1 = (E_2 - E_1)m = (1365.9 - 868.9)5 = 2485$ Btu (14)
$\Delta S = (S_2 - S_1)m = (1.7854 - 1.2589)5 = 2.6325$ Btu per F (15)
$\qquad W_2 = 0 \qquad \Delta V = 0 \qquad \Delta E = Q_1$ (16)

We must now find the state at $T_2 = 1000$ F and $V_2 = 2.6150$ cu ft per lb in the tables. Look down the 1000 F columns in the superheat table, and find that this lies between 325 and 330 psia. Equations (8) and (9) list properties for these two pressures at 1000 F. To find the final pressure we must interpolate between v values as in Eq. (10).

Since there is only 0.1-Btu difference between the two pressures, we can easily estimate h as in Eq. (11). In Eq. (12) we interpolate to find final entropy. Since the table does not list internal energy we find it from H_2 and P_2V_2 as in Eq. (13). Adding heat to a constant-V process affects only the internal energy as figured in Eq. (14). The total entropy change must take into account the total steam mass $m = 5$ lb, as in Eq. (15). By definition W and ΔV must be zero as shown in Eq. (16). Notice the curvatures of the constant-volume line on the T-S chart.

25·10 Constant-temperature process.

(See Fig. 25·9.) In the example below, Eq. (1) comes from ST 2 corresponding to 1,200 psia, as do the factors in Eqs. (2) to (5). When we look up 300 psia, we find saturation temperature of 417.33 F; so we know the steam must be superheated in state 2. In ST 3 we find temperatures listed at 560 and 580 F; so we must interpolate between these two states at a pressure of 300 psia, as listed in Eqs. (7) and (8). Equation (9) finds the H_2, Eq. (10) the S_2, and Eq. (11) the V_2. In Eq. (12) we find E_2 from the enthalpy equation. Q_1 is most easily found from the change in entropy as in Eq. (13). Remember to put temperature in degrees Rankine absolute by adding 460. The work output W_2 is found from the nonflow energy equation as in Eq. (16). Changes in S and V are found in Eqs. (17) and (18).

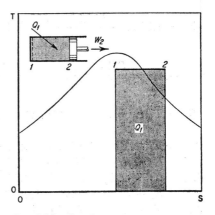

Fig. 25·9 Figuring a constant-temperature process.

Example

Given: $T = C$ expansion of 6 lb of wet steam initially at 1,200 psia and 50% moisture ending at 300 psia.

Find: T_1, H_1, E_1, V_1, S_1, T_2, H_2, E_2, V_2, S_2, Q_1, W_2, ΔE, ΔV, ΔS.

Solution: *State 1.* $P_1 = 1{,}200$ psia; $y_1 = 0.5$.

$T_1 = 567.22$ F (1)

$H_1 = h_g - y_1 h_{fg} = 1183.4 - 0.5 \times 611.7 = 877.5$ Btu per lb (2)

$E_1 = u_g - y_1 u_{fg} = 1103.0 - 0.5 \times 536.3 = 834.8$ Btu per lb (3)

$V_1 = v_g - y_1 v_{fg} = 0.3619 - 0.5 \times 0.3396 = 0.1921$ cu ft per lb (4)

$S_1 = s_g - y_1 s_{fg} = 1.3667 - 0.5 \times 0.5956 = 1.0689$ Btu per lb, F (5)

State 2. $P_2 = 300$ psia; $T_2 = T_1 = 567.22$ F (6)

At $T = 560$

$v = 1.9128 \quad h = 1292.5 \quad s = 1.6054$ (7)

At $T = 580$

$v = 1.9594 \quad h = 1303.7 \quad s = 1.6163$ (8)

$H_2 = 1292.5 + \dfrac{567.22 - 560}{580 - 560}(1303.7 - 1292.5) = 1296.5$ Btu per lb (9)

$$S_2 = 1.6054 + \frac{567.22 - 560}{580 - 560}(1.6163 - 1.6054)$$
$$= 1.6093 \text{ Btu per lb, F} \qquad (10)$$
$$V_2 = 1.9128 + \frac{567.22 - 560}{580 - 560}(1.9594 - 1.9128)$$
$$= 1.9296 \text{ cu ft per lb} \qquad (11)$$
$$E_2 = H_2 - \frac{P_2 V_2}{J} = 1296.5 - \frac{144 \times 300 \times 1.9296}{778}$$
$$= 1109.3 \text{ Btu per lb} \qquad (12)$$
$$Q_1 = T_1(S_2 - S_1)m = (567.22 + 460)(1.6093 - 1.0689)6 = 3330 \text{ Btu} \qquad (13)$$
$$\Delta E = E_2 - E_1 = 1109.3 - 834.8 = 274.5 \text{ Btu per lb} \qquad (14)$$
$$\Delta H = H_2 - H_1 = 1296.5 - 877.5 = 419.0 \text{ Btu per lb} \qquad (15)$$
$$W_2 = (Q_1 - \Delta E)m = (555 - 274.5)6 = 1683 \text{ Btu} \qquad (16)$$
$$\Delta S = S_2 - S_1 = 1.6093 - 1.0689 = 0.5404 \text{ Btu per lb, F} \qquad (17)$$
$$\Delta V = V_2 - V_1 = 1.7375 \text{ cu ft per lb} \qquad (18)$$

25·11 Constant-entropy process. (See Fig. 25·10.) In the example below we look at both a nonflow and steady-flow process between the same two end states. Items (1) to (4) we find in the superheat table corresponding to the pressure and temperature, except that E_1 must be figured from the enthalpy equation. The second state is wet steam, since the pressure is so low; so in Eq. (5) we list the saturated-liquid, saturated-vapor, and vaporization properties corresponding to 2 psia as found in ST 2. In Eq. (6) we set up the basic equation for entropy in the wet region and as the known quantity that stays constant in the process.

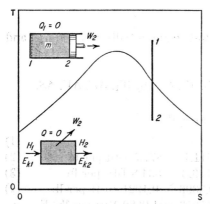

Fig. 25·10 Figuring a constant-entropy process.

Example

Given: Nonflow and steady-flow $S = C$ expansion of 10-lb steam from an initial pressure of 200 psia and a temperature of 800 F to a final pressure of 2 psia. In steady flow, assume $E_{k1} = E_{k2}$.

Find: H_1, E_1, V_1, S_1, T_2, y_2, H_2, E_2, V_2, S_2, ΔE, ΔH, ΔS, ΔV, Q_1, W_2.

Solution: *State* 1. $P_1 = 2000$ psia; $T_1 = 800$ F.

$$H_1 = 1335.5 \text{ Btu per lb} \qquad (1)$$
$$V_1 = 0.3074 \text{ cu ft per lb} \qquad (2)$$

$$E_1 = h - \frac{P_1 V_1}{J} = 1335.5 - \frac{144 \times 2000 \times 0.3074}{778}$$
$$= 1221.6 \text{ Btu per lb} \quad (3)$$
$$S_1 = 1.4576 \text{ Btu per lb, F} \quad (4)$$

State 2. $P_2 = 2$ psia; $T_2 = 126.08$ F.
At 2 *psia*,

$$\begin{aligned} s_f &= 0.1749 & s_{fg} &= 1.7451 & s_g &= 1.9200 \\ h_f &= 93.99 & h_{fg} &= 1022.2 & h_g &= 1116.2 \\ u_f &= 93.98 & u_{fg} &= 957.9 & u_g &= 1051.9 \\ v_f &= 0.016 & v_{fg} &= 173.71 & v_g &= 173.73 \end{aligned} \quad (5)$$

$$S_2 = S_1 = s_g - y_2 s_{fg} \quad (6)$$
$$y_2 = \frac{s_g - S_1}{s_{fg}} = \frac{1.9200 - 1.4576}{1.7451} = 0.265 \quad (7)$$
$$H_2 = h_g - y_2 h_{fg} = 1116.2 - 0.265 \times 1022.2 = 845.3 \text{ Btu per lb} \quad (8)$$
$$E_2 = u_g - y_2 u_{fg} = 1051.9 - 0.265 \times 957.9 = 798.0 \text{ Btu per lb} \quad (9)$$
$$V_2 = v_g - y_2 v_{fg} = 173.73 - 0.265 \times 173.71 = 127.7 \text{ cu ft per lb} \quad (10)$$
$$\Delta E = (E_1 - E_2)m = (1221.6 - 798.0)10 = 4236 \text{ Btu} \quad (11)$$
$$\Delta H = (H_1 - H_2)m = (1335.5 - 845.3)10 = 4902 \text{ Btu} \quad (12)$$
$$\Delta S = (S_1 - S_2)m = (1.4576 - 1.4576)10 = 0 \text{ Btu per F} \quad (13)$$
$$\Delta V = (V_1 - V_2)m = (0.3074 - 127.7)10 = -1274 \text{ cu ft} \quad (14)$$
$$Q_1 = 0 \text{ Btu (by definition)} \quad (15)$$
$$\text{Nonflow } W_2 = \Delta E = 4236 \text{ Btu} \quad (16)$$
$$\text{Steady-flow } W_2 = \Delta H = 4902 \text{ Btu} \quad (17)$$

In Eq. (7) we solve for y_2 and substitute factors from the steam table, (5). Knowing y_2, we solve for the other properties in Eqs. (8) to (10). In Eqs. (11) to (14) we find the total change in properties for the process for 10 lb of steam. By definition there is no transfer of heat in a constant-entropy process as in Eq. (15). Nonflow work depends on the change in internal energy, Eq. (16). The steady-flow work depends on the change in enthalpy, Eq. (17), and is larger than the nonflow work by the amount of change in the flow work.

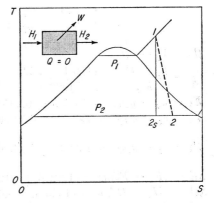

Fig. 25·11 Figuring an irreversible-adiabatic expansion process.

25·12 Irreversible-adiabatic expansion. (See Fig. 25·11.) For the example given below of this process, we find the answers to Eqs. (1) to (4) in the *Steam tables*, figuring item (3) as shown in the constant-entropy process of Art. 25·11. For this present process the expansion takes the steam into the wet region; so we know that the final pressure fixes the final temperature (saturation).

Example

Given: Steady-flow expansion of 10 lb of steam from an initial pressure and temperature of 2000 psia and 800 F to a final pressure of 2 psia at an expansion efficiency of 75%. In steady flow, assume $E_{k1} = E_{k2}$.

Find: $\Delta E, \Delta H, \Delta S, \Delta V, Q, W_2$.

Solution: *State* 1. $P_1 = 2000$ psia; $T_1 = 800$ F.

$$H_1 = 1335.5 \text{ Btu per lb} \tag{1}$$
$$V_1 = 0.3074 \text{ cu ft per lb} \tag{2}$$
$$E_1 = 1221.6 \text{ Btu per lb} \tag{3}$$
$$S_1 = 1.4576 \text{ Btu per lb, F} \tag{4}$$

State 2. $P_2 = 2$ psia; $T_2 = 126.08$ F.

$$H_{2s} = 845.3 \text{ Btu per lb} \tag{5}$$

$$\text{Expansion eff} = \frac{\text{actual work}}{\text{ideal work}} = \frac{H_1 - H_2}{H_1 - H_{2s}} \tag{6}$$

$$0.75 = \frac{1335.5 - H_2}{1335.5 - 845.3} \tag{7}$$

$$H_2 = 1335.5 - 0.75 \times 490.2 = 967.9 \text{ Btu per lb} \tag{8}$$

At 2 psia,

$$\begin{array}{lll} h_f = 93.99 & h_{fg} = 1022.2 & h_g = 1116.2 \\ s_f = 0.1749 & s_{fg} = 1.7451 & s_g = 1.9200 \\ u_f = 93.98 & u_{fg} = 957.9 & u_g = 1051.9 \\ v_f = 0.016 & v_{fg} = 173.71 & v_g = 173.73 \end{array} \tag{9}$$

$$H_2 = h_g - y_2 h_{fg} \tag{10}$$

$$y_2 = \frac{h_g - H_2}{h_{fg}} = \frac{1116.1 - 967.9}{1022.2} = 0.1451 \tag{11}$$

$$E_2 = u_g - y_2 u_{fg} = 1051.9 - 0.1451 \times 957.9 = 912.9 \text{ Btu per lb} \tag{12}$$

$$V_2 = v_g - y_2 v_{fg} = 173.73 - 0.1451 \times 173.71$$
$$= 148.5 \text{ cu ft per lb} \tag{13}$$

$$S_2 = s_g - y_2 s_{fg} = 1.9200 - 0.1451 \times 1.7451$$
$$= 1.6668 \text{ Btu per lb, F} \tag{14}$$

$$\Delta E = (E_1 - E_2)m = (1221.6 - 912.9)10 = 3087 \text{ Btu} \tag{15}$$
$$\Delta H = (H_1 - H_2)m = (1335.5 - 967.9)10 = 3676 \text{ Btu} \tag{16}$$
$$\Delta S = (S_2 - S_1)m = (1.6668 - 1.4576)10 = 2.092 \text{ Btu per F} \tag{17}$$
$$\Delta V = (V_2 - V_1)m = (148.5 - 0.3074)10 = 1{,}482 \text{ cu ft} \tag{18}$$
$$Q = 0 \text{ by definition} \tag{19}$$
$$W_2 = \Delta H = 3676 \text{ Btu} \tag{20}$$

STEAM PROCESSES

Before we can find state 2, we must first find the end state for a constant-entropy process between the same end pressures. We do this by the method shown in the $S = C$ process to find item (5), giving us the enthalpy H_{2s}. The subscript s means at constant entropy.

With this in hand we can use Eq. (6) to find the actual end state H_2 as in Eqs. (7) and (8). Next we look up the data in item (9) from the *Steam tables*, and by the moisture formulas we find the actual properties as in Eqs. (10) to (14). In Eq. (15) we find the change in internal energy ΔE for 10 lb of steam; this would be the mechanical work W developed in a nonflow expansion. Equations (16) to (18) give the change in other properties for 10 lb of steam. Equation (20) gives the mechanical work developed for the steady-flow process.

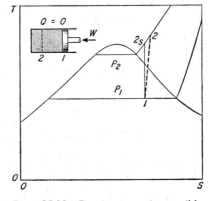

Fig. 25·12 Figuring an irreversible-adiabatic compression process.

25·13 Irreversible-adiabatic compression. This process, Fig. 25·12, can be used where a small amount of h-p steam is needed when the available boiler produces only l-p steam. An air compressor can be adapted for this work. Our example has a pressure range larger than usual.

Example

Given: Nonflow adiabatic compression of 2 lb of saturated steam at 120 psia with 80% quality to a final pressure of 1,700 psia at 75% compression efficiency.

Find: $T_2, \Delta E, \Delta S, W_1, Q$.

Solution: *State 1.* $P_1 = 120$ psia; $x_1 = 0.8$; $T_1 = 341.2$F; $E_1 = u_f + x_1 u_{fg} = 312.05 + 0.8 \times 795.6 = 948.5$ Btu per lb   (1)

$S_1 = s_f + x_1 s_{fg} = 0.4916 + 0.8 \times 1.0962 = 1.3686$ Btu per lb, F   (2)

$S_{2s} = S_1 = 1.3686$ Btu per lb, F   (3)

At 1,700 psia and $S_{2s} = 1.3686$,

$T_{2s} = 650$ F $H_{2s} = 1214.4$ Btu per lb $V_{2s} = 0.2755$ cu ft per lb
(4)

$$E_{2s} = H_{2s} - \frac{P_2 V_{2s}}{J} = 1214.4 - \frac{144 \times 1700 \times 0.2755}{778}$$
$$= 1127.8 \text{ Btu per lb} \quad (5)$$

$$\text{Ideal } W = E_{2s} - E_1 = 1127.8 - 948.5 = 179.3 \text{ Btu per lb} \quad (6)$$

$$\text{Compression eff} = \frac{\text{ideal } W}{\text{actual } W} = \frac{E_{2s} - E_1}{E_2 - E_1} \quad (7)$$

$$0.75 = \frac{1127.8 - 948.5}{E_2 - 948.5} \quad (8)$$

$$E_2 = \frac{1127.8 + 948.5(0.75 - 1.00)}{0.75} = 1187.6 \text{ Btu per lb} \quad (9)$$

$$\Delta E = (E_2 - E_1)m = (1187.6 - 948.5)2 = 478.2 \text{ Btu} \quad (10)$$

$$W = \Delta E = 478.2 \text{ Btu} \qquad Q = 0 \text{ by definition} \quad (11)$$

At 1,700 psia and 720 F,

$$E = 1288.4 - \frac{144 \times 1{,}700 \times 0.3283}{778} = 1185.1 \text{ Btu per lb} \quad (12)$$

At 1,700 psia and 740 F,

$$E = 1305.8 - \frac{144 \times 1{,}700 \times 0.3410}{778} = 1198.5 \text{ Btu, lb} \quad (13)$$

$$T_2 = 720 + \frac{1187.6 - 1185.1}{1198.5 - 1185.1}(740 - 720) = 723.7 \text{ F} \quad (14)$$

$$S_2 = 1.4333 + \frac{1187.6 - 1185.1}{1198.5 - 1185.1}(1.4480 - 1.4333)$$

$$= 1.4360 \text{ Btu per lb, F} \quad (15)$$

$$\Delta S = (S_2 - S_1)m = (1.4360 - 1.3686)2 = 0.1348 \text{ Btu per F} \quad (16)$$

Our process starts with wet steam; so Eqs. (1) to (3) find the initial properties. Again we must first figure a constant-entropy process. Searching the superheat tables at 1,700 psia, we find that at $T_{2s} = 650$ F, $s = 1.3686$; so we can read off the other properties easily as in item (4). Usually we do not have things this easy and must interpolate. We find the internal energy E_{2s} as in Eq. (5). Equation (6) figures the ideal work needed for compression between pressure levels.

With these data developed we use Eqs. (7) to (9) to find the actual final internal energy. Equations (10) and (11) figure the work and heat transfer for the actual process. Now we find it a little more complicated to figure the other final properties. We must interpolate between 720 and 740 F, using internal energies as proportioning factors. Since the superheat table does not list the internal energy, we must first figure it for each given temperature in Eqs. (12) and (13). Equation (14) gives us the final actual temperature, 73.7 F higher than the ideal compression. Equation (15) gives the final actual entropy and Eq. (16) the total increase in entropy for 2 lb of steam being compressed to 1,700 psia pressure.

25·14 Throttling process. Here, Fig. 25·13, we study the effects of throttling (a) superheated steam, (b) slightly wet steam, (c) very wet steam, and (d) saturated water between the same pressure levels. The

STEAM PROCESSES

definition of this process is $H_1 = H_2$. It is completely irreversible with zero work and heat transfer. In item (1), in the example below, we find H_1 from the superheat tables. Items (3) and (4) bracket the final enthalpy at 14.7 psia. Equation (5) interpolates this basic data to find the final temperature of the still superheated steam.

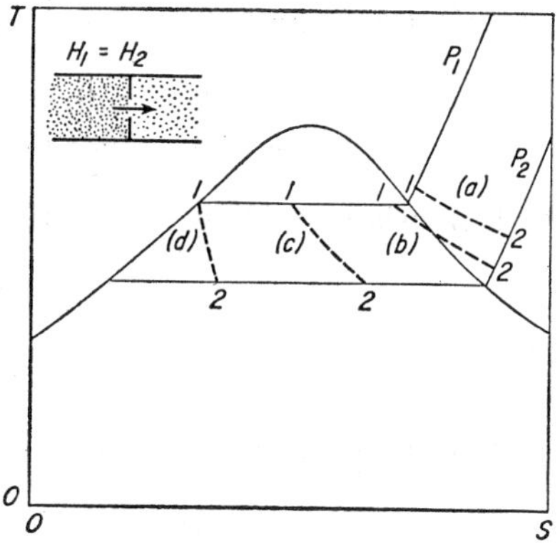

Fig. 25·13 Figuring a throttling process.

Example

Given: (a) Steam at 500 psia and 500 F, (b) steam at 500 psia with 4% moisture, (c) steam at 500 psia with 50% moisture, (d) saturated water at 500 psia. Each is throttled to a final pressure of 14.7 psia.

Find: H_2, T_2, and y_2 for each individual process.

Solutions:

(a) At 500 psia and 500 F:

$$H_1 = 1231.3 \text{ Btu per lb} \tag{1}$$
$$H_2 = H_1 = 1231.3 \text{ Btu per lb} \tag{2}$$

At 14.7 psia and 1230.5 Btu per lb

$$T = 380 \text{ F} \tag{3}$$

At 14.7 psia and 1239.9 Btu per lb

$$T = 400 \text{ F} \tag{4}$$

$$T_2 = 380 + \frac{1231.3 - 1230.5}{1239.9 - 1230.5} \times 20 = 381.7 \text{ F} \tag{5}$$

(b) At 500 psia and 4% moisture:

$$H_1 = h_g - y_1 h_{fg} = 1204.4 - 0.04 \times 755.0 = 1174.2 \text{ Btu per lb} \tag{6}$$
$$H_2 = H_1 = 1174.2 \text{ Btu per lb} \tag{7}$$

At 14.7 psia and 1173.8 Btu per lb

$$T = 260 \text{ F} \tag{8}$$

At 14.7 psia and 1183.3 Btu per lb

$$T = 280 \text{ F} \tag{9}$$

$$T_2 = 260 + \frac{1174.2 - 1173.8}{1183.3 - 1173.8} \times 20 = 260.8 \text{ F} \tag{10}$$

(c) At 500 psia and 50% moisture:

$$H_1 = 1204.4 - 0.5 \times 755.0 = 826.9 \text{ Btu per lb} \quad (11)$$
$$H_2 = H_1 = 826.9 \text{ Btu per lb} \quad (12)$$

At 14.7 psia and 826.9 Btu per lb,

$$y_2 = \frac{h_g - H_1}{h_{fg}} = \frac{1150.4 - 826.9}{970.3} = 0.3335 \quad (13)$$
$$T_2 = 212 \text{ F—saturation temp for } 14.7 \text{ psia} \quad (14)$$

(d) For 500-psia saturated water:

$$H_1 = H_f = 449.4 \text{ Btu per lb} \quad (15)$$
$$H_2 = H_1 = 449.4 \text{ Btu per lb} \quad (16)$$

At 14.7 psia and 449.4 Btu per lb,

$$y_2 = \frac{h_g - H_1}{h_{fg}} = \frac{1150.4 - 449.4}{970.3} = 0.723 \quad (17)$$
$$T_2 = 212 \text{ F—saturation temp for } 14.7 \text{ psia} \quad (18)$$

With slightly wet steam we find H_1 from the moisture formula as in Eq. (6). Then by interpolation in the superheat tables we find the final steam temperature in Eqs. (8) to (10). This example shows that, if we expand slightly wet steam through a great enough pressure range, we end up with superheated steam. This method is used experimentally to determine steam moisture in some testing methods.

When we start with really wet steam as in Eq. (11), we simply end up with somewhat drier steam at the lower pressure. Initial 50% moisture dries to 33.35%.

In Eq. (15) we list the enthalpy for saturated water; this can be looked up directly in the saturated-steam tables. When we expand it to 14.7 psia, Eq. (17) shows that we end up with a very wet steam—72.3% moisture. In other words, 27.7% of the original saturated water flashes to steam at the lower pressure and temperature. The remainder stays as saturated water at the lower end conditions. Note trend of the constant-enthalpy lines on the T-S chart. They tend to level off with increasing entropy.

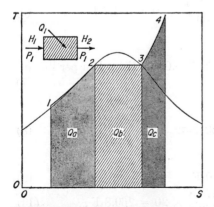

Fig. 25·14 Figuring a reversible heating process.

25·15 Reversible heating process. This is a process typical of a steam boiler and superheater, Fig. 25·14.

STEAM PROCESSES

Feedwater is usually a subcooled liquid. If the pressure is relatively high, we have to take the subcooling into account to get accurate results. Some authorities recommend that at pressures below 400 psia we can ignore subcooling and assume that the saturated-steam table will give us accurate enough data; that is, take enthalpies and other properties corresponding to actual temperature as good enough. To get accurate data at pressures above 400 psia we have to use ST 4—Compressed Liquid.

Example

Given: Subcooled water, state 1, at 1,500 psia and 140 F, heated at constant pressure to state 4 as superheated steam at 1000 F.

Find: Heat added (a) to raise compressed liquid to saturation temperature, (b) to vaporize saturated liquid to saturated steam, (c) to superheat steam to 1000 F. Also (d) find Q_1, ΔV, ΔS from state 1 to 4.

Solution: (a) At 140 F,

$h_f = 107.89$ Btu per lb $P_{1s} = 2.889$ psia $v_f = 0.01629$ cu ft per lb
$s_f = 0.1984$ Btu per lb, F   (1)

At 1,500-psia pressure			Interpolation	
Temp	100 F	200 F	140 F	
$(v - v_f)10^5$	−7.5	−8.1	−7.7	(2)
$(h - h_f)$	+3.99	+3.36	+3.74	
$(s - s_f)10^3$	−0.86	−1.79	−1.23	

$H_1 = h_f + 3.74 = 107.89 + 3.74 = 111.63$ Btu per lb   (3)

$V_1 = \dfrac{v_f - 7.7}{10^5} = 0.01629 - 0.000077 = 0.01621$ cu ft per lb   (4)

$S_1 = \dfrac{s_f - 1.23}{10^3} = 0.1984 - 0.00123 = 0.1972$ Btu per lb, F   (5)

At 1,500 psia saturation pressure,

$H_2 = H_f = 611.6$ Btu per lb   (6)
$Q_a = H_2 - H_1 = 611.6 - 111.6 = 500$ Btu per lb to heat water   (7)

(b) At 1,500 psia saturated pressure,

$H_3 = 1167.9$ Btu per lb   (8)
$Q_b = H_3 - H_2 = 1167.9 - 611.6 = 556.3$ Btu per lb to evaporate water   (9)

(c) At 1,500 psia and 1000 F,

$H_4 = 1490.1$ Btu per lb $V_4 = 0.5390$ cu ft per lb
$S_4 = 1.6001$ Btu per lb, F   (10)
$Q_c = H_4 - H_3 = 1490.1 - 1167.9 = 322.2$ Btu per lb to superheat steam   (11)

(d)
$$Q_1 = Q_a + Q_b + Q_c = H_4 - H_1 = 1490.1 - 111.63 = 1378.5 \text{ Btu per lb} \tag{12}$$
$$\Delta V = V_4 - V_1 = 0.5390 - 0.01621 = 0.5228 \text{ cu ft per lb} \tag{13}$$
$$\Delta S = S_4 - S_1 = 1.6001 - 0.1972 = 1.4029 \text{ Btu per lb, F} \tag{14}$$

Equation (1) in our example gives us the data we read from the saturated-steam table corresponding to saturation temperature of 140 F. Equation (2) extracts the pertinent data from ST 4 and interpolates to get the values at 140 F. Equations (3) to (5) apply these data to correct the saturation data to actual subcooled state 1 properties. Equation (6) comes directly from the saturated-steam table corresponding to the pressure. Equation (7) figures the heat needed to raise the subcooled water to saturated condition.

Equations (8) and (9) give the energy needed to evaporate saturated water to saturated steam; this can be read directly as h_{fg} from the saturated-steam pressure table.

Equations (10) and (11) give properties at the end state and figure the energy needed to heat the steam from saturated state to superheated state. Equations (12) to (14) give us the total energy added (and the change in specific volume and entropy) to heat the subcooled water to the final state as superheated steam.

REVIEW TOPICS

1. Draw corresponding constant-pressure lines on P-V, T-S, and Mollier charts for steam. Derive the equations for work and heat of reversible (and irreversible) constant-pressure processes.

2. Draw corresponding constant-volume lines on P-V, T-S, and Mollier charts for steam. Derive the equations for work and heat transferred in reversible constant-volume processes.

3. Draw corresponding constant-temperature lines on P-V, T-S, and Mollier charts for steam. Derive the equations for work and heat transferred in reversible constant-temperature processes.

4. Draw corresponding constant-entropy lines on P-V, T-S, and Mollier charts for steam. Derive the equations for work and heat transferred in reversible constant-entropy processes.

5. Draw corresponding irreversible-adiabatic expansion lines on P-V, T-S, and Mollier charts for steam. Derive the equations for work and heat transferred in irreversible-adiabatic expansion processes.

6. Draw corresponding irreversible-adiabatic compression lines on P-V, T-S, and Mollier charts for steam. Derive the equations for work and heat transferred in irreversible-adiabatic compression processes.

7. Draw corresponding throttling-process lines on P-V, T-S, and Mollier charts for steam.

PROBLEMS

1. One pound of water under 200 psia constant pressure and initially at 150 F is heated to 500 F. Find the work and heat transferred during the process.

2. One pound of steam at 350 psia and 550 F follows a nonflow constant-pressure process until the steam quality becomes 60%. Find (a) W, (b) change in enthalpy, (c) heat transferred, (d) change in internal energy, (e) change in entropy.

3. Two pounds of steam under a pressure of 2 in. Hg absolute with a quality of 89% condenses to a saturated liquid at constant pressure. Find the heat transferred, the change in entropy, and the change in total volume.

4. A tank holds 2 lb of steam at 215 psia and 500 F. When the steam is cooled until its pressure drops to 15 psia, find (a) the final moisture, (b) the change in enthalpy, (c) the change in internal energy, (d) the heat transferred, (e) the work transferred, (f) the change in entropy.

5. A saturated mixture of steam and water with 15% moisture is initially at 350 F. It is heated at constant temperature until its pressure drops to atmospheric. Find the heat transferred during the process and the work done.

6. Four pounds of steam at 600 psia and 500 F expands through a nonflow constant-entropy process to a final pressure of 300 psia. Find (a) the final moisture, (b) the final temperature, (c) the change in H, (d) the change in E, (e) the work transferred, (f) the heat transferred.

7. One pound of steam at 40 psia and 400 F is compressed at constant temperature until the moisture is 60%. Find the final pressure, work transferred, and heat transferred.

8. In steady flow, steam expands at constant entropy from 200 psia and 500 F to 65 psia with negligible change in kinetic energy. For 1 lb of steam find (a) the final temperature, (b) the final entropy, (c) the final specific volume, (d) the heat transferred, (e) the work transferred.

9. Five pounds of steam passes through an adiabatic steady-flow process from 1,500 psia and 700 F to a final pressure of 25 psia at an expansion efficiency of 60%. With zero change in kinetic energy, find (a) the change in enthalpy, (b) the change in specific volume, (c) the heat transferred, (d) the work transferred, (e) the change in entropy.

10. One pound of saturated steam with 70% moisture is compressed in a nonflow process adiabatically from 100 to 500 psia at a compression efficiency of 75%. Find (a) the final temperature, (b) the change in internal energy, (c) the change in enthalpy, (d) the work transferred, (e) the heat transferred.

11. Steam at 400 psia pressure with 5% moisture throttles to 25 psia. Find (a) H_2, (b) T_2, and (c) y_2.

CHAPTER 26

Steam Cycles

26·1 Rankine cycle. We have sharpened our understanding of steam processes in the last chapter. Now let us see how steam processes can be applied in power-production cycles. The simplest one and oldest is the Rankine cycle, Fig. 26·1. Here we have a simple boiler supplying steam

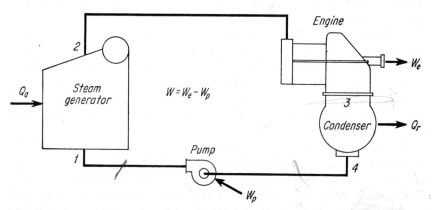

Fig. 26·1 Rankine cycle assumes the simplest type of power plant with a steam generator, engine or turbine, condenser, and feedwater pump. The ideal cycle assumes that there are no other auxiliaries.

to an engine (a turbine in this case); the turbine exhausts its steam to a condenser. Condensed steam (condensate) is then pumped back to the boiler as feedwater.

We must have many more auxiliaries to run an actual plant of this type, but as an ideal standard we consider only what happens to the water and steam acting as the working fluid in this steady-flow closed cycle. Reduced to this simple idea we find that energy enters and leaves the cycle at only four points.

To heat the water and vaporize it to steam at constant pressure, we have an energy input Q_a to the boiler. This successively changes the phase of the working fluid at state 1 from a subcooled liquid to a saturated liquid to a mixture of saturated liquid and vapor to a dry saturated vapor and perhaps to a superheated vapor at state 2. Then

$$Q_a = H_2 - H_1 \qquad (26\cdot1)$$

for this steady-flow constant-pressure process.

The steam with H_2 then flows as a wet or dry saturated steam or as superheated steam to the turbine. In the turbine the steam expands adiabatically at constant entropy to a lower pressure at state 3. During the expansion it produces mechanical work W_e, the useful output of the cycle. Then

$$W_e = H_2 - H_3 \qquad (26\cdot2)$$

for this steady-flow process.

At state 3 the steam as a wet saturated vapor enters the condenser to transfer its unavailable energy Q_r to the cooling water. In this way the energy leaves the closed cycle. The condensed vapor falls as a saturated liquid to state 4 in the bottom of the condenser, the condenser hot well. For this constant-pressure steady-flow cooling process

$$Q_r = H_3 - H_4 \qquad (26\cdot3)$$

The pump draws the condensate from the hot well at state 4 to pressurize it to state 1 for reinjection into the boiler as a subcooled liquid. The pump needs a mechanical energy input W_p to work on the water. This input to the cycle as a constant-entropy adiabatic compression is then

$$W_p = H_1 - H_4 \qquad (26\cdot4)$$

26·2 Thermal efficiency. Since this is a closed cycle, all energy entering must equal all energy leaving:

$$Q_a + W_p = W_e + Q_r \qquad (26\cdot5)$$

Rearranging, we get

$$W_e - W_p = Q_a - Q_r \qquad (26\cdot6)$$

The net work output of the cycle can then be written as

$$W = W_e - W_p = Q_a - Q_r \qquad (26\cdot7)$$

The efficiency of the Rankine cycle will then be

$$e = \frac{W}{Q_a} = \frac{Q_a - Q_r}{Q_a} = 1 - \frac{Q_r}{Q_a} = 1 - \frac{H_3 - H_4}{H_2 - H_1} \qquad (26.8)$$

The thermal efficiency of the cycle depends on the enthalpy of water and steam at the four states among the components of the plant. These can be chosen as needed for a particular situation of the power plant.

26·3 T-S graphs. These graphs, Figs. 26·2 to 26·4, are standard tools for describing steam-plant cycles. In Fig. 26·2 we have a simple Rankine cycle where the boiler produces dry saturated steam for the turbine. At state 1 the boiler receives the subcooled water and heats it to saturated water at state a. For low pressures, under about 400 psia, the constant-pressure line is practically coincident with the saturated-vapor locus. From a to state 2 the boiler vaporizes the saturated water to dry saturated steam. The area under 1–a is the heat added to bring the water to the saturated-liquid phase; the area under a–2 is the heat added to vaporize the saturated water to steam. The total gray area under 1–a–2 is Q_a, the heat added to the cycle.

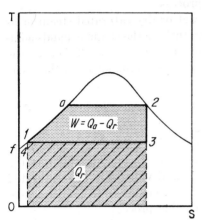

Fig. 26·2 Rankine cycle using dry saturated steam from boiler is now seldom used.

The constant-entropy process 2–3 is the steam expansion in the turbine to produce mechanical work W_e. Since it is adiabatic, no heat transfer is involved.

The constant-pressure cooling process 3–4 is the condensing of the steam in the condenser. The area under 3–4 measures the heat rejected by the working fluid, Q_r. This is the unavailable energy and is marked by the hatched area. The toe of the diagram marked 4–1 is the constant-entropy compression of the water by the pump input W_p. For low pressures this appears as a common point and W_p is almost zero.

The areas Q_a and Q_r have three common boundaries, and since this is an ideal cycle, the net remaining area is $Q_a - Q_r$. In Eq. (26·7) we have shown that this equals the net work output of the cycle for heat input Q_a.

Another way to interpret these charts is to remember that the total area under f–1–a–2 is H_2 and the area under f–1 is H_1. Similarly the total area under f–4–3 is H_3 and under f–4 is H_4. We showed before that $H_2 - H_1 = W_e$ and $H_3 - H_4 = Q_r$.

In Fig. 26·4 the cycle uses an h-p boiler; here g–1–2 is the boiler-pressure locus and the area underneath g–1 is H_1, the enthalpy of the subcooled water discharged by the pump entering the boiler. The area under f–4 is H_4, the enthalpy of saturated water entering the pump from the condenser hot well. Since $W_p = H_1 - H_4$, the net area f–g–1–4 measures W_p. Note particularly that in Figs. 26·2 to 26·4 all the net areas enclosed in the cycle processes represent the net work of the W of the cycle and not the gross work of the engine W_e. The latter equals the area of W plus the net pump work area f–g–1–4.

26·4 Superheating. Superheating, Fig. 26·3, of the steam in the boiler or steam generator brings the cycle processes out of the "steam dome" on the chart. The area under 1–a again measures the heat to raise the subcooled liquid to the saturated state; the area under a–b measures the heat to vaporize the saturated liquid; the area under b–2 measures the heat to superheat the saturated steam; the total gray area, of course, measures the heat input Q_a to the boiler, and therefore the cycle.

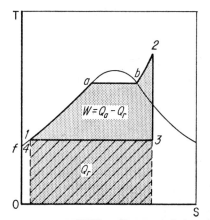

Fig. 26·3 Rankine cycle using superheated steam has a drier exhaust leaving the engine.

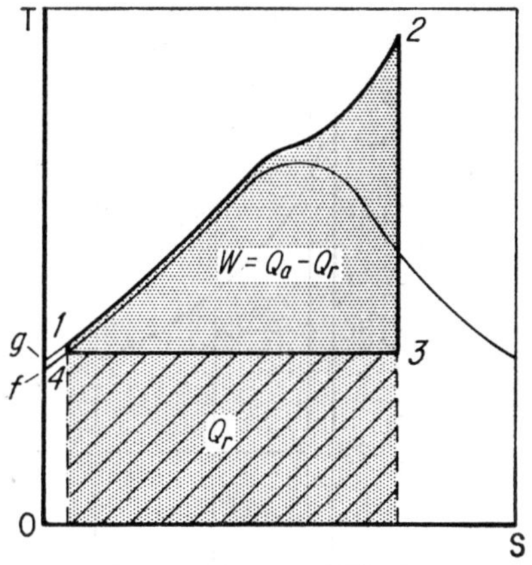

Fig. 26·4 Supercritical-pressure Rankine cycle has a triangular-shaped net work output area.

26·5 Superpressure. Superpressure, Fig. 26·4, means that the boiler pressure or engine-inlet pressure is above the critical 3,206.2 psia. The constant-pressure heating process 1–2 lies entirely above the saturation "steam dome." To be strictly consistent the steam at state 2 should not be called superheated because there is no saturation temperature corresponding to a superpressure. But it is becoming common practice to call any steam temperature superheated when it is above the critical 705.4 F.

The net work area begins to assume the shape of a triangle for this pressure level. The pump work area, as we discussed before, is outside the net work area.

The thermal efficiency of the Rankine cycle as in all thermodynamic cycles depends on the temperatures of heat addition and heat rejection. High-temperature addition and low-temperature rejection aid in increasing efficiency. The heat rejection is usually at constant pressure and temperature, the temperature primarily depending on the pressure that can be maintained at exhaust by the condenser.

The heat addition, however, is usually at a variable temperature, so it is the *average* temperature that determines the efficiency. For instance, in Figs. 26·2 to 26·4 all the Q_a's start at about the same low temperature T_1, but they progressively have higher end temperatures T_2. Too, all three have the same Q_r and $T_3 = T_4$. So the thermal efficiencies of the

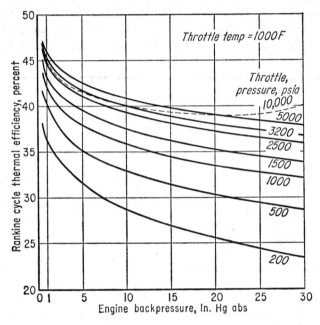

Fig. 26·5 Thermal efficiency of the Rankine cycle with a constant throttle temperature drops with increasing back pressure on the engine exhaust.

three cycles increase from Figs. 26·2 to 26·4. The increasing areas of the W's show this for these ideal cycles.

26·6 Back pressure. Back pressure, Fig. 26·5, has a marked effect on Rankine cycle thermal efficiency. The magnitude, though, varies with the throttle pressure, that is, the inlet pressure to the turbine or engine. With a constant throttle temperature of 1000 F and 200-psia throttle pressure, cycle thermal efficiency drops from 36.3 to 23.3% as back pressure rises from 1 to 30 in. Hg absolute. But at 5,000 psia the thermal efficiency drops from 46 to 37.7%. The drops are 13.0 against 8.3%. Back-pressure change has less effect with higher throttle pressures.

The increasing slopes of the constant-pressure curves with lowering back pressure show the high advantage of maintaining low back pressures on steam engines and turbines. This can be achieved by minimizing the temperature differential between cooling water and exhaust steam.

26·7 Throttle *P-T*. This, Fig. 26·6, has a controlling effect on Rankine-cycle efficiency. Low back pressures were achieved early in modern power-plant development, minute gains being made in the last decade. But continuing efforts are being made to harness higher and higher throttle pressures and temperatures. Figure 26·6 shows why.

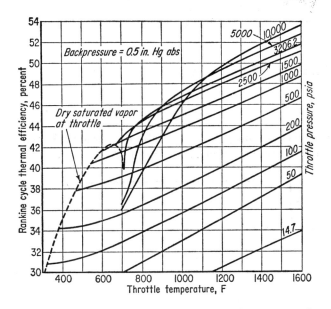

Fig. 26·6 Rankine-cycle thermal efficiency varies with throttle pressure and temperature. It generally rises with increasing temperature.

The dashed line shows how efficiency varies for cycles using dry saturated steam as the throttle temperature increases. The efficiency reaches a maximum at around 660 F and then drops sharply as it approaches the critical temperature of 705.4 F.

All the constant-pressure lines show that increasing temperatures raise the thermal efficiency. However, we cannot say that increasing pressures also consistently raise the efficiency unless we specify throttle temperatures above 1130 F. In the higher temperatures, note the decreasing incremental return in efficiency with rising pressure.

26·8 Steam rate. An operating convenience giving a quick performance index of a steam engine is the steam rate SR. Knowing the total pounds of steam that passed through an engine to produce a given kilowatt hour output, we get

$$SR = \frac{\text{total steam flow, lb}}{\text{total shaft output, kwhr}} \tag{26.9}$$

Since there are 3413 Btu in each kwhr, the steam rate is related to the work done per pound W as

$$SR = \frac{3413}{W} \tag{26.10}$$

26·9 Heat rate. Another way of measuring cycle performance is the heat rate which states the Btu input for each unit of output. This is

$$HR = SR(H_2 - H_1) \tag{26.11}$$

The heat rate and thermal efficiency are related by

$$e = \frac{3413}{SR(H_2 - H_1)} = \frac{3413}{HR} \tag{26.12}$$

For practice, get out your steam tables and slide rule and check some of the figures on the curves of Figs. 26·5 and 26·6. Do not forget to account

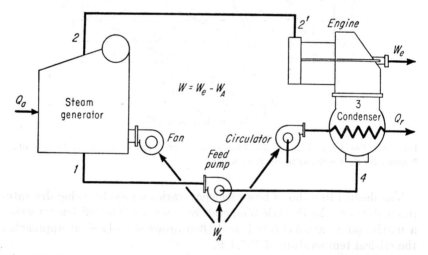

Fig. 26·7 Practical Rankine cycle needs an array of auxiliaries that consume energy and reduce the net work output of the plant. The engine and boiler have losses larger than the ideal cycle indicates.

for the compressed-liquid correction wherever it amounts to more than 1 Btu. This you can get from Keenan and Keyes' Table 4. Your Mollier chart can be used to advantage to find H_3—some of them you will have to figure from the saturated steam tables because the chart is not wide enough to include the lower entropy values.

As you probably recognized, we seldom use the Rankine cycle in our modern plants, except for the tiny ones. The cycle is valuable, though, in giving a fundamental introduction to the factors that control the efficiency of a steam cycle. These must be understood before we study the more advanced forms used today.

26·10 The practical Rankine cycle. This needs many auxiliaries, each consuming energy. Figure 26·7 suggests just a few of the more important.

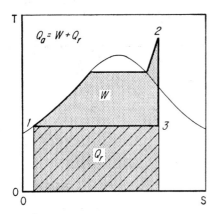 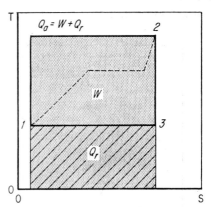

Fig. 26·8 Ideal Rankine cycle has a net W output that equals the difference of Q_a and Q_r.

Fig. 26·9 Ideal Carnot cycle with the same T_2 and Q_r as the Rankine cycle proves to be more efficient.

These reduce the net output W and so the thermal efficiency of the cycle. In addition to this, none of the cycle machinery works as well as the ideal. So all actual losses are greater than the theoretical, further depressing overall cycle thermal efficiency.

26·11 Carnot comparison. While the Rankine cycle is a theoretical ideal, let us not forget it is *not* the best ideal. The Carnot cycle still defines the best performance you can expect for given limits of heat added and rejected:

$$\text{Thermal eff} = \frac{T_a - T_r}{T_a} \tag{26·13}$$

The Rankine ideal falls considerably short of the Carnot. Figure 26·8 shows the T-S graph of an ideal Rankine cycle with heat addition at temperatures from T_1 to T_2 and heat rejection of unavailable energy at T_3.

An ideal Carnot cycle would operate with all heat addition at T_2, the highest in the cycle. Figure 26·9 shows how a Carnot cycle would compare with the ideal Rankine cycle of Figure 26·8. Both have the same T_2 and the same Q_r. The gray areas, of course, measure heat added Q_a. This is much greater for the Carnot cycle, but note that the additional Q_a for the Carnot all converts to useful work W. The dotted line in Fig. 26·9 shows the comparison of the two W's. This makes the Carnot more efficient thermally.

The advanced steam cycles we shall study later all strive to "square off" the T-S graph to approach Carnot thermal efficiency. We might ask again why an actual Carnot engine was not developed because of its superior theoretical performance. Most attempts in this direction showed that the engine would need an extraordinarily long stroke to achieve performance. Practical factors so far have made it uneconomical to develop, but attempts under way might just be successful eventually.

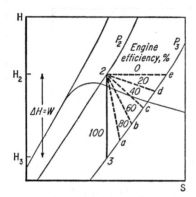

Fig. 26·10 Mollier chart shows the effect of engine efficiency on the exhaust enthalpy with constant initial state.

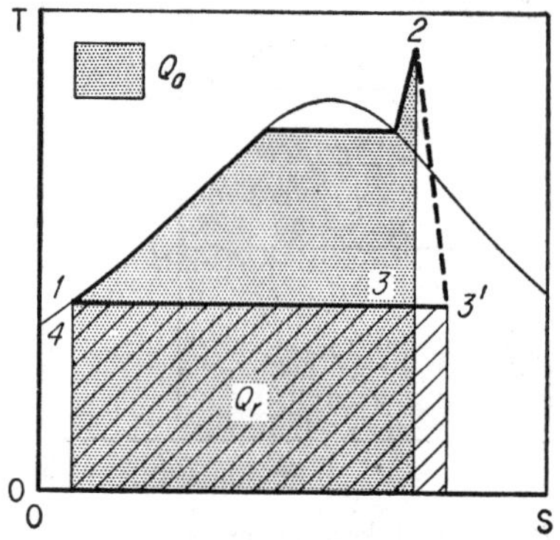

Fig. 26·11 Irreversible expansion of steam in the Rankine engine increases the heat rejected Q_r.

26·12 Engine efficiency. The efficiency of the Rankine cycle depends largely on the performance of the engine. The engine takes in steam with enthalpy H_2, converts some of the energy to work output W_e, and rejects the remainder to the condenser as enthalpy H_3 of the steam.

The Mollier chart, Fig. 26·10, shows that ideally the steam would expand at constant entropy S from state 2 to state 3. Then work output would be

$$W_e = H_2 - H_3 \quad \text{Btu per lb} \quad (26·14)$$

But because of irreversibilities in the engine, all this theoretical conversion is not realized and some internal energy remains in the steam exhausting from the engine. The actual H_3 might be H_a or H_b, H_c, etc., all higher than H_3.

The actual performance depends on many practical factors of design and loading. To know how close a machine runs to ideal, we use the engine efficiency:

$$e_e = \frac{H_2 - H_{3'}}{H_2 - H_3} \quad (26·15)$$

where H_3 is the ideal exhaust enthalpy of the steam and $H_{3'}$ the actual exhaust enthalpy. Figure 26·10 shows how engine efficiency varies the expansion line of the steam on the Mollier chart. Note at zero engine efficiency that $H_2 = H_e$. This is a throttling process where all the energy remains in the steam while it flows to a lower pressure.

26·13 Actual Rankine efficiency. Now let us study what engine inefficiency does to the performance of a cycle. Figure 26·11 shows the T-S graph for an ideal cycle excepting the engine, which has an irreversible expansion (at increasing entropy) from 2 to 3′. The ideal expansion would end at state 3.

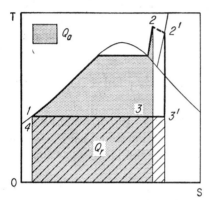

Fig. 26·12 Pressure drop of steam from the boiler to the engine raises the unavailable heat Q_r.

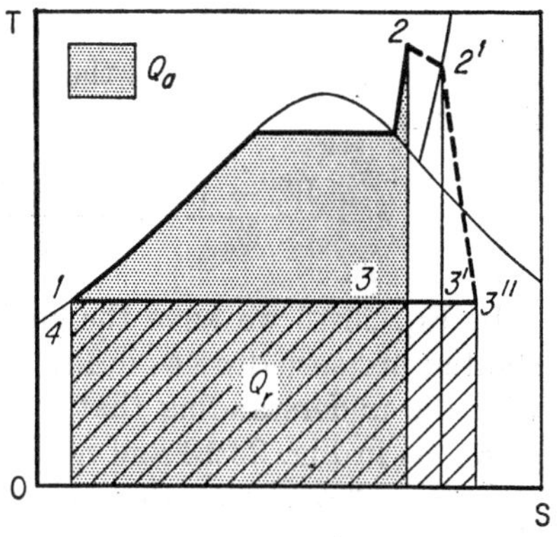

Fig. 26·13 Expansion irreversibility and pressure drop make the unavailable heat Q_r grow cumulatively.

Here we assume that heat is rejected reversibly at constant pressure in the condenser from state 3′ to state 4. The area underneath measures the heat rejected Q_r. The gray area under 1–2 measures the heat added to the cycle Q_a. The difference of the two areas measures the net work output W. But since the two areas do not coincide on three sides, there is no net definable area as in the ideal cycle that we can label W. The actual W is less than the upper gray area by the hatched area under 3–3′.

As engine efficiency drops, the hatched area Q_r grows and there is less and less net W produced. For zero engine efficiency, or pure throttling, $H_2 = H_3'$, Q_r would be equal to the gray area, and, of course, $W = 0$. This drops the cycle thermal efficiency to zero.

26·14 Steam-pressure drop. When steam flows from the steam generator to the engine inlet, it must overcome pipe friction and so drops in pressure. Pressure drop causes a loss in cycle thermal efficiency. In Fig. 26·12 we have steam leaving the steam generator at state 2. It drops in pressure to state 2′. Since this is a throttling process (usually with no heat loss) $H_{2'} = H_2$, there is a slight drop in temperature too. If the steam expanded in an ideal engine, it would exhaust with enthalpy $H_{3'}$. This is larger than the ideal enthalpy H_3 when steam expands from the steam generator outlet at state 2. Q_r increases, and the pressure drop lowers the thermal efficiency of the cycle.

Figure 26·13 shows the combination of pressure drop and irreversible expansion in the engine. Each acts independently on cycle efficiency—the effects of both add together, raising Q_r and correspondingly subtracting from net output W. These graphs show the need for careful design to achieve acceptable thermal efficiency.

26·15 Actual efficiency. For design needs, the theoretical efficiency is first computed and then efficiency factors applied to estimate actual performance. The theoretical thermal efficiency of the Rankine cycle can be figured as

$$e = \frac{W_e - W_p}{Q_a} = \frac{Q_a - Q_r}{Q_a} = 1 - \frac{Q_r}{Q_a} = 1 - \frac{H_3 - H_4}{H_2 - H_1} \quad (26\cdot16)$$

In Eq. 26·15 we have the engine efficiency e_e. But if we have a substantial pressure drop as in Fig. 26·13, the equation would have to be used with the understanding that, while $H_{2'} = H_2$, the ideal expansion ends at $H_{3'}$ as in Fig. 26·13 and not at H_3.

As Fig. 26·7 shows, there is additional, auxiliary energy consumption that must be accounted for to arrive at overall cycle thermal efficiency. Also, the steam generator does not get all the energy it releases by chemical combustion into the working fluid. Much of the energy disappears up the stack to the atmosphere. The ratio of the energy transmitted to the working fluid to the chemical energy in the fuel is the boiler or steam-generator efficiency:

$$e_b = \frac{H_2 - H_1}{F(HHV)} = \frac{Q_a}{F(HHV)} \quad (26\cdot17)$$

where e_b = boiler efficiency, decimal
H_2 = enthalpy of steam leaving boiler, Btu per lb
H_1 = enthalpy of feedwater entering boiler, Btu per lb
F = lb fuel burned per lb steam evaporated
HHV = higher heating value of fuel, Btu per lb fuel

Transposing Eq. 26·17 and recognizing that the fuel energy input per pound of working fluid is the overall cycle input, we get, in Btu per pound of steam,

$$\text{Station input} = \frac{Q_a}{e_b} \quad (26\cdot18)$$

Then if we let W_a represent all the auxiliary energy consumption of the plant cycle, we can figure the overall actual thermal efficiency for a plant as

$$e' = \frac{e_e W_e - W_a}{Q_a/e_b} = \frac{e_e(H_{2'} - H_{3'}) - W_a}{(H_2 - H_1)/e_b} \quad (26 \cdot 19)$$

To convert the efficiency from a decimal to per cent, multiply by 100.

We can compare energy flow for ideal and practical cycles as in Figs. 26·14 and 26·15. In Fig. 26·14 the ideal input Q_a loses its major portion to unavailable energy Q_r and a small part W_p goes to run the feed pump. This leaves W, the net output of the cycle.

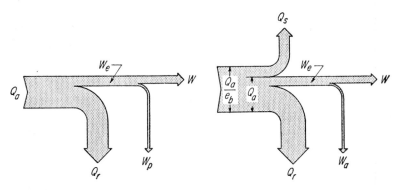

Fig. 26·14 Ideal energy-flow diagram shows the net W left after rejection of heat Q_r and feed pump input W_p.

Fig. 26·15 Actual energy flow shows losses up the stack, greater inherent plant losses, and auxiliary energy consumption greater than W_p only.

Figure 26·15 diagrams the flow for a practical cycle, where the plant input Q_a/e_b is diminished by the boiler losses (mostly stack loss) Q_s to leave Q_a, the energy entering the steam. In turn Q_a is disbursed to unavailable energy Q_r and auxiliary work W_a, leaving the net output W. In the actual cycle Q_r and W_a are larger than in the ideal by the amount caused by irreversibilities of actual equipment.

26·16 Availability loss. Since we release our input energy in the form of high-temperature fire, the Rankine cycle has more serious basic losses than we have studied so far. The true temperature of energy addition is at fire temperature in the furnace and may range from 2000 F upward. But the limitations of metal in strength and economics keep most practical steam temperatures to about 1050 F; 1200 F is considered highly experimental.

Figure 26·16 suggests the great loss we take in availability when the temperature drops from fire to working fluid. At the low end of the cycle we do pretty well. A temperature difference of about 10 F to transfer Q_r from steam to cooling water cannot be much improved.

26·17 Reheat cycle.

Engineering strives to make all its materials work to best advantage. In power generation this means trying to convert most of all heat inputs to mechanical-work output. We studied the Rankine cycle, Fig. 26·17, and found what we could expect in the way of performance. But as we pointed out, this is a basic cycle and could stand much improvement.

One way we can improve the cycle is by finding a way to raise the average temperature of heat addition Q_a. Figure 26·18 shows one means of doing this. Full-pressure throttle steam enters the engine as in the

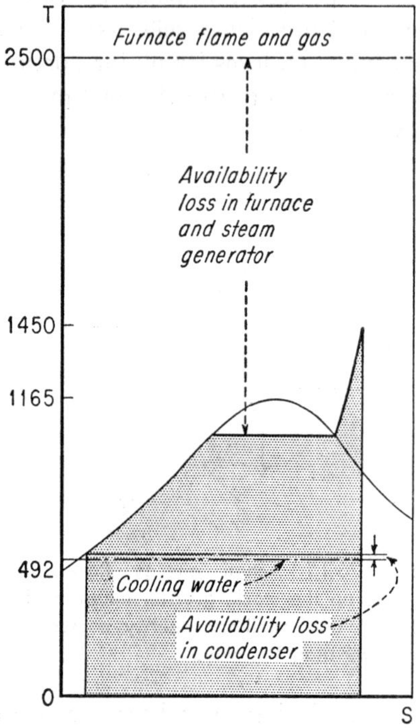

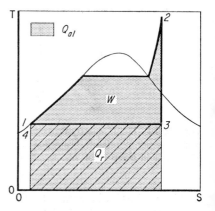

Fig. 26·16 High-temperature drop in the boiler furnace offers big cycle-efficiency gain for future development.

Fig. 26·17 Rankine cycle uses a minimum number of steam processes to produce work output.

Rankine cycle, but the steam is only partly expanded, 2–3. Steam at the lower pressure then flows back to the steam generator through a superheater to be reheated, usually back to its throttle temperature, 3–4. At 4, reheated steam enters the second or l-p section of the engine and expands to the back pressure held by the condenser at 5.

The cycle input consists of two parts, Q_{a1} and Q_{a2}. In most engines and turbines the output of the cycle will be through one shaft as W_e. In all other respects, the reheat cycle is the same as the basic Rankine cycle.

The pressure at which the steam is reheated can be varied. Figure 26·19 shows the steam expanded through a relatively small pressure drop 2–3. In the ideal cycle, steam is reheated at constant pressure from 3 to 4. It then expands at constant S to state 5 in the condenser.

The gray area under 1–2 measures the heat input to the boiler and main superheater, Q_{a1}. The darker area under 3–4 measures the heat input to the l-p reheat steam, Q_{a2}. The reheater is usually a second superheater, taking its heat from the furnace gases of the boiler that also supply Q_{a1}.

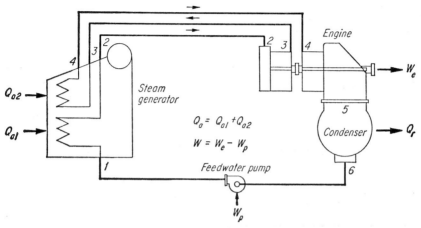

Fig. 26·18 Reheat cycle takes steam partly expanded in the first section of the engine back to the steam generator to be reheated. Then steam flows through the second engine section to the condenser to complete the expansion.

By varying the pressure of state 3 we can control the average temperature at which Q_{a2} is added to the cycle. For instance, in Fig. 26·20 state 3 is at a lower pressure and Q_{a2} is larger than in Fig. 26·19, but the average temperature of the addition is lower because of the wider temperature range during the addition.

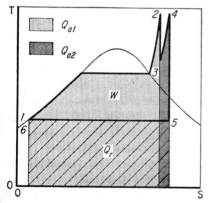

Fig. 26·19 Reheat cycle has partly expanded steam reheated in a steam generator from 3 to 4.

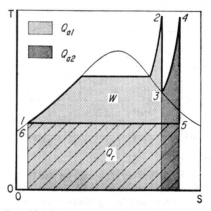

Fig. 26·20 Reheat cycle can be varied in the amount of expansion used in each of the engine sections.

To get the best reheating pressure we have to find the balance where the additional heat and its average temperature become optimum.

In Figs. 26·19 and 26·20 for the ideal reheat cycle, the net area enclosed by heavy process lines measures the net mechanical-work output W. As usual, the crosshatched area shows the entire unavailable heat Q_r rejected from the cycle. The area to the left of 6–1 paralleling the saturated-liquid line measures the pump work input W_p; this area becomes significant only at high pressure ranges for the cycle.

26·18 Cycle thermal efficiency. This is figured as the ratio of output to input as for other cycles. Referring to Figs. 26·18 to 26·20 we get

$$e = \frac{W}{Q_a} = \frac{W_e - W_p}{Q_{a1} + Q_{a2}} = \frac{(H_2 - H_3) + (H_4 - H_5) - W_p}{(H_2 - H_1) + (H_4 - H_3)} \quad (26\cdot20)$$

where the H's correspond to states given in the figures.

26·19 Reheat pressure. This influences the overall thermal efficiency of the cycle, as we have just discussed. Figure 26·21 shows how thermal efficiency varies with reheat pressure for five different throttle pressures. In all curves both the throttle and reheat temperatures are constant at 1000 F and exhaust at 1 in. Hg absolute. At unity pressure ratio of reheat to throttle pressure we have the same cycle as the Rankine; that is, reheating is actually zero.

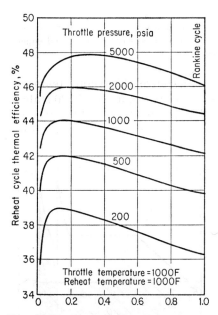

Fig. 26·21 Thermal-efficiency improvement by reheating depends on the initial and reheat pressures used in the cycle. Curves based on 1-in. Hg abs back pressure.

As we lower the reheat pressure (and H_3), the overall cycle thermal efficiency steadily rises until it reaches a maximum at pressure ratios varying from about 0.12 to 0.3, depending on the throttle-pressure level. Further pressure reduction drops the cycle efficiency sharply toward zero.

Theoretically, reheating has more to offer at the lower throttle pressures than at the higher, but significant gains can be made at all levels. In practice, in the more complicated cycles we shall discuss later, reheating is more often used at the higher pressures.

26·20 Multiple reheating. A few power-plant cycles running at superpressures are using two stages of reheating because of the gain in overall cycle thermal efficiency. Figure 26·22 shows a theoretical super-pressure cycle with two stages of reheating. Three separate inputs, Q_{a1}, Q_{a2}, and Q_{a3}, power the cycle. Adding "saw-tooth" edges to the top of the diagram adds additional heat at a high average temperature which raises the cycle efficiency.

The point of diminishing returns may be reached quickly with added reheat stages because of the complication of equipment. Added reheaters must be installed in the steam generator, the engine needs additional sectionalization, and more

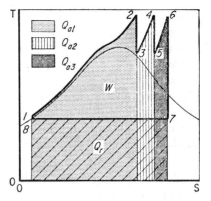

Fig. 26·22 Multiple reheat stages raise the theoretical cycle efficiency to higher levels.

piping must be installed to hook up the more complex cycle. Even the thermal-efficiency formula becomes more complicated:

$$e = \frac{(H_2 - H_3) + (H_4 - H_5) + (H_6 - H_7) - W_p}{(H_2 - H_1) + (H_4 - H_3) + (H_6 - H_5)} \quad (26·21)$$

26·21 Irreversibilities. As in all cycles, irreversibilities decrease thermal efficiency of the reheat cycle. The prime loss results from the irreversibility of steam expansion in the engine. Figure 26·23 shows its effects. Expansion through the h-p section of the engine is from 2 to 3' at increasing S. We assume that reheating takes place reversibly from 3' to 4. Then steam expands irreversibly from 4 to 5' in the l-p section of the engine. The total energy rejected Q_r is then the total hatched area under 5'–6.

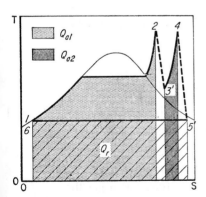

Fig. 26·23 Irreversible expansion in engine sections reduces the reheat-cycle efficiency performance.

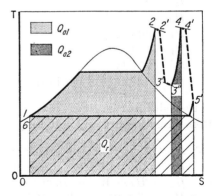

Fig. 26·24 Pressure drops in lines between the steam generator and the engine reduce the reheat-cycle efficiency.

The net work is not described by a single area but is the difference between the sum of the areas under 1–2 and 3′–4 less Q_r. This will be considerably less than the two gray areas above 5′–6.

Considerable pressure drop has to be built into reheaters to ensure that steam fills all the parallel tubes and prevent the metal from overheating. Pressure drop through connecting piping is unavoidable; the practical reheat cycle has these irreversibilities in addition to those of engine expansion. Figure 26·24 shows these effects. The net result increases Q_r at the expense of decreasing W.

26·22 Figuring reheat cycle performance. Find thermal efficiency of reheat cycle with:

Throttle pressure = 2,000 psia
Reheat pressure = 400 psia
Throttle and reheat temperatures = 1000 F
Condenser pressure = 1 in. Hg abs
Engine efficiencies of h-p and l-p engines = 80%

Using symbols of Fig. 26·23,

P_2 = 2,000 psia
T_2 = 1000 F
S_2 = 1.5603
H_2 = 1474.5 Btu per lb (from steam tables)

Enthalpy at 400 psia after constant-S expansion from H_2, read from Mollier chart = 1276.8 Btu per lb = H_3.

$H_2 - H_3$ = 1474.5 − 1276.8 = 197.7 Btu per lb
$H_2 - H_{3'}$ = 197.7 × 0.8 = 158.2 Btu per lb = W_{e1}
$H_{3'} = H_2 - W_{e1}$ = 1474.5 − 158.2 = 1316.3 Btu per lb
P_4 = 400 psia
T_4 = 1000 F
S_4 = 1.7623
H_4 = 1522.4 Btu per lb (from steam tables)

Enthalpy at 1 in. Hg absolute after constant-S expansion from H_4, read from Mollier chart = 947.4 Btu per lb.

$H_4 - H_5$ = 1522.4 − 947.4 = 575.0 Btu per lb
$H_4 - H_{5'}$ = 575.0 × 0.8 = 460.0 Btu per lb = W_{e2}
$H_{5'} = H_4 - W_{e2}$ = 1522.4 − 460.0 = 1062.4 Btu per lb
H_6 = 47.1 Btu per lb at 1 in. Hg abs
W_p = 5.5 Btu per lb (from compressed-liquid table)
$H_1 = H_6 + W_p$ = 47.1 + 5.5 = 52.6 Btu per lb

$$e = \frac{(H_2 - H_{3'}) + (H_4 - H_{5'}) - W_p}{(H_2 - H_1) + (H_4 - H_{3'})}$$

$$= \frac{(1474.5 - 1316.3) + (1522.4 - 1062.4) - 5.5}{(1474.5 - 52.6) + (1522.4 - 1316.3)}$$

$$= \frac{158.2 + 460.0 - 5.5}{1421.9 + 206.1} = \frac{612.7}{1628.0} = 0.3766$$

or a cycle thermal efficiency of 37.66%.

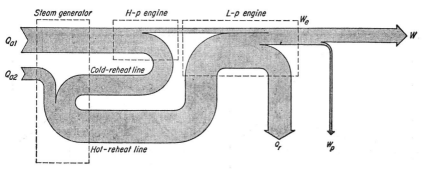

Fig. 26·25 Energy-flow diagram for the reheat cycle shows Q_{a1} as the largest part of the total input. The cold reheat line carries the major share of energy leaving the h-p engine. When engine sections have a common shaft, l-p engine output augments h-p engine output to make the total work output W_e.

26·23 Regenerative steam cycle. We have already learned of the limitations of the Rankine cycle in approaching the performance of the ideal Carnot cycle. A steady-flow steam cycle can duplicate Carnot performance, within limits, by adapting some of the tricks of regenerative reciprocating-engine cycles, see Art. 11-3.

Figure 26·26 shows the Rankine-cycle layout modified for regenerative

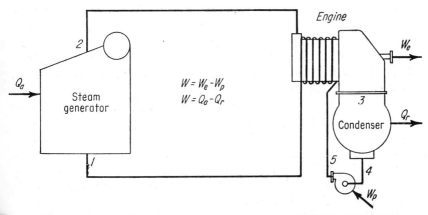

Fig. 26·26 Theoretical regenerative cycle features reversible heat exchange between condensate returning to steam generator and steam expanding reversibly in the engine.

feedwater heating. The modification features an ideal reversible heat-transfer surface between the steam expanding in the engine and the feedwater returning from the condenser to the steam generator. Ideally, this means that water entering the steam generator has the same temperature as steam entering the turbine.

26·24 Cycle processes. Figure 26·27 shows the effect of the heating process on feedwater and steam expansion in the engine. The chart

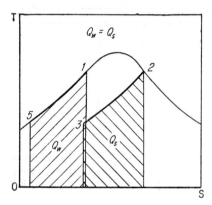 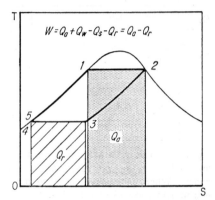

Fig. 26·27 Heat Q_w absorbed by water equals heat Q_s given up by expanding steam in the engine.

Fig. 26·28 Saturated regenerative steam cycle works at Carnot efficiency under given ideal conditions.

assumes that the steam leaving the generator is dry and saturated. The heat given up by the steam Q_s equals the heat absorbed by the water Q_w. So the entropy gain of the water equals the entropy loss of the steam. This means that the area under the water heat gain 5–1 or Q_w equals the area under the steam heat loss 2–3 or Q_s and has exactly the same shape. Process lines 5–1 and 3–2 are parallel.

Thus a sizable amount of energy continuously stays within and circulates in the cycle. This makes the steam exhausting to the condenser at state 3 very wet—about 75% moisture.

Figure 26·28 shows the complete cycle of processes. Water enters the generator at state 1 as a saturated liquid. Q_a simply adds the heat of vaporization without the need for subcooled-liquid heating, and it is all added at constant temperature, a requirement for achieving Carnot thermal efficiency. The steam expands polytropically to a very wet state, as we noted. Then heat is rejected as Q_r in the condenser; because of the wet state, Q_r is relatively small.

Taking into account all the areas under the processes, heat added to water and steam = $Q_w + Q_a$. Heat given up by steam = $Q_s + Q_r$. The difference, by the first law, must equal the shaft work W; then

$$W = Q_a + Q_w - (Q_s + Q_r) \tag{26·22}$$

But $Q_w = Q_s$, so we get

$$W = Q_a - Q_r \tag{26·23}$$

The net work area W is the slanted area enclosed by heavy process lines 1–2–3–4–5 in Fig. 26·28.

The thermal efficiency of the cycle is

$$e = \frac{Q_a - Q_r}{Q_a} = \frac{T_1(S_2 - S_1) - T_3(S_3 - S_4)}{T_1(S_2 - S_1)} \tag{26·24}$$

But $(S_2 - S_1) = (S_3 - S_4)$, so

$$e = \frac{T_1 - T_3}{T_1} \tag{26·25}$$

The net effect of the regenerative cycle raises the average temperature at which Q_a enters the cycle. The special case of using saturated steam at the engine inlet is met by completely eliminating external heating of the water to the saturated state.

Practical cycles would be seriously limited if we did not take advantage of superheating steam. Figure 26·26 would be arranged so the steam transfers heat to the water only after it has expanded to the saturation temperature of the water, Fig. 26·29. Here we recede from the conditions for Carnot efficiency because *all* the heat is not added at highest temperature, but we gain by having an overall higher average temperature than that for the saturated cycle. Partial regeneration offers a big gain in thermal efficiency.

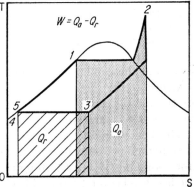

Fig. 26·29 Superheated regenerative steam cycle gains from the internal-cycle reheating of water returning to the steam generator.

26·25 Bleeding cycle. The wet exhaust steam of the ideal regenerative cycle proves a major practical defect. The engine is usually a turbine, and as expansion and cooling make the steam wetter, the moisture builds up as minute fog droplets.

The speed with which steam flows through the nozzles and buckets of a turbine gives these droplets a severe cutting ability; they can destroy the latter stages of the unit. So it becomes important to limit moisture to about 12% maximum. To achieve this, another approach to regenerative feedwater heating has been worked out, bleeding steam from the engine as in Fig. 26·30.

Here a part of the steam leaves the turbine at succeedingly lower pressures and flows to open mixing heaters below. Condensate from the condenser is pumped to the pressure of the lowest bleed point at state 5 and mixed with the steam in heater 3. Steam nominally fills the heater

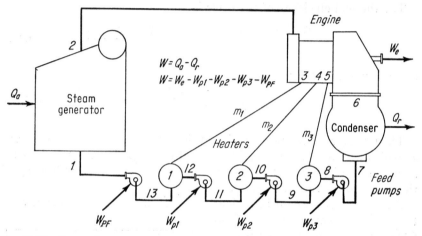

Fig. 26·30 Practical regenerative steam cycle bleeds steam from the engine to warm feedwater in mixing heaters. Condenser steam flow is diminished by the amount of steam bled for feedwater heating.

shell, and the condensate being sprayed in heats to the saturation temperature at the bleed-steam pressure by condensing some bleed steam. As the bleed steam condenses, more steam flows in automatically from the engine to maintain the pressure of the bleed point in the heater.

The heated condensate then gets pumped up to the pressure of the next higher pressure heater 2. The heating process draws bleed steam from point 4 at state 4 automatically. Next the feedwater is pumped to the last h-p heater 1 fed by bleed point 3 in the engine. Here it is heated by steam bled from point 3 at state 3. Finally the fourth or feedwater pump raises the water pressure to the boiler level at point 1.

Figure 26·31a shows the T-S chart for the bleed cycle. While the chart applies to 1 lb of working fluid, we have actually a varying amount of fluid flowing through different parts of the cycle. From state 2 to 3 one lb of steam expands at constant entropy through the engine. At point 3, m_1 lb of steam bleeds off to heater 1 (see Fig. 26·30), and $1 - m_1$ lb of steam expands in the engine to point 4 at state 4. At point 4, m_2 lb of steam bleeds off to heater 2, leaving $1 - m_1 - m_2$ lb expanding in the engine to point 5. Here m_3 lb of steam bleeds off to heater 3 at state 3, leaving $1 - m_1 - m_2 - m_3$ lb of steam to expand in the engine to exhaust at state 6 and condense to liquid at state 7. Condensate flow builds up as it passes through the heaters back to the boiler.

Now we are faced with interpreting the areas in Fig. 26·31a. One pound of steam expanding from 2 to 3 produces work in the engine, $W_1 = H_2 - H_3$. On the chart, W_1 is measured by the area 1–a–2–3.

One pound of steam expanding from 3 to 4 produces work

$$W_2 = H_3 - H_4$$

measured by the net area 1–3–4–12.

One pound of steam expanding from 4 to 5 produces work

$$W_3 = H_4 - H_5$$

and is measured by the net area 4–5–10–11.

Finally *one pound* of steam expanding from 5 to 6 produces work W_4, equals $H_5 - H_6$, and is measured by area 5–6–7–9. The area Q_r measures the heat given up by *one pound* of exhaust steam. Similarly the gray area marked Q_a measures the heat absorbed by 1 lb of water in the steam generator.

Q_a is true for our cycle, since 1 lb of water does flow through the steam generator and the first section of the engine. But Q_r is much too large; only $(1 - m_1 - m_2 - m_3)$ lb of steam flows through the condenser. Similarly the net area for W_2, W_3, and W_4 are all too large, because less than 1 lb of steam flows through the respective engine section. On the other hand, the area for W_1 is true.

We can get a true *proportionate-area* diagram by applying the factors of actual flow as in Fig. 26·31b. W_2 outlined by the heavy lines equals the similarly labeled area on Fig. 26·31a multiplied by $(1 - m_1)$. The states marked by 11' and 12' are not true state points because of the ratioing factor applied to the area for W_2. The true state points 11 and 12 of the liquid before and after the heater pump 3 stay as shown on Fig. 26·31a.

By applying $(1 - m_1 - m_2)$ to W_3 of Fig. 26·31a, we get the proportionate area of Fig. 26·31b; to get W_4 we multiply by $(1 - m_1 - m_2 - m_3)$. Multiplying by $(1 - m_1 - m_2 - m_3)$ also gives us Q_r. Then in Fig. 26·31b we have all areas in proper proportion for 1 lb of steam entering the engine throttle or inlet, but less in other parts of the cycle.

In Fig. 26·31b the work can be measured by the difference of the gray area Q_a and the crosshatched area Q_r. There is no simple net area left because the areas coincide on only two sides. But the area enclosed by the heavy lines *is* total net W for the cycle, equal to the sum of the work produced in the various sections of the engine. Then Q_a can be shown as the alternate area $Q_r + W_1 + W_2 + W_3 + W_4$ as shaded in Fig. 26·31c.

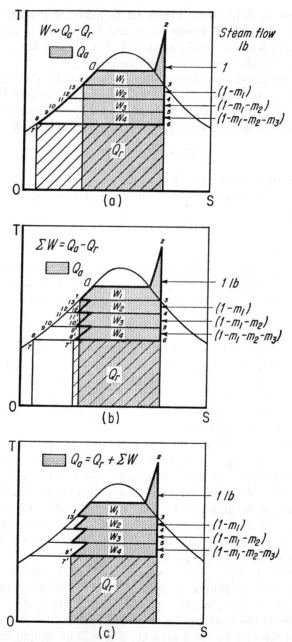

Fig. 26·31 (a) T-S chart for the bleeding regenerative cycle shows the state of the fluid at each point in the cycle for 1 lb of fluid, but the weight of working fluid varies during both steam expansion and water heating. (b) The T-S chart proportioned to actual fluid flow shows net work W as the difference of Q_a and Q_r with no net integral area remaining. (c) An alternative plot shows true Q_a as the sum of Q_r and net W of engine sections.

The saw-tooth appearance of the liquid-heating line suggests that, as the number of heaters in the cycle increases, it approaches a line of constant entropy. So an infinite number of heaters will achieve maximum efficiency for the bleed regenerative cycle. Actually the best number of heaters depends on the steam state at the engine inlet; many medium-pressure and -temperature cycles use five to six heaters. High-pressure and -temperature cycles use as many as nine.

26·26 Energy flow. Figure 26·32 clearly shows the amount of energy recirculating in the regenerative bleeding cycle. This chart assumes that

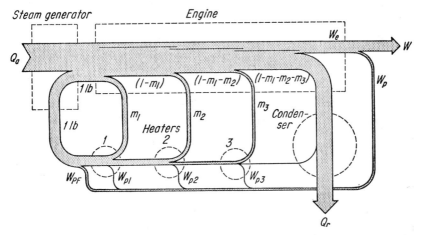

Fig. 26·32 Energy-flow chart shows the distribution of energy flow in engine sections to work, bleed steam for water heating, and downstream sections of engine. Energy in bleed or heating steam recirculates in the cycle.

energy in the condensate is taken as datum or zero; so we just show a single line leaving the condenser on its way back to the steam generator. Energy in the feedwater and heat Q_a added to the cycle enter the steam generator to combine as energy input to the engine.

In the first engine section, part of the energy converts to W_e and part leaves in the bleed steam m_1 to heater 1. The $(1 - m_1)$ lb of main steam with its energy flows to the second section of the engine. Here another part converts to W_e; another leaves in bleed m_2 to heater 2. In the third engine section more W_e is produced, some energy leaves in m_3 bleed to heater 3, and $(1 - m_1 - m_2 - m_3)$ lb flows through the last section of the engine on its way to the condenser. Here the final part of W_e is produced, leaving heat Q_r to be rejected from the cycle.

Following the condensate back to the steam generator we see the bleed energy in m_3 added to heat the liquid. Pump work W_{p3} also adds energy. The energy load of the condensate is augmented in each heater by the incoming bleed steam and the pump work needed to raise the condensate pressure to the heater pressure. The final boost of energy comes from the last or feedwater pump, raising water pressure to the steam-generator pressure. We have shown here the recirculating nature of pump work in a steady-flow cycle. This is also true of the Rankine cycle; in earlier

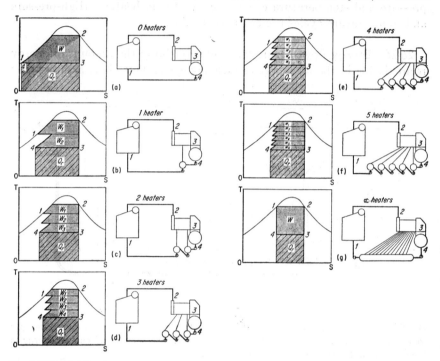

Fig. 26·33 Adding heaters at proper places raises the thermal efficiency of the regenerative bleed-steam cycle. Each additional heater adds a smaller increment of efficiency, but theoretically an infinite number of heaters raises the cycle to work at the ultimate Carnot efficiency.

diagrams we simply showed W_p as withdrawn from W_e without indicating that it really recirculates through the cycle.

The evolution of the steady-flow steam cycle from the simple Rankine to the infinite-heater bleed regenerative is shown in Fig. 26·33. Cycle diagrams show the growing complication of the plant as heaters and pumps are added. Proportionate-area T-S charts show that the net work areas shrink with added heaters. But the heat-rejected areas Q_r shrink at a faster rate; so the net effect raises the thermal efficiency. For an infinite number of heaters we get Carnot-cycle diagram of Fig. 26·33g.

26·27 Heater placement. In Fig. 26·33b, the one-heater cycle, we can bleed the engine anywhere along its length from the inlet or throttle down to the exhaust for feedwater heating. If we tap steam from the exhaust, we actually would take no bleed steam because the saturation temperature of the exhaust steam and the condensed liquid are the same and there would be no heat transfer between them. But as we move the bleed-steam tap toward the engine inlet, we would get higher pressure and temperature steam and deliver higher temperature feedwater to the

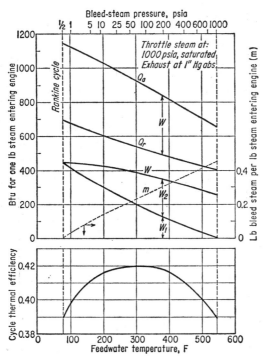

Fig. 26·34 Performance of a one-heater bleed-regenerative cycle depends on the pressure of bleed steam taken from the engine for heating.

steam generator. Figure 26·34 shows how the cycle factors change with the feedwater temperature and bleed-steam pressure as we do this. As the feedwater temperature is raised, Q_a drops. It reaches a minimum when we bleed live steam from the engine throttle—this steam has not passed through the engine at all.

Diminishing Q_a means that Q_r also drops, but not at so fast a rate. This results in W shrinking as the feedwater temperature rises. But W has two components, W_1 generated by the steam flowing between the throttle and bleed point and W_2 generated by steam flowing between the bleed point and exhaust. W_1 is the total output at minimum feedwater temperature but drops to zero when we bleed live steam from the engine inlet. W_2 grows as W_1 shrinks to zero output.

The bleed-steam flow m steadily rises as we raise the feedwater temperature by tapping the higher pressure bleed steam. The most important factor is effect on thermal efficiency. We get the same thermal efficiency either at Rankine-cycle conditions or by heating feedwater by bleeding live steam. At just half the total potential feedwater-temperature rise we get a maximum gain in thermal efficiency. For the conditions shown

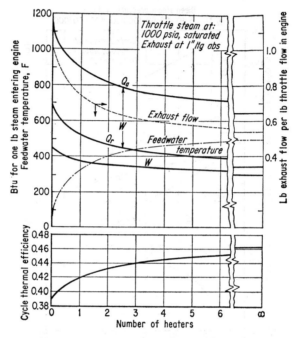

Fig. 26·35 Adding heaters to the bleed cycle improves overall thermal efficiency at diminishing increments of gain.

this amounts to three percentage points, a rise from 39 to 42% in round figures.

26·28 Multiple heating. The one-heater study shows the importance of properly placing the tap for bleed steam. Now let us study the effect of adding more heaters for feedwater heating, Fig. 26·35. We can use a rule of thumb in placing n heaters in series by tapping the bleed steam so the temperature rise per heater is $1/(1 + n)$ of the temperature difference between the boiler saturated temperature and condensate temperature. This gives about the best thermal efficiency for each number of heaters.

Figure 26·35 shows that, as we add heaters to the Rankine cycle, Q_a drops, but with a sharply diminishing rate. Q_r follows the same general trend. W shrinks sharply with the first two heaters but then drops slightly for more heaters. The exhaust flow of the engine, another way of measuring the total bleed steam, parallels Q_a closely. The feedwater temperature rises sharply at first and then reaches a maximum at infinite number of

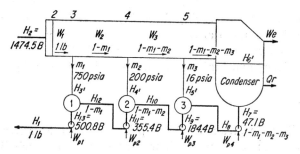

Fig. 26·36 Steam and water states in the bleed regenerative cycle used in the example. Feedwater enthalpies correspond to saturated values for the bleed pressures.

heaters. With these changes we steadily gain in cycle thermal efficiency, but after 10 heaters the gain becomes quite small per added heater. The variation of Q_a, Q_r, and W can be seen in the T-S charts of Fig. 26·33.

26·29 Practical cycle. In the calculations below and Figs. 26·36 to 26·38 we look at a simplified irreversible, regenerative bleed cycle. We have given steam conditions at throttle and exhaust of the engine and specified engine efficiencies of individual sections of the engine and the pressures at which steam is bled from the engine.

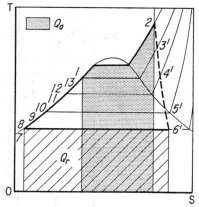

Fig. 26·37 T-S chart for an actual bleed cycle gives true-state points for steam and water in the example.

With these data we first find the state of the steam at each engine bleed point. After establishing the engine expansion line on a Mollier chart, we figure the amounts of steam bled to each heater. This calculation starts by recognizing that 1 lb of water enters the boiler for heating and this same pound of fluid enters the engine as steam. Where bled steam is superheated, we assume that the feedwater temperature equals the saturation temperature corresponding to the bleed pressure of the extracted steam.

We know all the factors about heater 1 except the amount of steam bled from the engine m_1 and the amount of feedwater entering it from

410   BASIC THERMODYNAMICS

heater 2. But the feedwater entering must be $(1 - m_1)$ lb, so we have just one unknown. Applying the first law we can set up an energy balance for the heater and solve for m_1.

After finding m_1 we can figure the balance about heater 2 to get m_2 and then do this for heater 3 to find m_3. The amount of steam extracted tells us the amount of steam flowing through each section of the engine; then we can calculate amount of W produced in each section. With these data established, we can figure heat input and rejection, engine and plant thermal efficiency and heat rates, and an engine steam rate.

Example

Given: Throttle steam at 2,000 psia and 1000 F; steam-generator efficiency = 0.88; station auxiliary consumption (excluding pump work) = 6% of W_e; engine efficiency of engine sections each = 0.8; engine cycle with three feedwater heaters and bleed-steam pressures as Fig. 26·36 shows; 1 in. Hg absolute exhaust pressure to condenser.

Solution: From steam tables find $H_2 = 1474.5$ Btu per lb and $S_2 = 1.5603$ Btu per lb, F.

First find the actual enthalpy $H_{3'}$ of steam at 750 psia bleed pressure in engine. As in Fig. 26·38, on a Mollier chart find H_2 and read at con-

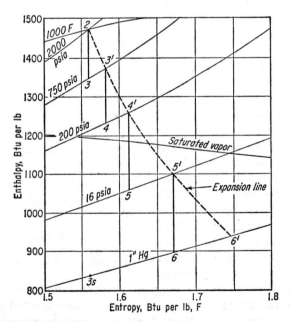

Fig. 26·38 Mollier chart shows the expansion line of steam flowing through a turbine at 80% engine efficiency for individual turbine sections.

stant-S expansion to 750 psia, $H_3 = 1346.7$ Btu per lb. Since

$$e_e = \frac{H_2 - H_{3'}}{H_2 - H_3} = 0.8 = \frac{1474.5 - H_{3'}}{1474.5 - 1346.7} \quad (1)$$

$$H_{3'} = 1474.5 - 0.8(1474.5 - 1346.7) = 1372.2 \text{ Btu}$$

From Mollier chart find $S_{3'} = 1.5819$ Btu per lb, F at 750 psia and 1372.2 Btu per lb.

At constant-S expansion from $H_{3'}$ find $H_4 = 1230.0$ Btu per lb at 200 psia. From an equation similar to (1) find

$$\begin{aligned} H_{4'} &= H_{3'} - e_e(H_{3'} - H_4) \\ &= 1372.2 - 0.8(1372.2 - 1230.0) \\ &= 1258.4 \text{ Btu} \end{aligned} \quad (2)$$

From Mollier chart find $S_{4'} = 1.613$ Btu per lb, F.

At constant-S expansion from $H_{4'}$ find $H_5 = 1059.5$ Btu per lb at 16 psia. Next find

$$\begin{aligned} H_{5'} &= H_{4'} - e_e(H_{4'} - H_5) \\ &= 1258.4 - 0.8(1258.4 - 1059.5) \\ &= 1099.2 \text{ Btu per lb} \end{aligned} \quad (3)$$

From Mollier chart find $S_{5'} = 1.6712$ Btu per lb, F.

At constant-S expansion from $H_{5'}$ find $H_6 = 898.2$ Btu per lb at 1 in. Hg absolute exhaust pressure.

$$\begin{aligned} H_{6'} &= H_{5'} - e_e(H_{5'} - H_6) \\ &= 1099.2 - 0.8(1099.2 - 898.2) \\ &= 938.4 \text{ Btu per lb} \end{aligned} \quad (4)$$

The Mollier chart shows that moisture at the exhaust is 15.1%.

The overall engine efficiency is better than the engine-section efficiency because of partial available-energy recovery between sections. Constant-S expansion from the throttle to the 1 in. Hg absolute exhaust gives $H_{3s} = 838.3$ Btu per lb, assuming that all the steam went to the condenser. Then

$$\begin{aligned} \text{Overall } e_e &= \frac{H_2 - H_{6'}}{H_2 - H_{3s}} \\ &= \frac{1474.5 - 938.4}{1474.5 - 838.3} = 0.8425 \end{aligned} \quad (5)$$

compared with 0.8 for individual engine sections.

For each heater, energy in = energy out; also note that the heated condensate leaving each heater is a saturated liquid at the heater bleed-steam pressure. Zero pressure drop is assumed between engine bleed point and heater inlet. Pump work comes from the chart with the Compressed Liquid Table in Keenan and Keyes' *Steam tables*.

No. 1 heater balance

$$H_{3'}m_1 + H_{12}(1 - m_1) = H_{13}$$
$$H_{3'}m_1 + (H_{11} + W_{p2})(1 - m_1) = H_{13}$$
$$1372.2m_1 + (355.4 + 1.7)(1 - m_1) = 500.8$$
$$m_1 = 0.1416 \text{ lb per lb throttle flow}$$
$$H_1 = H_{13} + W_{p1} = 500.8 + 4.7 = 505.5 \text{ Btu per lb}$$

No. 2 heater balance

$$H_{4'}m_2 + H_{10}(1 - m_1 - m_2) = H_{11}(1 - m_1)$$
$$H_{4'}m_2 + (H_9 + W_{p3})(1 - m_1 - m_2) = H_{11}(1 - m_1)$$
$$1258.4m_2 + (184.4 + 0.5)(0.8584 - m_2) = 355.4 \times 0.8584$$
$$m_2 = 0.1365 \text{ lb per lb throttle flow}$$

No. 3 heater balance

$$H_{5'}m_3 + H_8(1 - m_1 - m_2 - m_3) = H_9(1 - m_1 - m_2)$$
$$H_{5'}m_3 + (H_7 + W_{p4})(1 - m_1 - m_2 - m_3) = H_9(1 - m_1 - m_2)$$
$$1099.2m_3 + (47.1 + 0.1)(0.7219 - m_3) = 184.4 \times 0.7219$$
$$m_3 = 0.0942 \text{ lb per lb throttle flow}$$

Engine-section W_e per lb throttle flow

$$W_1 = H_2 - H_{3'} = 1474.5 - 1372.2 = 102.3 \text{ Btu}$$
$$W_2 = (H_{3'} - H_{4'})(1 - m_1)$$
$$= (1372.2 - 1258.4)(1 - 0.1416) = 97.7 \text{ Btu}$$
$$W_3 = (H_{4'} - H_{5'})(1 - m_1 - m_2)$$
$$= (1258.4 - 1099.2)(1 - 0.1416 - 0.1365) = 115.0 \text{ Btu}$$
$$W_4 = (H_{5'} - H_{6'})(1 - m_1 - m_2 - m_3)$$
$$= (1099.2 - 938.4)(1 - 0.1416 - 0.1365 - 0.0942) = 100.9 \text{ Btu}$$
$$\text{Total } W_e = 102.3 + 97.7 + 115.0 + 100.9 = 415.9 \text{ Btu}$$
$$\text{Total } W_p = W_{p1} + W_{p2} + W_{p3} + W_{p4}$$
$$\text{Total } W_p = 4.7 + 1.7 + 0.5 + 0.1 = 7.0 \text{ Btu}$$
$$\text{Station auxiliaries} = 415.9 \times 0.06 = 25.0 \text{ Btu}$$
$$\text{Net station } W = 415.9 - 7.0 - 25.0 = 383.9 \text{ Btu}$$

Check:

$$Q_a = H_2 - H_1 = 1474.5 - 505.5 = 969.0 \text{ Btu}$$
$$Q_r = (H_{6'} - H_7)(1 - m_1 - m_2 - m_3)$$
$$= (938.4 - 47.1)0.6277 = 559.5 \text{ Btu}$$
$$W_e - W_p = Q_a - Q_r = 969.0 - 559.5 = 409.5 \text{ Btu}$$

This compares with $415.9 - 7.0 = 408.9$ Btu from the work calculations. This is a good check; the difference of 0.6 Btu comes from errors in Mollier-chart and slide-rule readings. We shall use 408.9 Btu as correct.

$$\text{Plant energy input} = \frac{Q_a}{e_b} = \frac{969.0}{0.88} = 1101.0 \text{ Btu}$$

$$\text{Plant thermal efficiency} = \frac{W}{Q_a/e_b} = \frac{383.9}{1101.0} = 0.3486$$

$$\text{Engine thermal efficiency} = \frac{W_e}{Q_a} = \frac{415.9}{969.0} = 0.4292$$

$$\text{Plant heat rate} = \frac{3413}{0.3486} = 9790 \text{ Btu per kwhr}$$

$$\text{Engine heat rate} = \frac{3413}{0.4292} = 7950 \text{ Btu per kwhr}$$

$$\text{Engine steam rate (throttle)} = \frac{\text{engine } HR}{H_2 - H_1}$$

$$= \frac{7950}{1474.5 - 505.5} = 8.21 \text{ lb per kwhr}$$

The methods of achieving high thermal efficiency center on adding all heat input to a cycle at the highest average temperature possible and rejecting the unavailable energy at the lowest average temperature. No other fundamental methods have been proposed outside direct electric-energy generation now being researched so vigorously.

26·30 Combinations. The ingenuity of our plant designers, under the pressure of rising fuel costs, has not let them rest with the basic cycles. They have found that by combining these they can raise overall plant thermal efficiency without too much complication. We look at these combinations: (1) reheat-regenerative cycle, (2) mercury-steam cycle, and (3) gas-turbine–steam-turbine cycle.

26·31 Reheat-regenerative cycle. This cycle, Fig. 26·39, uses both bleed heaters for feedwater heating and one or more stages of reheating. For cycle analysis in Fig. 26·39 we divide the heat input into two parts, Q_{a1} and Q_{a2}. The first evaporates and superheats the full feedwater flow entering the boiler and steam generator. Superheated steam flows through the first part of the engine (turbine), part of the steam going to feedwater heater 1 and the remainder back to the reheater to receive the heat input Q_{a2}.

The reheated steam then enters the second part of the engine, and part bleeds off to the heaters at 5 and 6. The remainder exhausts to the condenser. The condensate returns to the steam generator joined on the way by the bleed condensate in the heaters to start the whole cycle over again at point 1.

Figure 26·40 shows the energy flow through this cycle; it can be cross referred to Fig. 26·39. Note that the work output W_1 of the first part of the engine (h-p engine) flows along the shaft while steam with diminished energy returns to be resuperheated in the steam-generator reheater.

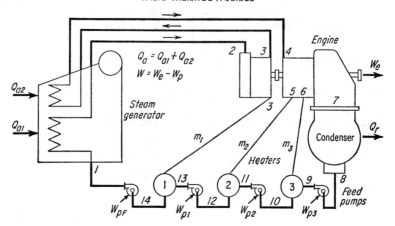

Fig. 26·39 Combined reheat and bleed-regenerative cycle makes use of economy features of each cycle type. Reheated-steam flow is less than the flow through initial superheater in the steam generator.

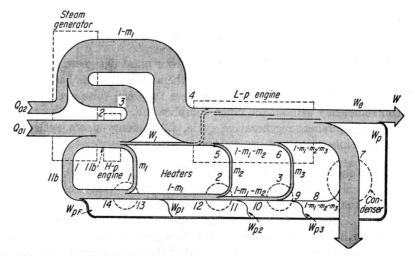

Fig. 26·40 Energy flow of combined reheat and bleed-regenerative cycle: W_1 is mechanical output of the first or h-p engine section flowing through the shaft, while most of the main steam flow returns for reheating in the steam generator.

Figure 26·41a shows a proportionate-area T-S chart for an ideal cycle. Regenerative heating of the feedwater diminishes the amount of Q_{a1} by raising the feedwater temperature to the boiler and so boosting the average temperature of the heat addition to increase the overall availability of the energy input. Q_{a2} input for reheating also raises the average temperature of the total input for an overall gain in cycle efficiency.

Figure 26·41b shows an actual-state-point T-S chart of an actual combined cycle with irreversible engine expansion of the steam. The inherent high efficiency of the cycle offsets the deteriorating effect of irreversible expansion processes. As usual, the net areas on an actual-cycle chart do not measure the net work produced.

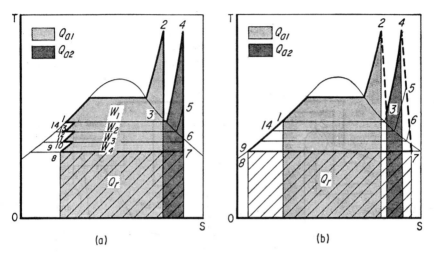

Fig. 26·41 (a) Proportionate-area T-S chart for the ideal reheat-regenerative-bleed cycle shows separate heat inputs, common heat rejection, and net work output. (b) Actual state points for cycle with an irreversible engine must be interpreted in the light of diminishing flow through parts of the cycle as steam is bled for feedwater heating.

Despite the relative complexity of this cycle, its thermal efficiency can be figured easily as

$$e = \frac{W}{Q_a} = \frac{Q_a - Q_r}{Q_a} = 1 - \frac{Q_r}{Q_{a1} + Q_{a2}} \qquad (26\cdot26)$$

Based on 1 lb of working fluid entering the boiler and turbine throttle, we can write for each of the terms from Fig. 26·41b

$$Q_r = (1 - m_1 - m_2 - m_3)(H_7 - H_8) \qquad (26\cdot27)$$
$$Q_{a1} = (H_2 - H_1) \qquad (26\cdot28)$$
$$Q_{a2} = (1 - m_1)(H_4 - H_3) \qquad (26\cdot29)$$

The bleed-flow terms m will have to be calculated by a first-law energy balance about each of the heaters, as we did for the regenerative cycle in Art. 26·29.

26·32 Steam rate of the cycle. This refers to the pounds of steam entering the turbine inlet per unit of cycle output. For example,

$$SR = \frac{3413}{W} = \frac{3413}{Q_{a1} + Q_{a2} - Q_r} \qquad (26\cdot30)$$

This gives us the steam rate in pounds per kilowatthour ($=3413$ Btu).
The cycle heat rate is simply

$$HR = \frac{3413}{e} \quad \text{Btu per kwhr} \quad (26\cdot 31)$$

26·33 Mercury-steam cycle. This cycle, Fig. 26·42, has a common furnace to supply heat energy to two separate fluid circuits in the mercury-boiler steam-superheater unit. The heat input Q_{am} vaporizes liquid

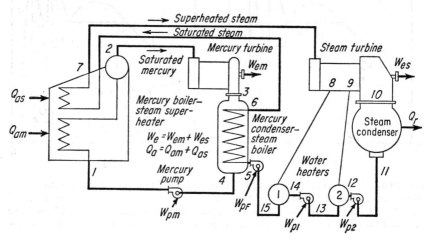

Fig. 26·42 Mercury-steam cycle has two separate heat inputs from a common furnace and two separate engines—a mercury turbine and a steam turbine—but only one heat rejection in the steam condenser.

mercury entering the unit at 1. The saturated mercury vapor enters the mercury turbine to generate shaft energy W_{em}. The wet vapor exhausts to the mercury-condenser steam-boiler unit, which it leaves as a saturated mercury liquid to be pumped back to the mercury boiler. The mercury part of the plant works as a simple Rankine cycle at relatively high temperature levels.

The energy rejected by the mercury cycle is part of the heat input to the steam cycle which the mercury cycle "tops." Water enters the mercury-condenser steam-boiler unit as a compressed liquid at 5 to be evaporated to a saturated vapor at 6. It then goes to the steam superheater to receive Q_{as} and enters the steam turbine to generate shaft work W_{es}. Exhausting to the condenser at 10 to reject Q_r, the saturated liquid leaves at 11 to flow through the feedwater heaters and repeat the cycle from point 5. Figure 26·43 shows the energy flow of this binary-vapor cycle with two points of energy addition and one point of rejection but two points of shaft-work output.

Figure 26·44 is a combined T-S chart of the mercury-steam cycle. The part for mercury refers to 10 lb of mercury, while that for steam refers

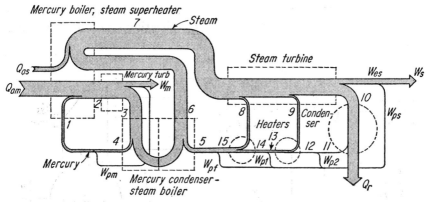

Fig. 26·43 Energy flow of the mercury-steam cycle. Here the heat rejected by the mercury cycle is part of heat input to the steam cycle with additional heating Q_{as} from the common furnace in mercury boiler-steam superheater.

to 1 lb of water. The chart shows the advantage of the binary-vapor idea. In vaporizing mercury at 940 F, the saturation pressure is only about 125 psia, minimizing the weight of equipment needed. Mercury lets the cycle take in most of its heat at a constant high temperature. It rejects heat at a high temperature, too, 475 F at 1.3 psia pressure, and this is absorbed by the steam cycle at about 430 F and 350 psia. Separate steam superheating helps raise the overall efficiency of the cycle by absorbing energy that may be otherwise wasted.

Several mercury-steam-cycle plants are running in the United States, but the vast improvement in performance of high-temperature reheat-regenerative cycles now used makes it unlikely that more binary-vapor plants will be built in the future.

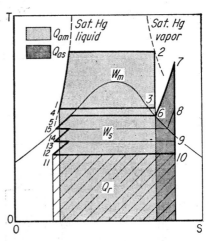

Fig. 26·44 Combined T-S chart for the mercury-steam cycle shows constant-temperature heat input to evaporate mercury liquid to saturated vapor.

26·34 Gas-turbine–steam-turbine cycle. This cycle, Fig. 26·45, is newly being applied in the United States. Several plants are now running.

The two types of cycles can be combined in a variety of ways; Fig. 26·45 shows only one of them. Here air enters a compressor at 1 to be pressurized for delivery to the pressurized furnace of a steam generator at 2. Fuel burned in the furnace furnishes Q_{a1} for the gas to the gas turbine at 3 and also Q_{a2} for evaporating and superheating steam for the steam cycle at 8.

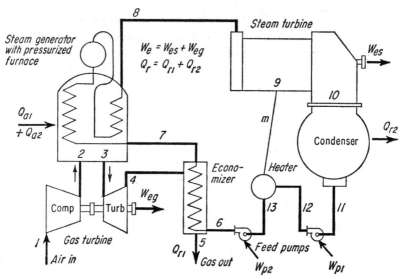

Fig. 26·45 Combined gas-turbine–steam-turbine cycle has a common gas-turbine-cycle combustor and steam-generator furnace. The economizer salvages some of the heat rejected by the gas turbine.

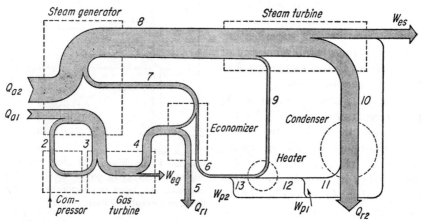

Fig. 26·46 Energy-flow chart of the gas-turbine–steam-turbine cycle shows two separate engines and two separate heat rejections from cycle. A pressurized combustor-furnace features the cycle.

STEAM CYCLES

Hot gas leaves the furnace at 3 to pass through the gas turbine and generate the shaft output W_{eg}. Exhaust gas then leaves at 4 to flow through the economizer, giving up part of its heat to the feedwater, and then leaves at 5 to the atmosphere to carry out the rejected heat Q_{r1}.

Feedwater in the steam cycle enters the economizer at 6 to be preheated at 7 before entering the steam generator, where it absorbs the input Q_{a2}. The steam and its condensate flow through a normal regenerative-bleed cycle to produce the shaft output W_{es} and reject the heat Q_{r2}, returning the heated feedwater to point 6.

Figure 26·46 shows the energy-flow diagram for the gas-turbine–steam-turbine cycle. This features two heat additions, two shaft-work outputs, and two heat rejections. Part of the heat rejected by the gas turbine is waste-heat-recovery input to the steam cycle.

Figure 26·47 is a pair of T-S charts showing the individual gas and steam cycles and how they are tied through heat transferred in the economizer Q_T. This assumes an irreversible transfer of this heat from the gas to the water. The chart is drawn for 2.6 lb of gas and 1 lb of water.

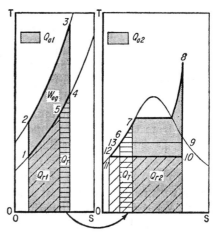

Fig. 26·47 T-S charts for the gas-turbine–steam-turbine cycle has irreversible heat transfer Q_T from the gas-turbine exhaust to the feedwater of the steam-turbine cycle, which helps shrink the amount of Q_{a2}.

Thermal efficiency of this combined cycle can be figured as

$$e = \frac{W}{Q_a} = \frac{Q_a - Q_r}{Q_a} = 1 - \frac{Q_{r1} + Q_{r2}}{Q_{a1} + Q_{a2}} \qquad (26·32)$$

The relative weight of the gas w_g to 1 lb of water must be figured by energy balance about the economizer:

$$H_7 - H_6 = w_g(H_4 - H_5) \qquad (26·33)$$

Then the other factors for efficiency will figure as

$$Q_{r1} = w_g(H_5 - H_1) \qquad (26·34)$$
$$Q_{r2} = (1 - m)(H_{10} - H_{11}) \qquad (26·35)$$
$$Q_{a1} = w_g(H_3 - H_2) \qquad (26·36)$$
$$Q_{a2} = H_8 - H_7 \qquad (26·37)$$

The bleed steam flow m must be calculated from an energy balance about the feedwater heater.

Steam rate for this cycle would have no meaning, since two different working fluids are involved. The heat rate for the cycle would simply be

$$HR = \frac{3413}{e} \quad \text{Btu per kwhr} \tag{26·38}$$

REVIEW TOPICS

1. Sketch the principal elements of a Rankine cycle, and show the points of energy input and departure. Evaluate the energy flows in terms of the enthalpy of the working fluid.
2. Derive the thermal-efficiency equation for a Rankine cycle.
3. Draw the T-S graph for a typical Rankine cycle using (*a*) saturated steam from the boiler, (*b*) superheated steam, (*c*) supercritical pressure steam. Designate the meanings of the areas on the graphs for reversible cycles.
4. What is the effect of raising the initial pressure of the Rankine cycle with respect to thermal efficiency? What about temperature rise?
5. What effect does lowering the back pressure of a Rankine cycle have on the thermal efficiency?
6. What is the steam rate of a cycle? How is it related to the work output per pound of working fluid?
7. What is the heat rate of a cycle? How is it related to the thermal efficiency of a cycle?
8. How do the Carnot and Rankine cycles compare in thermal efficiency for the same temperatures of heat addition?
9. Define engine efficiency with the aid of a Mollier chart.
10. What do the areas on a T-S chart mean for a Rankine cycle with an irreversible expansion in the engine?
11. Write the equation for boiler efficiency using the enthalpy added to the working fluid, the fuel burned, and its higher heating value.
12. Derive the equation for overall plant thermal efficiency including engine efficiency, boiler efficiency, and total auxiliary energy consumption.
13. What is the greatest availability loss in an actual power plant cycle?
14. Draw the simple reheat cycle, and show the principal points of energy input and departure. Draw the comparable T-S chart for this cycle.
15. Why does the reheat cycle have better thermal efficiency than the Rankine cycle?
16. Derive the equation for thermal efficiency of the reheat cycle.
17. What effect does reheat pressure have on cycle thermal efficiency?
18. List the irreversibilities that an actual reheat-cycle power plant may have.

STEAM CYCLES

19. Draw the diagram for the ideal regenerative feedwater heating cycle. Show the main points of energy input and departure.

20. Draw the T-S graph for the ideal regenerative feedwater heating cycle.

21. Derive the thermal efficiency of the ideal regenerative feedwater heating cycle.

22. Draw the bleeding regenerative feedwater heating cycle, and describe its operation.

23. Draw the equivalent T-S graph for a bleeding regenerative feedwater heating cycle, and explain its meaning.

24. Draw a proportionate-area T-S chart for the bleeding regenerative feedwater heating cycle, and explain its meaning.

25. Draw and explain an energy-flow diagram for a three-heater bleeding regenerative feedwater heating cycle.

26. What effect does placement of the bleed steam tap have on the overall thermal efficiency of a single-heater bleeding regenerative cycle. Discuss.

27. Outline the main steps in figuring the performance of a practical bleeding regenerative cycle with four heaters.

28. What effect does the addition of bleed points and heaters have on the thermal efficiency of a bleeding regenerative feedwater heating cycle.

29. Draw a sketch of a combined reheat-regenerative bleeding cycle, and describe its operation. Draw the corresponding T-S chart for this combined cycle.

30. How are the steam rate and heat rate figured for reheat, regenerative, and reheat-rengenerative cycles?

31. Draw the cycle and T-S chart for a combined mercury-steam regenerative cycle, and describe its operation. Draw an approximate T-S chart for the combined cycle.

32. Draw a sketch of a combined gas-turbine–steam-turbine cycle and explain its operation. Draw a corresponding T-S chart for the combined cycle.

PROBLEMS

1. In a Rankine cycle, steam from the steam generator enters the engine at 1,500 psia and 900 F and then expands to a condenser pressure of 1.5 in. Hg absolute. Using the compressed-liquid chart in Keenan and Keyes' *Steam tables* to figure pump work, find (*a*) the work done per pound of steam, (*b*) the heat added per pound of steam, (*c*) the unavailable energy per pound of steam, (*d*) the thermal efficiency of the cycle, (*e*) the steam rate of the cycle, (*f*) the heat rate of the cycle. Draw the T-S graph.

2. Compute the thermal efficiency of the ideal Carnot cycle working between the same temperature limits as the Rankine cycle in Prob. 1. Draw the T-S graph to compare with Prob. 1.

3. Steam entering the engine of a Rankine cycle is at 2,400 psia and 1000 F and then expands to 0.75 in. Hg absolute pressure in the condenser. If the engine efficiency is 85%, find (a) the thermal efficiency of the cycle, (b) the steam rate of the cycle, (c) the heat rate of the cycle. Draw the T-S graph of the cycle.

4. Steam leaves the boiler of a Rankine cycle at 2,000 psia and 1050 F, but the pressure drops to 1,800 psia before entering the engine. The steam expands to 1 psia in the condenser after flowing through the engine having 87% engine efficiency. Find (a) the thermal efficiency of the cycle, (b) the steam rate of the cycle, (c) the heat rate of the cycle. Draw the T-S graph for the cycle.

5. Steam leaves the boiler of a Rankine cycle at 2,800 psia and 1000 F, but the pressure drops to 2,600 psia before entering the engine. The condenser pressure is 1.5 in. Hg absolute, the engine efficiency is 0.83, the boiler efficiency is 86%, and the auxiliaries take 8% of the gross engine output. Find (a) the thermal efficiency of the cycle, (b) the steam rate of the cycle, (c) the heat rate of the cycle.

6. In an ideal reheat cycle steam at 800 psia and 900 F enters the inlet of the turbine. The steam expands to 160 psia in the turbine and is then reheated to 900 F. The steam enters the reheat inlet of the turbine and expands to 2.0 in. Hg absolute. Find (a) W in each section of the turbine, (b) total Q_a for the cycle, (c) the heat rejected by the cycle, (d) the thermal efficiency of the cycle, (e) the steam rate of the cycle, (f) the heat rate of the cycle. Draw the T-S graph for the cycle.

7. Find the thermal efficiency of a reheat cycle with a throttle pressure of 2,400 psia, reheat pressure of 450 psia, throttle and reheat temperatures of 850 F, condenser pressure of 2.0 in. Hg absolute, and engine efficiencies of both h-p and l-p engines of 85%. Draw the T-S diagram.

8. An actual bleeding regenerative cycle has throttle steam at 2,400 psia and 1000 F, steam generator efficiency of 86%, station auxiliary consumption of 7% (exclusive of pump work), engine efficiency of 83% in all sections of the turbine, and three bleed points and heaters with steam pressures of 850, 240, and 20 psia. Steam exhausts at 2 in. Hg absolute. Compare the overall engine efficiency for a straight-condensing operation to the sectional engine efficiency. Find the cycle thermal efficiency, steam rate, and heat rate.

CHAPTER 27

Steam-cycle Components

The steam turbine, heart of the steam cycle, converts the internal energy of the working fluid into mechanical shaft work. In Chap. 19 we studied the functions of the nozzles and buckets of gas turbines.

The action of the working fluid in passing through buckets or blades of turbines is the same whether the working fluid is gas or vapor, so we shall not go any further into these processes for steam turbines. But we encounter some interesting actions when steam changes phase in flowing through a nozzle. Let us study nozzle steam flow and its peculiarities.

27·1 Supersaturation. When steam enters and leaves a nozzle in superheated states, it acts much like a gas. We can use either the steam tables or $PV^n = C$ to figure the change in states and flows. But when superheated steam enters a nozzle and exhausts in a wet state, we find that it does not act smoothly while flowing in the nozzle.

In Fig. 27·1, superheated steam enters a nozzle at pressure P_a and exhausts at P_2. As it flows through the nozzle, we can measure its static pressure. Assuming a constant-entropy expansion, we can figure its state for each pressure, Fig. 27·2.

Such a calculation shows that at section b of the nozzle, Fig. 27·1, the steam should start condensing. But instead it remains a vapor until section c,e, when it suddenly starts condensing and forms an entrained fog or mist made up of saturated-water droplets. This plane is a condensation shock—a standing shock wave (Art. 18·11). Studying the expansion process on T-S and H-S charts, Fig. 27·2, shows some interesting actions. As the static temperature drops along the constant-entropy expansion below b, the actual static pressure is higher than the corresponding saturation pressure. At the condensation shock c, actual static temperature is T_c but actual static pressure is P_e instead of the lower saturation pressure P_c corresponding to T_c. As the steam passes through the shock wave, its static temperature and pressure rise to state e at the corresponding saturation temperature and pressure T_e and P_e.

Between states b and c the steam is in the *supersaturated* or *metastable* state. If the steam passed through all corresponding saturated pressures and temperatures while it expanded, it would be in *equilibrium*. The

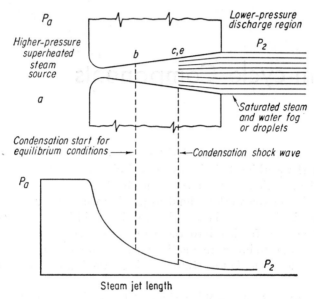

Fig. 27·1 When steam passing through a nozzle enters its saturation state, it becomes supersaturated before entering a condensation shock wave.

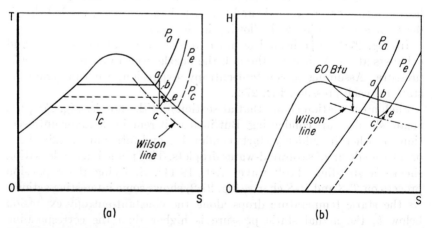

Fig. 27·2 Steam expanding into a saturation region from a becomes supercooled at c, though its actual pressure is P_e. After the steam passes through the condensation shock wave, the temperature rises to T_e at e.

degree of supersaturation is measured by the so-called *supersaturation ratio* P_b/P_e.

C. T. R. Wilson discovered this behavior in 1897 while experimenting with mixtures of air and water vapor. He found that air free of dust but saturated with water vapor could be expanded by 25% in volume before the vapor condensed to a fog, that is, 25% more than the specific volume corresponding to saturation.

Flow of condensing steam was studied by John Yellott and others during 1934 to 1937. They found that condensation shocks formed on the Mollier and T-S charts at a curve that was about 60 Btu below the saturated vapor locus; this is the Wilson line.

As Fig. 27·1 shows, static pressure jumps as the jet passes through the condensation shock wave. Liquid drops expanding beyond the shock probably stay hotter than the vapor, which cools rapidly with further expansion until it reaches the exhaust pressure. This is a *two-phase flow* with the saturated-steam molecules moving faster than the saturated-water molecules. The steam jet probably does not reach equilibrium until it is well out of the exit section of the nozzle.

27·2 Molecular action. To maintain equilibrium conditions we would slowly expand steam trapped in a cylinder by a piston. As we expand the volume by letting the steam do work adiabatically on the piston, all molecules of the steam move in all directions with a wide variety of speeds (Maxwell distribution, Art. 5·11). They slow down as they do work on the piston, and their helter-skelter directions and speeds make them collide billions of times a second with their neighbors. The slower ones, when colliding, quickly coagulate into drops of saturated water.

We have different conditions for flow through a nozzle as in Figs. 27·1 and 27·3. Even for small pressure differences across a nozzle, steam jets acquire a high speed. But more significantly the jet is made up of steam molecules traveling essentially in the same common direction. For a diverging nozzle, Fig. 27·3, we see that the initial random directions of molecules entering the nozzle are rearranged to move almost axially out of the nozzle exit. Paths *a*, *b*, and *c* show how the flared nozzle walls do this, assuming that individual lone molecules travel without interference and have equal angles of impact and rebound from the walls.

Trillions of molecules streaming through the nozzle continuously jostle one another, but the walls have the same aligning effect. The half star of vectors at the nozzle entrance shows the range of molecular direction and average speed of the entering molecules. The narrow fan of vectors at the exit shows the comparable direction range of the molecules leaving the nozzle. The drop in static pressure along the nozzle indicates fewer collisions of molecules with the walls and with one another because of the aligning effect.

With fewer collisions it becomes more difficult for the slower molecules to meet and coagulate into drops of water. The steam becomes supersaturated. Then, what causes the condensation shock wave? It might be the backflow effect of the exhaust-region molecules where the leaving jet is surrounded by vapor molecules traveling in all directions. These act on the jet and probably generate a backflow element of molecules in the jet that are effective up to the condensation shock wave.

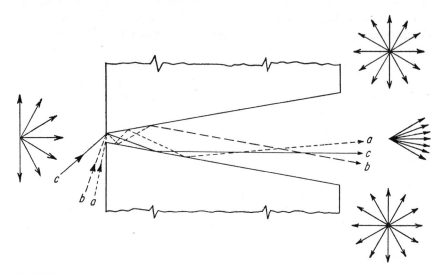

Fig. 27·3 Diverging part of a nozzle reflects molecules of steam into an axial path and makes the static pressure drop. The exhaust-region molecules tend to backflow into the steam jet and probably help form condensation shock wave.

At the shock wave, the momentum of the advancing jet molecules probably balances the momentum of the backflow molecules, so they have no effect further upstream. This makes the density of the steam higher downstream of the shock wave. Backflow molecules at and after the shock wave produce an increased mixing effect; so slower molecules in the jet can now meet more easily to form water drops. This is pure speculation, not bolstered by mathematical analysis, but it does seem plausible.

27·3 Nozzle performance. Like all thermodynamic machines, nozzles do not perform perfectly despite their simplicity. They convert the random motion of the molecules (internal energy) of the working fluid into an organized motion of a jet that we measure as kinetic energy of the body of fluid. In one sense the latter is also internal energy, since a thermometer can translate this into stagnation temperature, ideally the same as source temperature (Art. 18·4).

Steam flowing through the nozzle is retarded by friction with the wall of the nozzle. This irreversible effect returns internal energy to the steam jet at the expense of decreasing the kinetic energy gained by the jet. Actually there is a slight loss of heat by radiation, but this is usually assumed negligible; so the expansion is considered adiabatic. Straight nozzles will usually have fewer losses than curved ones; larger nozzles usually perform better than smaller ones.

From the steady flow energy equation for a nozzle

$$H_1 + \frac{v_1^2}{2gJ} = H_2 + \frac{v_2^2}{2gJ} \quad (27\cdot1)$$

but $$H_o = H_1 + \frac{v_1^2}{2gJ} \quad (27\cdot2)$$

then $$v_2 = \sqrt{2gJ(H_o - H_2)} \quad (27\cdot3)$$

where H_o is the stagnation enthalpy of the steam entering the nozzle. If the entering velocity $v_1 = 0$, then $H_o = H_1$ and $P_o = P_1$.

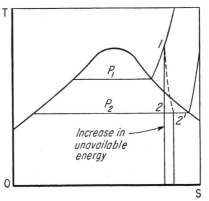

Fig. 27·4 Irreversibility of steam flow in a nozzle increases the unavailable energy of the steam jet.

The ideal leaving velocity v_2 depends on a constant-entropy expansion from state 1 to state 2 as in Fig. 27·4. The actual leaving velocity $v_{2'}$ will be lower because the actual state $2'$ will have a higher enthalpy $H_{2'}$ than H_2. The irreversible flow will increase the unavailable energy by the amount of the rectangular area under 2–2' in Fig. 27·4.

Nozzle performance is measured by nozzle efficiency defined as

$$e_N = \frac{v_{2'}^2/2gJ}{v_2^2/2gJ} = \frac{H_o - H_{2'}}{H_o - H_2} \quad (27\cdot4)$$

That is, it is the ratio of actual to ideal leaving kinetic energies. Another measure of nozzle performance is the nozzle velocity coefficient, defined as

$$C_v = \frac{v_{2'}}{v_2} = \frac{\sqrt{2gJ(H_o - H_{2'})}}{\sqrt{2gJ(H_o - H_2)}} = \sqrt{e_N} \quad (27\cdot5)$$

This becomes simply the square root of the nozzle efficiency.

Another measure of nozzle performance is the coefficient of discharge, defined as

$$C_d = \frac{\text{actual mass flow rate}}{\text{ideal mass flow rate}} \quad (27\cdot6)$$

This would be figured from tables or property formulas and known and ideal end states of the expansion process. It can vary from about 0.9 to over unity for supersaturated conditions.

27·4 Diffusers. A reversed nozzle is called a *diffuser*, Fig. 27·5. A diffuser takes in a high-speed jet of steam or gas at a relatively lower pressure, slows it down, and discharges the flow to a higher pressure region at a lower velocity, usually near zero. In other words, the diffuser is a pump or compressor without a mechanical impeller. Usually the diffuser works at less efficiency than a pump or compressor. The advantage of the diffuser lies in its simplicity. As a static device with no moving mechanical parts it is usually much less expensive.

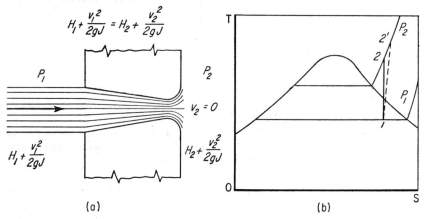

Fig. 27·5 Diffuser acts as a reversed nozzle by taking in a high-speed steam jet emerging from an l-p region and slowing it up by converting the kinetic energy of jet to random internal energy of steam at higher pressure.

The steady flow energy equation for a diffuser is the same as for the nozzle when using subscripts 1 and 2 for initial and final states, Eq. (27·1). But now the initial velocity is of major importance and the final velocity of little importance, since it is usually relatively low. We need to know the final enthalpy H_2 that can be developed for a given initial enthalpy H_1 and velocity v_1. By transposing Eq. (27·1) we get

$$H_2 = H_1 + \frac{1}{2gJ}(v_1^2 - v_2^2) \tag{27·7}$$

If v_2 is negligible, this reduces to

$$H_2 = H_1 + \frac{v_1^2}{2gJ} \tag{27·8}$$

27·5 Diffuser performance. Often we want to deliver working fluid from a lower to a higher pressure. Ideally this would be as a constant-entropy compression from 1 to 2 as in Fig. 27·5b. But because of frictional irreversibilities the entropy increases during the process and we find that the fluid must be delivered at a higher temperature (and enthalpy) at the same pressure P_2. For an actual diffuser, then,

$$H_{2'} = H_1 + \frac{v_{1'}^2}{2gJ} \qquad (27 \cdot 9)$$

Since H_1 is the same for the ideal and actual compression and $H_{2'}$ is larger than H_2, it follows that the actual initial velocity $v_{1'}$ must be higher than the ideal v_1. We have to use more kinetic energy initially than we would in the ideal diffuser to make the actual diffuser work.

Diffuser efficiency can be defined as

$$e_D = \frac{H_2 - H_1}{H_{2'} - H_1} \qquad (27 \cdot 10)$$

assuming that the final velocity at higher pressure $v_2 = 0$. If this is not so, then H_2 and $H_{2'}$ should be the stagnation enthalpies for the two states.

27·6 Diffuser application. This, Fig. 27·6, is most common in ejectors, low-cost pumps for removing gases and vapors from l-p spaces. Here the steam jet forms in a converging-diverging nozzle, the steam being supplied by a line. The steam jet is aimed into a converging-diverging section, like a venturi tube, which is the diffuser.

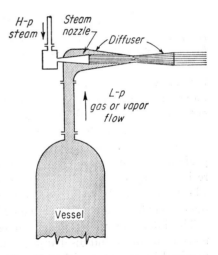

Fig. 27·6 Steam ejector uses h-p steam to build a vacuum in the vessel by removing gases and vapors.

The vessel's gases or vapors must be removed so its pressure can be kept low. These expand to fill the vessel and the pipeline connecting to the diffuser. Gases or vapors get entrained in the fast-moving steam jet and are carried through the diffuser to a higher pressure region, usually the atmosphere. As the gases and vapors are removed from the diffuser entrance, more of them flow up from the vessel by expansion, a continuous removal or purging process.

Another form of this application is water injection into small boilers. Here, Fig. 27·6, we substitute a feedwater line for the vessel and connecting pipe. The steam nozzle and diffuser are usually much smaller physically for water injection than gas removal. But the water gets entrained in the steam jet and is pumped or carried into the boiler drum at a higher pressure than the initial feedwater pressure to the injector.

27·7 Steam engines. Supreme prime mover of the nineteenth-century industrial world, the reciprocating steam engine now serves only low-capacity special applications. Despite its limited use today, the steam engine raises special thermodynamic problems that will be of interest to us—so let us take a brief look at its theory and methods of control.

Operation. The reciprocating steam engine has some features in common with the reciprocating air engines we studied in Chap. 12. Figure

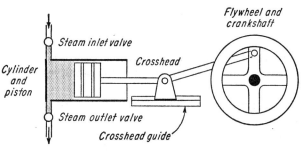

Fig. 27·7 Separate inlet and outlet valves control the admission and exhaust of steam from the engine cylinder. Flywheel smooths the rate of energy output.

27·7 shows elements of the simple single-acting steam engine. A piston moves back and forth in a cylinder under the force developed by steam entering and leaving the cylinder.

Force on the piston varies during the working stroke as the steam expands. To smooth the rate of shaft-work output and store work to push out spent steam, the piston drives a heavy flywheel through a reciprocating crosshead and crankshaft. Mechanical energy to the flywheel goes mostly to drive the load through the shaft, but part of it removes exhaust steam from the cylinder.

The inlet and outlet steam valves must be synchronized with piston motion to admit and discharge steam as needed and hold the shaft speed required by the load, either variable or constant.

Engine Cycle. Figure 27·8a and b shows idealized P-V variations of a piston and cylinder with zero clearance. At a in Fig. 27·8b the piston is at the head end of the cylinder with zero volume of steam in the cylinder. The inlet valve opens and admits steam which fills the expanding cylinder volume and moves the piston to point 1; this is called *admission.* Here the inlet valve closes and the steam begins expanding at constant entropy to the exhaust pressure at 2. At 2 the outlet or exhaust valve opens and the piston reverses direction, moved by the energy stored in the flywheel.

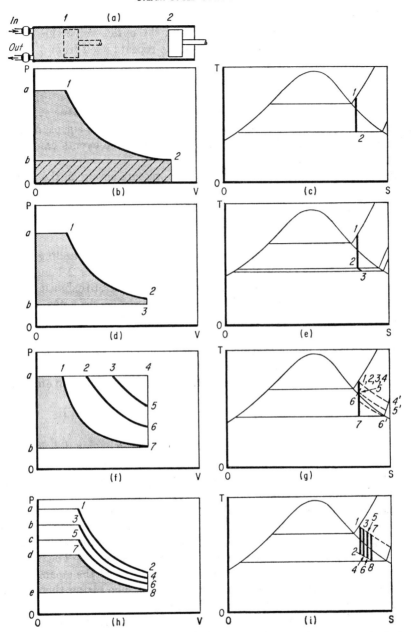

Fig. 27·8 Output of a zero-clearance engine depends on two points: the method of controlling flow in and out of cylinder and the extent of steam expansion. Throttling lowers the efficiency of the engine.

The piston pushes the l-p steam out of the cylinder until it reaches b, where cylinder volume reduces to zero and the outlet valve closes. The inlet valve then opens, and the engine cycle repeats itself from a.

What do the areas on the P-V chart mean? To begin with, this is not a thermodynamic-cycle chart because all the lines are not processes in the usual sense. Line a–1 applies to a varying mass of steam entering the cylinder, and 2–b to a varying mass leaving the cylinder. However, 1–2 is a process applying to a constant mass of steam. This corresponds to 1–2 on the T-S chart in Fig. 27·8c. The other engine events cannot be shown on Fig. 27·8c.

On the P-V chart, line a–1 measures the pressure acting on the piston. Its length (admission) measures the distance through which the force acts; so the product PV measures the *work* done by the entering steam *on the piston*, the area under line a–1. This equals the flow work of the entering steam.

The area under 1–2 measures the *work* done by the expanding steam at constant entropy *on the piston*. The hatched area under 2–b measures the *work done by the piston* on the steam in forcing it out to exhaust, which is also the flow work of the leaving steam. Summing up the areas, we find that the net gray area a–1–2–b measures the net work output of the piston or net work done by the steam flowing through the engine. Again note that this is not a thermodynamic cycle but a measure of engine events. This type of P-V chart is called an *indicator card* of the engine.

Assuming that 1 lb of steam flows through the reciprocating engine,

$$W = W_1 + W_2 - W_3 = \frac{P_1 V_1}{J} + E_1 - E_2 - \frac{P_2 V_2}{J} = H_1 - H_2 \quad (27 \cdot 11)$$

This shows that theoretically the ideal reciprocating engine develops the same work as a steady-flow engine like the turbine.

Engine Factors. The toe of the indicator card at state 2 develops little work. The additional stroke needed to realize this small increment does not pay for itself in the practical engine; friction losses may more than eat up the small work produced. As a compromise, the *incomplete expansion* engine is used, Fig. 27·8d and e. Here the piston completes its stroke at 2 with the pressure higher than the exhaust pressure at 3. The exhaust valve opens at 2, and the excess pressure makes part of the steam in the cylinder blow down to the exhaust pressure at 3. The piston then forces the remaining steam out of the cylinder along 3–b.

Throttling the steam from P_2 to P_3 raises the entropy and thus the unavailable energy, the area under 2–3 in Fig. 27·8e. This is less efficient than the complete expansion engine but a good practical compromise.

Most engines run at essentially constant speed. How, we may ask, can output be varied? Figure 27·8f and 27·8g shows how this can be done by changing the *cutoff* to vary W per cycle. The point where the inlet valve closes, 1, 2, 3, or 4 (or in between), is the cutoff. This is the end of the admission of steam to the cylinder and the start of expansion 1–7, 2–6, and 3–5. The engine with cutoff at 4 is a *full-admission* engine and does not expand the steam at all. Expansion shortens with increasing admission; Fig. 27·8g shows the increasing throttling that takes place as the inlet valve closes later. We get increasing work per cycle at the expense of poorer efficiency. In Fig. 27·8g, 4′ represents the steam state at exhaust pressure P_7, as do 5′ and 6′ from the corresponding expansion ends, 4, 5, and 6.

Figure 27·8h and i shows inlet-pressure throttling to control the engine output. This engine has a fixed cutoff, but as the load drops, the inlet steam pressure is throttled to reduce W per cycle. This engine proves rather inefficient because there is throttling at both ends of the expansion process, but its valve gearing is usually relatively inexpensive.

Indicated Horsepower. In practice, the net area on the indicator cards is called the indicated horsepower of the engine and can be figured as

$$Ihp = \frac{P_m LAN}{33,000} \tag{27·12}$$

where P_m = mean effective measure, psf
  = W/V_d
W = work, ft-lb per cycle
V_d = cylinder volumetric displacement, cu ft
L = piston stroke, ft
A = piston area, sq ft
N = number of engine cycles per min

Figure 27·9 compares an actual indicator card with an ideal conventional card. As expected, the net area of the actual card is less than the ideal—meaning less work per cycle. At 1 the piston reaches the end of its head-end stroke and the inlet steam valve opens, admitting steam to the cylinder and raising the pressure rapidly. The actual engine has a clearance volume V_1, depending on the design of valve, ports, and connecting passages. As the piston recedes on its working stroke, steam flows into the cylinder. Pressure fluctuates because of the sudden admission and uneven flow. At 2, the cutoff, inlet valve closes and steam expands to a lower pressure. Expansion is not at constant S because of heat transfer between cylinder walls and the steam.

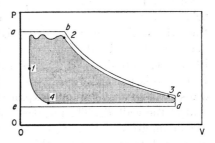

Fig. 27·9 Actual output of an engine with clearance is markedly less than the output of an ideal zero-clearance engine because of various losses.

At 3, before the end of the piston working stroke, the steam-exhaust valve opens and the steam begins blowing down to the exhaust pressure. On the exhaust stroke, the piston forces the steam out of the cylinder at about constant pressure. At 4, the exhaust valve closes and the piston compresses the entrapped cushion or clearance steam to point 1, where inlet valve opens and the cycle repeats.

Note that the admission pressure is less than the steam pressure P_a, expansion lies below the ideal b–c, and the cylinder pressure P_4 is higher than the exhaust pressure P_e. Irreversible processes of friction, heat

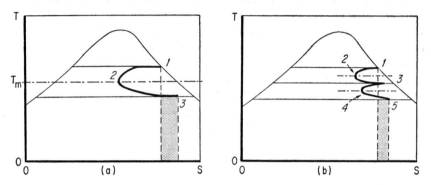

Fig. 27·10 Initial condensation of steam entering the engine distorts the expansion path of the steam and raises the unavailable energy of the steam.

transfer, and pressure loss contribute to efficiency losses in the steam engine.

Cylinder Condensation. The expansion process of an engine often lies close to $PV = C$, but it is not a constant-temperature process. Figure 27·10a shows its nature. As steam enters the cylinder at 1, it is hotter than the walls, and if the steam is saturated, a considerable part condenses as shown by the horizontal process line. But as the steam expands, its temperature drops to T_m at state 2 where it is the same as the walls. During this part of the expansion, heat transfer to the wall dwindles, dropping to zero. Continued expansion puts the steam temperature below that of the wall T_m. Then heat retransfers from the wall to the steam as the steam grows colder. Since the area under 1–2 is the same as the area under 2–3, the net effect of this complex expansion increases the entropy and the unavailable energy as shown by the gray area in Fig. 27·10a.

To raise engine efficiency, cylinder condensation can be reduced by several methods: superheating, valving, compounding, a uniflow engine.

Superheating. Figure 27·10a assumes steam initially saturated and ready to condense on the slightest cooling. If steam is initially superheated, initial cooling will drop its temperature but not form the moisture which aids heat transfer. Superheating, of course, also offers the thermodynamic advantage of higher thermal head. The generally higher temperature of the whole cycle also reduces exhaust moisture.

Valving. Designing the valves to shorten the contact of the steam with the metal walls of passages will reduce heat losses. Therefore the cylinder condensation caused by loss of internal energy will be reduced as well.

Compounding. A compound engine, Fig. 27·11, splits steam expansion into two processes and helps reduce initial condensation. Here steam first expands in an h-p cylinder, exhausts to an intermediate receiver R, flows

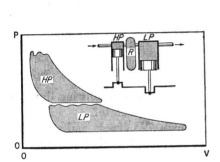

Fig. 27·11 Compound steam engine may have two or more cylinders receiving successively lower pressure steam. Compound engines can work through a larger pressure range and develop a larger capacity and higher efficiency.

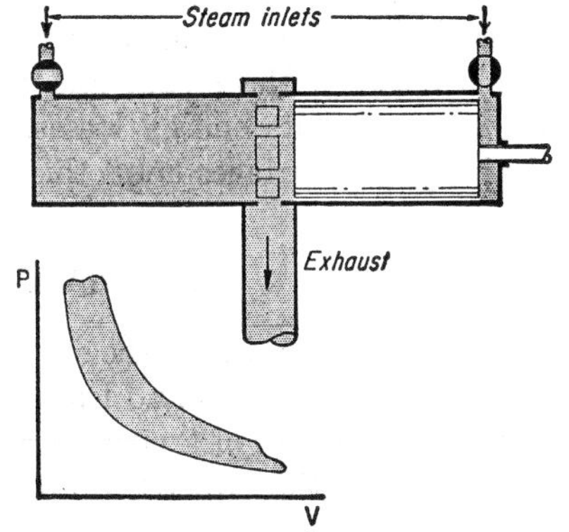

Fig. 27·12 Uniflow steam engine has the piston filling about half the cylinder and exhaust ports at the cylinder center. A double-acting engine, it takes in steam at both ends but exhausts only through the center ports.

into the l-p cylinder for final expansion then to exhaust. Both pistons work on a common crankshaft.

Compounding shortens the temperature range within each cylinder, reducing the opportunity for heat transfer as shown in Fig. 27·10b. The gray area measuring rise in unavailable energy is much less for the double expansion than the single expansion between the same pressure levels. Compounding also helps in building engines with large capacity when work output is split between two or more pistons for the same steam flow.

Uniflow Engine. This engine, Fig. 27·12, achieves low cylinder condensation essentially by making steam "flow" in one direction through the cylinder. The double-acting engine has a piston that fills one-half of the cylinder. In the position shown, the inlet valve on the right has just opened and steam enters the right-hand clearance volume. During the early part of the leftward piston admission stroke, the right inlet valve closes and the right-hand steam starts expanding. During the right-hand admission stroke the piston closes off the center exhaust ports and starts compressing the left-hand cylinder cushion or clearance steam.

While the right-hand steam expands, the left-hand steam compresses. Before the leftward stroke ends, the piston uncovers the exhaust ports for the right-hand steam and it blows down to exhaust. Just before completion of the stroke the left-hand admission valve opens to admit steam to the left-hand piston face. Actions in both ends of the cylinder then take place in the reverse order.

Figure 27·12 shows the indicator card for the left-hand end of the cylinder; the right-hand end has a similar one.

In this engine the inlet ends of the cylinder generally remain hotter than the center exhaust area. This reduces the average temperature differentials between the steam and the cylinder and thus the initial condensation with its energy losses. Avoiding the *flow* of hot steam over relatively cold surfaces helps this engine achieve its high efficiency.

REVIEW TOPICS

1. How can we figure the state of superheated steam flowing through a nozzle when it enters and leaves in a superheated state?

2. With the aid of a nozzle cross section, a T-S graph, and a Mollier chart, explain the supersaturated state of superheated steam flowing through a nozzle when it starts condensing.

3. What is meant by the equilibrium flow of a saturated vapor through a nozzle?

4. What is the Wilson line on a T-S graph and Mollier chart?

5. Give a qualitative explanation of why supersaturation takes place in a nozzle.

6. What happens at the shock wave in a steam nozzle?

7. Write and explain the steady-flow energy equation for a nozzle.

8. Derive the equation for exit steam velocity from a nozzle.

9. Define and evaluate nozzle efficiency.

10. Define and evaluate the nozzle-velocity coefficient. Relate it to the nozzle efficiency.

11. Define the coefficient of discharge for a nozzle.

12. What is a diffuser? How does it work?

13. Derive the steady-flow energy equation for a diffuser.

14. Define the equation for diffuser efficiency.

15. Explain the application of a diffuser in two forms.

16. Describe the processes of a reciprocating steam engine with the aid of a cylinder cross section, a P-V graph, and a T-S graph.

17. What precautions must be observed in describing the events in a steam-engine cylinder on a P-V graph.

18. Derive the energy equation for a reciprocating steam engine under ideal conditions.

19. How does the incomplete expansion engine differ from the ideal

complete expansion engine? What advantage does the former have over the latter?

20. Describe cutoff governing of a reciprocating steam engine.
21. Describe throttle governing of a reciprocating steam engine.
22. Derive the indicated horsepower for a reciprocating steam engine.
23. Compare an actual indicator card with the ideal events in a reciprocating steam engine.
24. Describe the effects of cylinder condensation on the expansion of steam in an engine cylinder.
25. How can cylinder condensation be reduced in a reciprocating steam engine?
26. Describe the operation of a uniflow steam engine.

PROBLEMS

1. Steam enters a nozzle at 100 psia and 340 F and leaves at atmospheric pressure. If the entering velocity is 500 fps, find (a) the leaving velocity for ideal expansion, (b) the critical pressure (see Art. 18·8), (c) the velocity at the throat, (d) the specific volume at the throat. (e) If the throat area is 1 sq in., how many lb per hr of steam flows through the nozzle?

2. If the nozzle in Prob. 1 has a nozzle efficiency of 97%, find the leaving velocity, assuming that the entering velocity is negligible. What is the quality of the leaving steam?

3. Steam enters a nozzle at 200 psia and 420 F and leaves at 50 psia. If the nozzle-velocity coefficient is 0.98, find the leaving velocity, assuming negligible entering velocity.

4. A diffuser receives saturated steam at 50 psia and having an entering velocity of 2500 fps. If the diffuser efficiency is 86% and exit pressure is 100 psia, what is the leaving velocity?

5. A steam engine works with a mean effective pressure of 25 psi; it has a stroke of 54 in. and piston diameter of 14 in. If the engine is single acting, find the indicated horsepower when it runs at 600 rpm.

CHAPTER 28

Vapor-compression Refrigeration

It will pay us to review briefly the basics of refrigeration before studying the widely used vapor cycles; see Chap. 9 also.

28·1 Reversed Carnot cycle. Our ideal Carnot engine becomes a refrigerating machine when we reverse the order of its processes. The purpose of a refrigerating cycle is to produce a fluid with a temperature lower than that of the space to be cooled. This induces a flow of internal energy or heat transfer from the cooled space to the refrigerating fluid. Next, the fluid is compressed by work input to raise its temperature above that of the atmosphere and let part of the internal energy of the fluid transfer to the air. Then the temperature of the fluid is reduced by making it do mechanical work, and it is ready to receive more heat from the lower level.

Figure 28·1a shows the P-V chart and Fig. 28·1b the T-S chart for a Carnot refrigerator. At state 1, the working (refrigerating) fluid is at the upper temperature level of the cycle. It expands at constant entropy by doing mechanical work on the piston to state 2; this drops its temperature to the lower level of the cycle. From 2 to 3 the fluid expands at constant temperature, doing work on the piston and absorbing heat from the fluid or space being cooled.

From 3 to 4 the piston works on the fluid, raising its temperature and pressure by a constant-entropy compression. In the last process, from 4 to 1, the piston does more work on the fluid by a constant-temperature compression that rejects heat to the atmosphere.

Ideally, heat transfers 2–3 and 4–1 would be reversible; that is, there would be zero temperature difference between the two fluids involved. Actually, T_1 is 10 F or more higher than atmospheric temperature and T_2 is 10 F or more lower than the cooled space or fluid.

The net area in Fig. 28·1a measures work input, in foot-pounds per pound of fluid, needed to make the Carnot refrigerating machine transport internal energy from the lower to the higher temperature level. The net area 1-2-3-4 in Fig. 28·1b also measures net work input W, but this

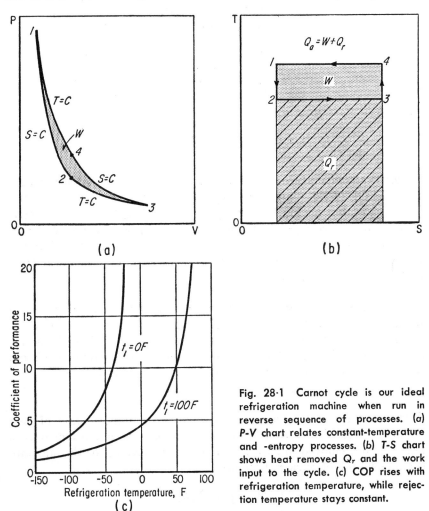

Fig. 28·1 Carnot cycle is our ideal refrigeration machine when run in reverse sequence of processes. (a) P-V chart relates constant-temperature and -entropy processes. (b) T-S chart shows heat removed Q_r and the work input to the cycle. (c) COP rises with refrigeration temperature, while rejection temperature stays constant.

time in Btu per pound of fluid. The lower hatched area measures Q_r, the heat removed from the refrigerated or cooled space, in Btu per pound of fluid. Total gray area measures the heat rejected to atmosphere Q_a, in Btu per pound of fluid.

28·2 Performance. We measure refrigerating-cycle performance by the ratio of energy output to input, but instead of being called an efficiency, it is named the *coefficient of performance* (COP). The useful output is the heat removed from the low-temperature level. The COP is always greater than one. For the Carnot refrigerator in Fig. 28·1b we have

$$\text{COP} = \frac{\text{energy output}}{\text{energy input}} = \frac{Q_r}{W}$$

$$= \frac{Q_r}{Q_a - Q_r}$$

$$= \frac{T_r(S_3 - S_2)}{T_a(S_4 - S_1) - T_r(S_3 - S_2)}$$

Since $S_3 - S_2 = S_4 - S_1$,

$$\text{COP} = \frac{T_r}{T_a - T_r} \tag{28.1}$$

Remember, temperatures are absolute. Figure 28·1c shows the variation in ideal COP of Eq. (28·1) for two atmospheric temperatures: $T_a = 0$ F (460°R) and 100 F (560°R). The curves show that the smaller the temperature difference $T_a - T_r$, the higher the COP, that is, the greater the units of refrigeration realized per unit of work input. To maintain a very-low-temperature region we have to expend a lot of mechanical work, that is, available energy.

28·3 Rating. A refrigerating system is rated by *tons of refrigeration*. A ton is defined as the energy that must be removed from a ton of saturated water at 32 F to freeze it to saturated ice at 32 F. While the actual enthalpy of fusion is 143.35 Btu per lb, the definition rounds this out to 144 Btu per lb, so that a *ton of refrigeration* is equal to

$$144 \times 2000 = 288{,}000 \text{ Btu}$$

The capacity of a refrigerating system is the uniform rate of heat removal, specified as a *standard commercial ton*—the number of tons of refrigeration removed in 24 hr. Equivalent expressions for the standard commercial ton of refrigeration are

$$\frac{288{,}000}{24} = 12{,}000 \text{ Btu per hr}$$

$$\frac{288{,}000}{24 \times 60} = 200 \text{ Btu per min}$$

Another practical measure of refrigerating performance of a cycle is the horsepower input per ton of refrigeration. This comes from

$$\text{COP} = \frac{\text{refrigeration per hr}}{\text{work input per hr}} = \frac{12{,}000N}{2{,}544Hp} \tag{28.2}$$

where N is the number of standard commercial tons of refrigeration and Hp is the horsepower input. Then

$$\frac{2{,}544Hp}{12{,}000N} = \frac{1}{\text{COP}}$$

$$\frac{Hp}{N} = \frac{12{,}000}{2{,}544 \times \text{COP}} = \frac{4.72}{\text{COP}} \tag{28·3}$$

28·4 Vapor cycle. Figure 28·2a shows the flow diagram for a steady-flow vapor-compression cycle of a refrigerating system; Fig. 28·2b gives a corresponding ideal T-S chart.

Here a refrigerating fluid like ammonia (NH_3) leaves the condenser as a liquid (usually slightly supercooled) in state 1, at the upper pressure level of the cycle. From 1 to 2 the pressurized liquid flows through an expansion or throttling valve to the lower pressure of the cycle.

During the expansion a small part of the liquid vaporizes as shown by throttling process on the T-S chart. More importantly, the temperature drops to the refrigerating level. Note that the vapor acts differently from a gas, which changes very little in temperature when throttled. The air-cycle refrigerator must make the expanding air do work to drop its temperature (see Art. 9·3).

In a liquid, however, the molecules are close together as they move past one another at different speeds (Maxwell velocity distribution). When the liquid flows into a lower pressure space through a narrow opening (throttle), the faster molecules have a chance to escape from their slower moving neighbors to form a vapor. But in doing so they must work against the mutual forces of attraction among all the molecules and so are slowed to a lower speed. Also, at the lower pressure, all the liquid molecules move slightly farther from one another, so they all work against the restraining forces of mutual attraction. This slows them down, too, and changes part of the molecular kinetic energy to potential energy. Decrease in kinetic energy shows as a lower temperature (Art. 5·10).

In a gas, on the other hand, the molecules are relatively far apart to begin with. So the additional separation after throttling does little to slow them down and drop the temperature. Only by making them impact a moving piston and do mechanical work can the molecules be slowed down appreciably.

Returning to Fig. 28·2a and b: after expansion at 2 the refrigerating fluid is a saturated liquid with entrained bubbles of saturated vapor at the lower temperature T_2. Fluid enters the evaporator to absorb heat Q_r from the liquid being chilled. Ideally, the refrigerating fluid is boiled off to a saturated vapor state at 3, which means that Q_r is absorbed at constant temperature $T_2 = T_3$.

The saturated vapor then enters the compressor to be pressurized and superheated to state 4 at constant entropy. The superheated vapor next flows into the condenser, where it is desuperheated, condensed, and slightly supercooled at constant pressure to state 1 while giving up heat Q_a to the cooling water. The cycle then repeats.

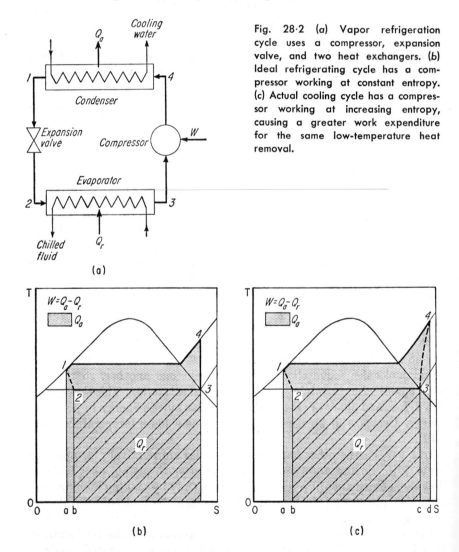

Fig. 28·2 (a) Vapor refrigeration cycle uses a compressor, expansion valve, and two heat exchangers. (b) Ideal refrigerating cycle has a compressor working at constant entropy. (c) Actual cooling cycle has a compressor working at increasing entropy, causing a greater work expenditure for the same low-temperature heat removal.

As in the Carnot refrigerator, there are only three points of energy transition in and out of the cycle. They can be related by $Q_a = W + Q_r$ or, transposing, $W = Q_a - Q_r$. Since Q_a is measured by the total gray area in Fig. 28·2b and Q_r lies wholly within area Q_a, the work input is the difference between the two areas: area a–b–2–3–4–1–a.

VAPOR-COMPRESSION REFRIGERATION 443

For this cycle we find that

$$\text{COP} = \frac{Q_r}{W} = \frac{H_3 - H_2}{H_4 - H_3} \qquad (28\cdot4)$$

We can also figure this as

$$\text{COP} = \frac{Q_r}{Q_a - Q_r} = \frac{H_3 - H_2}{(H_4 - H_1) - (H_3 - H_2)}$$

But since $H_1 = H_2$ for the throttling process, this equation reduces to the same form as Eq. (28·4).

28·5 Actual cycle. The ideal vapor cycle of Fig. 28·2b has one irreversible process 1–2, throttling the vapor. Despite this, the T-S diagram can be used to demonstrate the net work input W by the remaining unique area. An actual cycle, of course, has all irreversible processes; the compression process 3–4 will be at increasing entropy as in Fig. 28·2c.

Here we have an interesting situation again. Despite two irreversible processes, a net remaining area a–b–2–3–c–d–4–1–a measures the net work input W. Comparing this with Fig. 28·2b we see the increase in work input caused by the compressor irreversibility. For the same top pressure, the actual compressor produces higher superheat of the vapor. This places greater duty on the condenser in the larger Q_a that must be transferred, even though the same refrigeration Q_r is performed.

In actual cycles as well, the pressure of the fluid drops as fluid flows through the evaporator and condenser. These are not shown in Fig. 28·2, but note that Eqs. (28·2), (28·3), and (28·4) apply to both ideal and irreversible cycles.

28·6 Refrigerating fluids. These fluids obviously must stay in their liquid and vapor phases throughout the temperature range of a refrigerat-

Table 28·1 Thermodynamic properties of saturated ammonia[a]

Temp, F t	Press, psia p	Volume, cu ft per lb		Enthalpy, Btu per lb			Entropy, Btu per lb, F	
		Liquid v_f	Vapor v_g	Liquid h_f	Evap h_{fg}	Vapor h_g	Liquid s_f	Vapor s_g
−60	5.55	0.0227	44.73	−21.2	610.8	589.6	−0.0517	1.4769
−40	10.41	0.0232	24.86	0.0	597.6	597.6	0.0000	1.4242
−20	18.30	0.0237	14.68	21.4	583.6	605.0	0.0497	1.3774
0	30.42	0.0242	9.116	42.9	568.9	611.8	0.0975	1.3352
20	48.21	0.0247	5.910	64.7	553.1	617.8	0.1437	1.2969
40	73.32	0.0253	3.971	86.8	536.2	623.0	0.1885	1.2618
60	107.6	0.0260	2.751	109.2	518.1	627.3	0.2322	1.2294
80	153.0	0.0268	1.955	132.0	498.7	630.7	0.2749	1.1991
100	211.9	0.0272	1.419	155.2	477.8	633.0	0.3166	1.1705
120	286.4	0.0284	1.047	179.0	455.0	634.0	0.3576	1.1427

[a] Abstracted from National Bureau of Standards *Circular 142*, Apr. 16, 1923.

Table 28-2 Thermodynamic properties of superheated ammonia[a]

Temp, F	15 psia (−27.29-F sat)			50 psia (21.67-F sat)			100 psia (56.05-F sat)		
	v	h	s	v	h	s	v	h	s
−20	18.01	606.4	1.4031	—	—	—	—	—	—
0	18.92	617.2	1.4272	—	—	—	—	—	—
20	19.82	627.8	1.4497	—	—	—	—	—	—
40	20.70	638.2	1.4709	5.988	623.4	1.3046	—	—	—
60	21.58	648.5	1.4912	6.280	641.2	1.3399	2.985	629.3	1.2409
80	22.44	658.9	1.5108	6.564	652.6	1.3613	3.149	642.6	1.2661
100	23.31	669.2	1.5296	6.843	663.7	1.3816	3.304	655.2	1.2891
120	24.17	679.6	1.5478	7.117	674.7	1.4009	3.454	667.3	1.3104
140	25.03	690.0	1.5655	7.387	685.7	1.4195	3.600	679.2	1.3305
160	25.88	700.5	1.5827	7.655	696.6	1.4374	3.743	690.8	1.3495
180	26.74	711.1	1.5995	7.921	707.5	1.4548	3.883	702.3	1.3678
200	27.59	721.7	1.6158	8.185	718.5	1.4716	4.021	713.7	1.3854
220	28.44	732.4	1.6318	8.448	729.4	1.4880	4.158	725.1	1.4024
240	—	—	—	8.710	740.5	1.5040	4.294	736.5	1.4190
260	—	—	—	8.970	751.6	1.5197	4.428	747.9	1.4350
280	—	—	—	9.230	762.7	1.5350	4.562	759.4	1.4507
300	—	—	—	9.489	774.0	1.5500	4.695	770.8	1.4660
	150 psia (78.81-F sat)			200 psia (96.34-F sat)			300 psia (123.21-F sat)		
80	2.001	631.4	1.2025	—	—	—	—	—	—
100	2.118	645.9	1.2289	1.520	635.6	1.1809	—	—	—
120	2.228	659.4	1.2526	1.612	650.9	1.2077	—	—	—
140	2.334	672.3	1.2745	1.698	665.0	1.2317	1.058	648.7	1.1632
160	2.435	684.8	1.2949	1.780	678.4	1.2537	1.123	664.7	1.1894
180	2.534	696.9	1.3142	1.859	691.3	1.2742	1.183	679.5	1.2129
200	2.631	708.9	1.3327	1.935	703.9	1.2935	1.239	693.5	1.2344
220	2.726	720.7	1.3504	2.009	716.3	1.3120	1.294	706.9	1.2546
240	2.820	732.5	1.3675	2.082	728.4	1.3296	1.346	720.0	1.2736
260	2.912	744.3	1.3840	2.154	740.5	1.3467	1.397	732.9	1.2917
280	3.004	756.0	1.4001	2.225	752.5	1.3631	1.447	745.5	1.3090
300	3.095	767.7	1.4157	2.295	764.5	1.3791	1.496	758.1	1.3257

[a] Abstracted from National Bureau of Standards *Circular 142*, Apr. 16, 1923.

ing cycle. The fluid must stay well above its solid phase at the low-temperature end of the cycle.

Refrigerants that are used commercially include ammonia, butane, CO_2, carrene, Freon, methyl chloride, sulfur dioxide, and propane. These offer varying advantages—low temperature range, modest pressure range, heat capacity, and specific volume—and these factors all affect the design and size of the cycle components.

Table 28·1 lists the thermodynamic properties of saturated ammonia and Table 28·2 of superheated ammonia. Let us run through a simple refrigerating-cycle problem using ammonia as the refrigerant.

Example: For the cycle in Fig. 28·2a, ammonia leaves the evaporator as a wet vapor with 95% quality and a temperature of -20 F. The compressor raises the pressure at constant entropy to 100 psia. Ammonia leaves the condenser as a subcooled liquid at 50 F. Find (a) the heat removed by the evaporator in Btu per pound of ammonia, (b) the work input to the compressor, (c) the heat rejected in the condenser, (d) the COP of the ideal cycle, (e) horsepower per ton of refrigeration, (f) quality at state 2.

Solution: Using data from Table 28·1 in

$$H_3 = h_f + xh_{fg}$$
$$= 21.4 + 0.95 \times 583.6$$
$$= 575.8 \text{ Btu per lb}$$
$$S_3 = s_f + xs_{fg}$$
$$= 0.0497 + 0.95 \times 1.3277$$
$$= 1.3110 \text{ Btu per lb, F}$$
$$S_4 = S_3 = 1.3110 \text{ Btu per lb, F}$$

Interpolating in the 100-psia superheat section of Table 28·2,

$$T_4 = 130 - \frac{1.3206 - 1.3110}{1.3206 - 1.3104}(130 - 120)$$
$$= 120.6 \text{ F}$$
$$H_4 = 673.3 - \frac{1.3206 - 1.3110}{1.3206 - 1.3104}(673.3 - 667.3)$$
$$= 667.7 \text{ Btu per lb}$$

From Table 28·1 at 50 F,

$$H_1 = 97.9 \text{ Btu per lb} = H_2$$

(a)
$$Q_r = H_3 - H_2$$
$$= 575.8 - 97.9$$
$$= 477.9 \text{ Btu per lb}$$

(b)
$$W = H_4 - H_3$$
$$= 667.7 - 575.8$$
$$= 91.9 \text{ Btu per lb}$$

(c)
$$Q_a = H_4 - H_1$$
$$= 667.7 - 97.9$$
$$= 569.8 \text{ Btu per lb}$$

(d)
$$\text{COP} = \frac{Q_r}{W}$$
$$= \frac{477.9}{91.9}$$
$$= 5.20$$

(e)
$$\frac{\text{Hp}}{N} = \frac{4.72}{5.20} = 0.908$$

(f) From Table 28·1 at -20 F,

$$x = \frac{H_2 - h_f}{h_{fg}}$$
$$= \frac{97.9 - 21.4}{583.6}$$
$$= 0.1311 \text{ or } 13.11\% \text{ quality}$$

Energy, being costly, should be used as economically as possible. So we try to run refrigerating cycles at least cost. Studying compressed-air cycles, we learned that by splitting gas compression into steps and using intercooling we can reduce the work input. This same idea fits into the vapor-compression refrigerating cycle to raise the COP.

28·7 Simple intercooled cycle. This cycle, Fig. 28·3a, has two compressors in series with an interposed intercooler. Cooling water carries off some of the internal energy of the compressed vapor leaving the l-p compressor 1. This gives an additional point of energy removal from the closed cycle. Its energy balance is

$$Q_r + W_1 + W_2 = Q_a + Q_i \quad (28\cdot5)$$
$$W_1 + W_2 = Q_a + Q_i - Q_r \quad (28\cdot6)$$

Figure 28·3b shows the T-S graph for this cycle. The temperature at state 5 leaving the intercooler cannot be any lower than that of the cooling water entering the intercooler. This limits the advantage of intercooling and may mean that the work of the l-p compressor may be considerably more than that of the h-p compressor 2. Equal work in both compressors usually is the most economical division. The graph assumes that each compressor produces an irreversible adiabatic compression of the vapor.

The hatched area measures the heat Q_r absorbed by the evaporator and the gray area measures the heat Q_i removed by the intercooler. The area under process 6–1 (not outlined) would measure the heat rejected by the condenser, Q_a. The dotted process line 3–4–A shows the path for an irreversible single-stage compression between the top and bottom pressures of the cycle, without intercooling. The work saved by intercooling can be *roughly* estimated as area 4–5–6–A, since

$$W_1 = H_4 - H_3$$
$$W_2 = H_6 - H_5$$

and the work of single-stage compression is

$$W_s = H_A - H_3$$
$$\text{Work saved} = W_s - (W_1 + W_2)$$
$$= H_A - H_3 - (H_4 - H_3 + H_6 - H_5)$$
$$= H_A - H_4 + H_5 - H_6 \qquad (28\cdot7)$$

28·8 Regenerative intercooling. This, Fig. 28·4, improves on the simple cycle. Relatively high-temperature cooling water is not needed to remove Q_i. The cycle has two expansion valves in series, two compressors in series, and a flash tank connected between. Adjusting compressors and expansion valves controls the pressure of the refrigerant fluid in the flash tank.

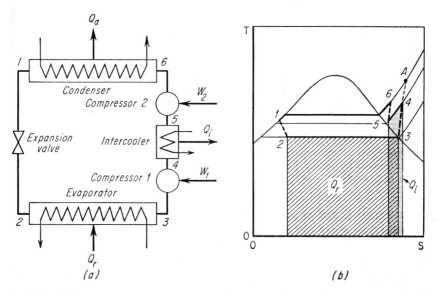

Fig. 28·3 (a) Splitting the compression of the refrigerant vapor into two steps and adding intercooling reduces the work input to the cycle and improves the COP. (b) Crosshatched area shows the heat removed and the gray area the heat rejected by the refrigerant in the intercooler.

Assume that we have 1 lb of vapor leaving the evaporator at state 6; the l-p compressor pressurizes and superheats it to state 7. Here m lb of dry saturated vapor from the flash tank at state 3 joins the 1 lb from the l-p compressor.

Mixing the two vapors produces $(1 + m)$ lb of vapor at state 8 (superheat less than at state 7) that enters the h-p compressor. Irreversible adiabatic compression pressurizes and superheats the $(1 + m)$ lb of vapor to state 9. The condenser then removes Q_a from $(1 + m)$ lb of vapor to change it to a subcooled liquid at state 1.

The $(1 + m)$ lb of liquid then throttles through the h-p expansion valve to state 2 at the intermediate pressure in the flash tank. One pound

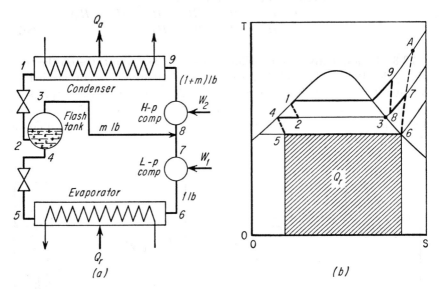

Fig. 28·4 (a) Regenerative intercooling refrigeration cycle uses two stages of vapor expansion with a flash tank in between. The flashed refrigerant cools the vapor entering h-p compressor by mixing. (b) T-S chart shows the states of the vapor for 1 lb; states 8, 9, 1, and 2 are for $(1 + m)$ lb of vapor.

of saturated liquid at state 4 throttles through the l-p expansion valve to state 5 to enter the evaporator. At the same time m lb of saturated vapor at state 3 leaves the flash tank to join with 1 lb of superheated vapor at state 7 leaving the l-p compressor. From here the cycle repeats, with the 1 lb of saturated vapor at stage 6 leaving the evaporator after having absorbed the refrigeration load Q_r.

Since the T-S graph of Fig. 28·4b is drawn for 1 lb of fluid, we can directly show Q_r as an area. Processes 8–9–1–2 all have $(1 + m)$ lb of fluid flowing in the cycle, so true Q_a referred to Q_r cannot be shown in proper proportion. But using the same reasoning as in Fig. 28·3, irreversible adiabatic compression 6–7–A shows the path for a single compression process between cycle pressure limits. *Roughly*, then, area 7–8–9–A shows the work input saved by regenerative intercooling.

A variation of intercooling uses a single-stage compressor. During compression liquid refrigerant is injected to reduce superheat of the discharged pressurized vapor. Depending on their refrigerant, most compressors have jacket cooling at the h-p end of the cylinder to help reduce final superheat and work input.

28·9 Cascade system. This system, Fig. 28·5, is used where very low temperatures are wanted for refrigeration. Here the low-temperature

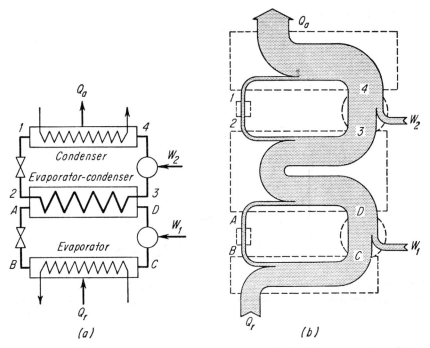

Fig. 28·5 (a) Cascading two cycles using different refrigerants can be arranged to achieve extra-low-temperature refrigeration; condenser of the lower cycle is evaporator of the upper cycle. (b) The energy flow through a cascaded cycle is augmented by the work inputs of the compressors.

cycle A–B–C–D may use one refrigerant to absorb the refrigerating load Q_r. It rejects heat

$$Q_{a1} = Q_r + W_1$$

to the high-temperature cycle 1–2–3–4 in the evaporator-condenser, connecting the two cycles by the heat-transfer surface.

In the high-temperature cycle the condenser finally rejects cascade-cycle heat to atmosphere as

$$Q_a = Q_{a1} + W_2 = Q_r + W_1 + W_2 \tag{28·8}$$

The COP in an actual cycle suffers from the additional temperature drop that must be carried in the combined evaporator-condenser.

Figure 28·5b is the energy-flow diagram for the cascade system. Refrigerants in each cycle would be chosen to gain the best design factors, taking into account specific heats, specific volume, P-T relations, and other properties.

28·10 Absorption refrigeration. This, Fig. 28·6, uses a system with a different method of raising the refrigerant from the l-p to the h-p level

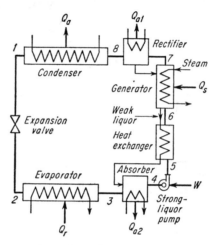

Fig. 28·6 Ammonia-water absorption refrigerating system uses four heat exchangers and a liquid pump instead of a vapor compressor to raise the vapor to upper temperature level.

of the cycle. Processes 8–1–2–3 are the same as in Fig. 28·3, but in place of a compressor we have four heat exchangers and one liquid pump. Their advantage is that they take much less valuable mechanical work input (for the pump); so most of the energy input goes into the generator in the form of heat Q_s.

Saturated ammonia vapor leaving the evaporator enters an absorber filled with water. Ammonia goes into solution with the water at various concentrations. High concentration, called *strong liquor*, rises to the top of the absorber while lighter concentration, *weak liquor*, drops to the bottom.

Heat released by the process of dissolving ammonia in the water is carried away by cooling water. Cooling the water raises its ability to absorb ammonia. The pump forces the strong liquor through a heat exchanger, where it preheats weak liquor returning from the generator to the absorber. From the heat exchanger at 6, strong liquor enters the generator where the steam coils carrying Q_s heat it and drive the ammonia out of solution with the water.

A mixture of ammonia and water vapor then enters the rectifier. It leaves behind it, in the generator, a weak liquor that returns to the absorber through the heat exchanger. The rectifier further separates water and ammonia vapors, returning the condensed water vapor to the generator for disposal.

Cooling water through the rectifier carries off some of the heat of vaporization of the water and ammonia. Ammonia vapor then enters the condenser to reject heat Q_a. The heat balance of the entire cycle is

$$Q_r + W + Q_s = Q_a + Q_{a1} + Q_{a2} \qquad (28\cdot9)$$

In many systems, the sum of Q_{a1} and Q_{a2} is of the same order as Q_a.

This cycle is most likely to be used where heating steam for the generator is available as a byproduct of some other process.

28·11 Vacuum refrigeration. This method, Fig. 28·7, develops moderately low temperatures, such as used in space air conditioning for comfort. The evaporator develops a supply of chilled saturated water that the chilled-water pump forces through cooling coils to absorb the refrigerating load Q_r.

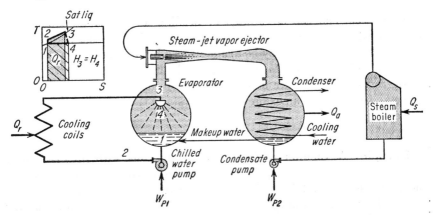

Fig. 28·7 Vacuum refrigeration system develops moderate cooling temperature and uses pumps and a steam-jet vapor ejector instead of a vapor compressor for cycle work input. This cycle is frequently used for air conditioning.

Warmed water returns to the evaporator, where it sprays into the evacuated space in a throttling process. In this constant-enthalpy process the warmed, pressurized water evaporates to a very wet vapor at the lower pressure and correspondingly lower temperature held in the evaporator. Drops of liquid fall to the bottom of the evaporator, while vapor fills the space above the liquid pool.

The small T-S chart in Fig. 28·7 shows the processes of the chilled water; 1–2 is the ideal constant-entropy pressurizing in the pump; 2–3 is the constant-pressure warming in the cooling coils by absorption of heat Q_r (gray area); 3–4 is the throttling in the spray head of the evaporator. Process 4–1 shows the equivalent constant-pressure cooling needed to remove the heat Q_r (hatched area) from the chilled-water circuit.

In the T-S graph, the total gray area under 2–3 measures Q_r. But since $H_4 = H_3$, the hatched area under 4–1 also measures Q_r. With warmed chilled water returning to the evaporator, an accumulation of dry saturated vapor builds up. To maintain the low pressure, the steam-jet vapor ejector must continuously remove this vapor. Steam from the boiler powers the ejector.

A mixture of evaporator vapor and jet steam discharges into a closed condenser with cooling coils or tubes. Cooling water flowing through the tubes absorbs the heat of vaporization and condenses the mixed vapors by carrying off heat Q_a. The condensate pump returns the jet part of the condensate to the boiler. The small remainder goes back to the evaporator, where it replaces vapor withdrawn by the ejector.

The heat balance of the cycle in Fig. 28·7 is

$$Q_r + Q_s + W_{p1} + W_{p2} = Q_a$$

This cycle has the advantage of needing a very small amount of mechanical work input. Steam for the vapor ejector need be only at a moderate pressure and may be available as a byproduct from some other process. The disadvantage of this cycle lies in the relatively high refrigeration temperature it develops, which limits the range of its usefulness. Let us study an example to understand the factors of this cooling cycle.

Example: A vacuum refrigeration system runs with an evaporator temperature of 45 F and chilled water returning at 60 F. The condenser holds a pressure of 1.135 psia, and the steam-jet ejectors need 3.1 lb of boiler steam per pound of vapor removed from the evaporator. Find the pounds of boiler steam needed per ton of refrigeration.

Solution: First, let us refer to the *Steam tables* and list needed enthalpies for our calculations.

At 45 F, $P = 0.14752$ psia; $h_f = 13.06$ Btu per lb; $h_{fg} = 1068.4$ Btu per lb.

At 60 F, $h_f = 28.06$ Btu per lb.

At 1.135 psia, $h_f = 73.95$ Btu per lb.

Each pound of chilled water picks up $28.06 - 13.06 = 15.0$ Btu.

VAPOR-COMPRESSION REFRIGERATION

Chilled-water flow for 1 ton of refrigeration

$$= \frac{12{,}000 \text{ Btu per hr}}{15.0 \text{ Btu per lb}}$$
$$= 8{,}000 \text{ lb per hr per ton}$$

Figure 28·7 shows the three fluid cycles involved: (1) chilled-water flow from evaporator to the cooling coils and back, (2) chilled-water vapor flow from evaporator through the ejector to the condenser and back as makeup, (3) boiler steam flow from the boiler to the ejector to the condenser and back to the boiler as condensate.

We shall base our calculations on 1 lb of chilled water flowing through the cooling coils. For the throttling process from 3 to 4 in the evaporator the enthalpy remains constant but part of the chilled water vaporizes at the lower pressure; so

$$H_3 = H_4 = h_f + xh_{fg}$$
$$28.06 = 13.06 + x1068.4$$
$$x = \frac{15.0}{1068.4} = 0.01405 \text{ lb vapor per lb of chilled water entering}$$

Chilled water remaining in the evaporator at 1

$$= 1.0 - 0.01405 = 0.98595 \text{ lb per lb of chilled water recirculating}$$

Since makeup from condenser also throttles into the evaporator, part of it evaporates; so

$$H_m = h_f + x_m h_{fg}$$
$$73.95 = 13.06 + x_m \, 1068.4$$
$$x_m = 60.89/1068.4 = 0.0570 \text{ lb of makeup vaporized per}$$
lb of makeup water entering

Makeup vapor simply recirculates between the evaporator and the condenser, so the total makeup water entering the evaporator must replace both chilled-water vapor and makeup vapor formed by the two throttling processes.

Pounds of makeup vapor per pound of makeup water remaining in evaporator = $0.0570/(1.0 - 0.0570) = 0.0604$.

The total makeup water to the evaporator needed to replace vapors = $0.01405 \times 1.0604 = 0.01491$ lb per pound of chilled water circulating. This is also the vapor removed from the evaporator by the ejector.

The total vapor removed from the evaporator

$$= 8{,}000 \text{ lb} \times 0.01491 = 119.3 \text{ lb per ton refrigeration}$$

Boiler steam = $119.3 \times 3.1 = 370$ lb per ton refrigeration.

REVIEW TOPICS

1. With the aid of P-V and T-S graphs describe the Carnot-cycle application as a refrigerator.
2. How is the performance of a refrigerating cycle computed?
3. What is the useful output of a refrigerating cycle?
4. Derive the COP for a reversible Carnot refrigerating cycle.
5. Plot the variation of COP with refrigeration temperature for given atmospheric temperatures.
6. What is a ton of refrigeration? What does it mean?
7. With the aid of a flow diagram and a T-S graph describe the operation of a vapor-compression refrigeration cycle.
8. Why does the temperature of a liquid drop when it throttles to a lower pressure region?
9. Why does not the temperature of a gas throttling to a lower pressure region drop as much as that of a liquid?
10. Draw the T-S graph for an irreversible cycle of vapor-compression refrigeration using an adiabatic irreversible compression process. Explain the meaning of the areas.
11. Describe the operation and energy balance of an intercooled vapor-compression refrigeration cycle with the aid of a flow diagram and a T-S graph.
12. With the aid of a flow diagram and T-S graph show the operation of a regeneratively intercooled vapor-compression refrigeration cycle.
13. With the aid of a flow diagram explain the operation of a cascade system of vapor-compression refrigeration. Why is this type of system used?
14. With the aid of a flow diagram explain the operation of an absorption refrigerating system using ammonia.
15. With the aid of a flow diagram explain the operation of a vacuum refrigeration cycle.

PROBLEMS

1. In a simple vapor-compression refrigerating system working under ideal conditions with ammonia as the refrigerant the evaporator pressure is 15.72 psig. The ammonia leaves the evaporator with 100% quality to enter the compressor suction. The compressor raises the pressure of the vapor at constant entropy to 140 psia. The ammonia leaves the condenser as a subcooled liquid at 70 F. Find (*a*) the heat rejected in the condenser, (*b*) the work input of the compressor, (*c*) the heat absorbed in the evaporator, (*d*) the COP of the cycle, (*e*) the pounds of ammonia to be circulated per ton of refrigeration, (*f*) the horsepower per ton of refrigeration needed to run the cycle.

2. A simple vapor-compression refrigerating cycle using ammonia has the fluid leaving the evaporator at a pressure of 15 psia and a temperature of 0 F. The compressor with 85% compression efficiency raises the pressure to 100 psia where the ammonia enters the condenser. The ammonia leaves the condenser as a saturated liquid to throttle to the evaporator. Find (a) the heat rejected in the condenser, (b) the work input to the compressor, (c) the heat absorbed in the evaporator, (d) the COP of the cycle, (e) the pounds of ammonia circulation needed per ton of refrigeration, (f) the horsepower per ton of refrigeration needed to operate the cycle.

3. A vacuum refrigerating system runs with an evaporator temperature of 40 F with chilled water returning at 58 F. The condenser holds a pressure of 1.1 psia, and the steam-jet ejectors need 3.2 lb of boiler steam per pound of vapor removed from the evaporator. Find the pounds of boiler steam needed per ton of refrigeration.

CHAPTER 29

Gas and Vapor Mixtures

29·1 Atmospheric humidity. "It's not the heat, it's the humidity," sums up the oppressive action superheated steam in the atmosphere can have on mankind. Our common experience points up the variable nature of the air and steam mixture as the earth's atmosphere changes with the seasons.

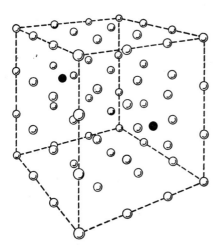

Fig. 29·1 This model shows the ratio of air molecules (white balls) to vapor molecules (black balls) for highly humid air; this could be 100% humidity at about 77 F dry-bulb temperature.

Humidity measurement has been reduced to an exact science and its theory proved. Figure 29·1 shows a hypothetical mixture of air (white balls) and steam (black balls). Strange as it may seem, this low ratio of vapor molecules may represent a debilitating humidity—100% relative humidity at about 77 F dry-bulb temperature.

Though the steam in the atmosphere is superheated, its temperature is the same as that of the air. A glance at the *Steam tables* (Table 24·1) shows that the partial pressure of the steam must be very low, well below 1 psia.

As preparation for this chapter it would be well to review Chap. 3, on perfect-gas laws and Chap. 22 on gas mixtures. Important laws and equations are summarized in Table 29·1.

Fortunately, water vapor at very low pressures behaves essentially as a

Table 29·1 Basic laws predict behavior of mixtures

Avogadro's Law

Equal volumes of perfect gases held under exactly the same conditions of temperature and pressure each have equal numbers of molecules (see Chap. 3).

Gibbs-Dalton Law

Gases occupying a common volume each fill that volume and behave as if the others were not present (see Chap. 22).

The pressure of a gas mixture is the sum of the pressures that each gas would exert if it alone occupied the volume.

The properties of a gas mixture are equal to the sums of those properties for each component when it occupies the total volume by itself; these properties are internal energy, enthalpy, and entropy.

Mixture equations

$$V'_{\text{total}} = V'_1 = V'_2 = V'_3 = \cdots$$
$$= m_1V_1 + m_2V_2 + m_3V_3 + \cdots$$
$$T_{\text{total}} = T_1 = T_2 = T_3 = \cdots$$
$$P_{\text{total}} = P_1 + P_2 + P_3 + \cdots$$
$$R_u = M_1R_1 = M_2R_2 = M_3R_3$$
$$= \cdots = 1545$$

$$D_{\text{total}} = D_1 + D_2 + D_3 + \cdots$$
$$m_{\text{total}} = m_1 + m_2 + m_3 + \cdots$$
$$\%m_1 = \frac{m_1}{\Sigma m}\, 100, \text{ etc.}$$
$$\%m_1 + \%m_2 + \%m_3 + \cdots = 100$$
$$\%V_1 = \frac{m_1/M_1}{\Sigma(m/M)}\, 100, \text{ etc.}$$
$$\%V_1 + \%V_2 + \%V_3 + \cdots = 100$$

perfect gas; so we can use the perfect-gas equation $PV = RT$. Figure 29·1 shows the reason: Steam molecules are so far apart that their mutual forces of attraction are negligible; so they behave like a perfect gas. This means that the temperature of the steam determines its enthalpy and not the pressure, since $H_2 - H'_1 = c_p\,(T_2 - T_1)$.

Figure 29·2 shows superheated atmospheric steam (at very low partial pressure) at state 1. The T-S graph then measures the enthalpy of the steam above 32 F, the total gray area under the constant-pressure line for state 1. Now suppose we have saturated steam at the same temperature, state 2; its pressure will be higher and its enthalpy is measured by the hatched area under the higher pressure line. Compare the two areas, and you will see that they are equal, bearing out the gaslike behavior of l-p steam.

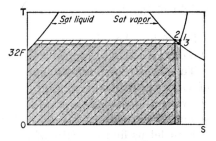

Fig. 29·2 Partial pressure of water vapor in atmospheric air is very low at ambient temperatures. Luckily, vapor can be figured as a perfect gas. At atmospheric temperatures steam enthalpy for any level is constant whether it is saturated or a superheated vapor.

29·2 Specific humidity. We can go about measuring the humidity of air or any gas and vapor mixture, for that matter, in several ways. Looking at one method we find:

Specific humidity is the ratio of pounds of water vapor per pound of dry air in a mixture, or

$$\overline{SH} = \frac{D_s}{D_a} = \frac{V_a}{V_s} \qquad (29 \cdot 1)$$

where $\overline{SH}$ = specific humidity, lb steam per lb air
D_s = steam density, lb per cu ft
D_a = air density, lb per cu ft
V_a = air specific volume, cu ft per lb
V_s = steam specific volume, cu ft per lb

According to Dalton's law, the steam and air occupy the same volume and each exerts its own partial pressure to add up to the total; so we can consider 1 cu ft of mixture for convenient calculation.

From the perfect-gas equation

$$V = \frac{R_u T}{MP}$$

where $R_u/M = R$, M being the molecular weight of the gas or vapor. Substituting in Eq. (29·1),

$$\overline{SH} = \frac{R_u T_a / M_a P_a}{R_u T_s / M_s P_s} = \frac{M_s P_s}{M_a P_a}$$

since $T_a = T_s$ = temperature of mixture. Then

$$\overline{SH} = \frac{18 P_s}{29 P_a} = 0.622 \frac{P_s}{P_a} \qquad (29 \cdot 2)$$

where we take 18 as the molecular weight of H_2O and 29 as the molecular weight of the dry-air mixture of O_2 and N_2.

Example: An air-steam mixture has 4% of its volume in steam at standard atmospheric pressure and 90 F. Find its specific humidity.

To solve this problem we must first establish a relation between ratios of volumes, ratios of pressures, and mass of components. Figure 29·3a shows the mixture of steam and air for a total volume V'_t having a mass of steam m_s and mass of dry air m_a for a total mass $m_t = m_s + m_a$.

First let us find the ratio of the steam partial pressure P_s to the total mixture pressure P_t. Starting with Dalton's law,

$$P_t = P_s + P_a = \frac{R_u T_t}{M_s (V'_t / m_s)} + \frac{R_u T_t}{M_a (V'_t / m_a)}$$

$$= \frac{R_u T_t}{V'_t} \left(\frac{m_s}{M_s} + \frac{m_a}{M_a} \right)$$

Then taking the ratio P_s/P_t and substituting, we get

$$\frac{P_s}{P_t} = \frac{m_s/M_s}{m_s/M_s + m_a/M_a} \quad (29\cdot3)$$

But when we separate the steam and air for a volumetric analysis as in

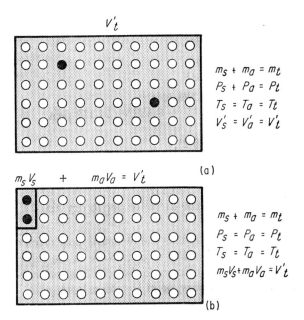

Fig. 29·3 (a) In a mixture of air and steam, both occupy the same total volume at their individual partial pressures but at the same temperatures. (b) In a volumetric analysis of an air-steam mixture, both are assumed to be at the same pressure and temperature.

Fig. 29·3b, with each component at the pressure and temperature of the original mixture, we start with the basic relation

$$m_s V_s + m_a V_a = V'_t$$
$$V'_t = m_s \frac{R_u T_t}{M_s P_t} + m_a \frac{R_u T_t}{M_a P_t}$$
$$= \frac{R_u T_t}{P_t} \left(\frac{m_s}{M_s} + \frac{m_a}{M_a} \right)$$

Then the volumetric proportion of steam in the mixture is

$$\frac{m_s V_s}{V'_t} = \frac{m_s/M_s}{m_s/M_s + m_a/M_a} = \frac{\%V_s}{100} \quad (29\cdot4)$$

(See Table 29·1 for the general form of this equation.) But this has the same value as Eq. (29·3); so we can immediately write

$$\frac{\%V_s}{100} = \frac{P_s}{P_t} \tag{29·5}$$

Then for our problem

$$\frac{\%V_s}{100} = 0.04 = \frac{P_s}{P_t}$$

Substituting,

$$0.04 = \frac{P_s}{14.7}$$
$$P_s = 0.588 \text{ psia}$$
$$P_a = 14.7 - 0.588 = 14.112 \text{ psia}$$

Then substituting in Eq. (29·2),

$$\overline{SH} = 0.622 \frac{0.588}{14.112}$$
$$= 0.0259 \text{ lb vapor per lb dry air}$$

29·3 Relative humidity. This is the measure most often quoted and is defined as: *The relative humidity of a mixture is the ratio of the actual density of water vapor in the mixture to the density of saturated water vapor at the temperature of the mixture,* or in equation form

$$\overline{RH} = \frac{D_s}{D_g} = \frac{V_g}{V_s} \tag{29·6}$$

Relating this to Fig. 29·2, V_s would be at state 1, a superheated vapor, while V_g would be at state 2, a saturated vapor at the same temperature (but a higher pressure).

We can state this relation in terms of partial pressures. Substituting from the perfect-gas equation in (29·6), we get

$$\overline{RH} = \frac{RT_t/P_g}{RT_t/P_s} = \frac{P_s}{P_g} \tag{29·7}$$

By finding the value of P_s from Eqs. (29·2) and (29·7) and equating, we find that

$$\frac{\overline{SH}P_a}{0.622} = \overline{RH}P_g$$

$$\overline{RH} = \frac{\overline{SH}}{0.622}\frac{P_a}{P_g} \tag{29·8}$$

Carrying on with the previous example, we can find the $\overline{RH}$ by substituting in Eq. (29·8) to get

$$\overline{RH} = \frac{0.0259}{0.622} \frac{14.112}{0.6982}$$
$$= 0.8416 \text{ or } 84.16\%$$

Here 0.6982 psia is the steam saturation pressure for the 90 F mixture.

29·4 Dew point. The dew point of a mixture is another method of measuring its humidity. To define:

Dew point is the temperature of a mixture at which liquid water starts forming when the mixture cools at constant pressure.

We can explain this in Fig. 29·2. Here the temperature of the steam in the mixture (and of the mixture) is defined by state 1. As the mixture cools at constant total pressure, the steam state falls from 1 to 3, where the steam begins to condense with further cooling.

Note that this definition specifies a constant-total-pressure cooling process. Since the steam cools at its constant partial pressure from an initial temperature T_1, dry air must also cool at its constant partial pressure. Evaluating this,

$$P_s = \frac{R_s T}{V_s}$$

$$P_t = \frac{R_t T}{V_t}$$

$$\frac{P_s}{P_t} = \frac{R_s V_t}{R_t V_s}$$

Since the air and steam occupy the same total volume, it follows that

$$V'_t = m_s V_s = m_a V_a = (m_s + m_a) V_t$$

Then
$$\frac{P_s}{P_t} = \frac{R_s m_s}{R_t (m_s + m_a)}$$

Thus partial pressures and the total pressure remain unchanged during the cooling process providing the mass of steam and the mass of air stay constant (the total volume does change, however). Partial pressure of the steam in a mixture is the saturation pressure corresponding to the dew point.

Again referring to the previous example, $P_s = 0.588$ psia. Looking in the *Steam tables* we find a corresponding saturated temperature of 84.58 F; this is the dew point of the mixture.

If a mixture is cooled below the dew point, state 3 in Fig. 29·2, some vapor condenses to liquid and the pressure of the remaining vapor will be the saturation pressure corresponding to the temperature of the mixture.

29·5 Wet-bulb measurement.

The easiest method of measuring humidities uses the wet-bulb thermometer. This is a thermometer with its bulb covered by a soaking-wet wick, Fig. 29·4c. The air-steam mixture to be measured is made to flow rapidly over the wick. If the mixture is unsaturated (below 100% relative humidity), water evaporates from the wick to join the mixture and raise it to saturation.

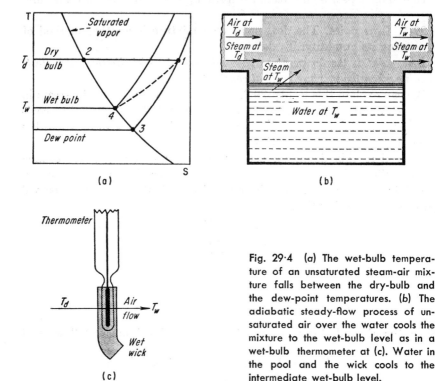

Fig. 29·4 (a) The wet-bulb temperature of an unsaturated steam-air mixture falls between the dry-bulb and the dew-point temperatures. (b) The adiabatic steady-flow process of unsaturated air over the water cools the mixture to the wet-bulb level as in a wet-bulb thermometer at (c). Water in the pool and the wick cools to the intermediate wet-bulb level.

No heat is supplied to the process for water evaporation, making this an adiabatic process. Usually the water can evaporate only by loss of its faster molecules, so it cools down to a lower temperature. The level to which it cools depends on the degree of saturation of the steam-air mixture. When the entering mixture is fully saturated, as many molecules from the steam enter the water in the wick as leave the wick to join the steam. Water in the wick and the air-steam mixture passing it do not change temperature, so the wet-bulb thermometer will read the same as a wickless dry-bulb thermometer.

But as the humidity of the entering air-steam mixture drops, the wet-bulb reading falls below the corresponding dry-bulb reading. Figure 29·4a shows the relation of the various steam temperatures on a section of a T-S chart. Point 1 shows state of the superheated steam in the mixture approaching the wet-bulb thermometer at its dry-bulb temperature. State 2 shows the corresponding saturation pressure, and state 3 the corresponding dew point. State 4 shows the wet-bulb temperature which lies between the dry-bulb and dew-point levels.

Process 1–4 indicates the change of state for the steam in the mixture. Note that the steam gives up energy as indicated by the area under the process line even though the overall mixture process is adiabatic. This goes toward evaporating water from the wick. In other words, heat is transferred between components of the system.

Figure 29·4b shows a water-removal or drying process that parallels the wet-bulb thermometer. The unsaturated air-steam mixture flows over water to be removed from a chamber. For stabilized steady-flow conditions the water will be at the wet-bulb temperature T_w, the same as the temperature of the air-steam mixture leaving.

Let us write the steady-flow energy equation for Fig. 29·4b in terms of 1 lb of dry air entering the chamber:

$$H_{ad} + \overline{SH}_d H_{sd} + Z H_{fw} = H_{aw} + \overline{SH}_w H_{sw}$$

where H_{ad} = enthalpy of 1 lb of dry air entering at dry-bulb temperature t_d, Btu per lb
$\overline{SH}_d$ = specific humidity of entering air-steam mixture at t_d, lb steam per lb air
H_{sd} = enthalpy of 1 lb steam in mixture entering, Btu per lb
Z = lb of water evaporated into mixture at t_w per lb of dry air entering
H_{fw} = enthalpy of water evaporated into mixture at t_w, Btu per lb
H_{aw} = enthalpy of one lb of dry air leaving at wet-bulb temperature t_w, Btu per lb
$\overline{SH}_w$ = specific humidity of saturated air-steam mixture leaving
H_{sw} = enthalpy of 1 lb steam in mixture leaving at t_w, Btu per lb

From the process we see that the specific humidity of the air-steam mixture increases with addition of the water; then $Z = \overline{SH}_w - \overline{SH}_d$. Making the substitution and solving,

$$\overline{SH}_d = \frac{\overline{SH}_w(H_{sw} - H_{fw}) + H_{aw} - H_{ad}}{H_{sd} - H_{fw}} \qquad (29\cdot9)$$

Knowing the mixture pressure and dry-bulb and wet-bulb temperatures we can figure the specific humidity of the incoming air-steam mixture.

Example: Atmosphere at 14.0 psia total pressure has a dry-bulb temperature of 70 F and a wet-bulb of 60 F. What is its specific humidity?

First let us remember that

$$H_{aw} - H_{ad} = c_p(t_w - t_d) = 0.24(60 - 70) = -2.4 \text{ Btu per lb}$$

From Eq. (29·2),

$$\overline{SH}_w = 0.622 \frac{P_{sw}}{P_{aw}}$$

P_{sw} is the saturation pressure corresponding to $t_w = 60$ F. (The steam tables list this as 0.2563 psia.) $P_{aw} = 14.0 - 0.2563 = 13.7437$ psia; then

$$\overline{SH}_w = 0.622 \frac{0.2563}{13.7437} = 0.0116 \text{ lb steam per lb dry air}$$

H_{sw} will be the same as the saturation H_g at 60 F, or 1088.0 Btu per lb. H_{fw} is the saturation H_f at 60 F, or 28.06 Btu per lb. H_{sd} is the same as the saturation H_g at 70 F, or 1092.3 Btu per lb. Substituting in Eq. (29·9),

$$\overline{SH}_d = \frac{0.0116(1088.0 - 28.1) - 2.4}{1092.3 - 28.1} = 0.00931 \text{ lb steam per lb air}$$

We can convert this to relative humidity by

$$\overline{RH}_d = \frac{\overline{SH}_d P_{ad}}{0.622 P_{gd}}$$

First let us find P_{ad} by Eq. (29·2):

$$P_{ad} = 0.622 \frac{P_{sd}}{\overline{SH}_d} = 0.622 \frac{P_t - P_{ad}}{\overline{SH}_d}$$

$$= \frac{(0.622/\overline{SH}_d)P_t}{1 + 0.622/\overline{SH}_d}$$

$$= \frac{0.622/0.00931 \times 14.0}{1 + 0.622/0.00931} = 13.78 \text{ psia}$$

Substituting in the above conversion equation,

$$\overline{RH}_d = \frac{0.00931}{0.622} \frac{13.78}{0.3631}$$
$$= 0.568 \text{ or } 56.8\%$$

The dew point of the air-steam mixture entering is the saturation temperature of $P_{sd} = 14.0 - 13.78 = 0.22$ psia, or 55.8 F.

29·6 Air conditioning. We developed formulas and demonstrated them in examples. Remember that the basic relations hold for all gas-vapor mixtures, although constants in the formulas for other mixtures will be different from those for air-steam. But since air-steam mixtures are the ones we meet most frequently in engineering, we shall concentrate on them.

We can make Eq. (29·9) more inclusive by substituting from (29·2) and noting that

$$H_{aw} - H_{ad} = 0.24(t_w - t_d)$$

We then get

$$\overline{SH}_d = \frac{0.622(P_{sw}/P_{aw})(H_{sw} - H_{fw}) + 0.24(t_w - t_d)}{H_{sd} - H_{fw}} \quad (29·10)$$

In air-conditioning work, two factors find frequent use: (1) the enthalpy of the air-steam mixture and (2) the total heat of a saturated air-steam mixture. For the first we have

$$H_m = H_{ad} + \overline{SH}_d H_{sd} \quad (29·11)$$

But convenience dictates measuring air enthalpy from 0 F and steam enthalpy from its usual 32 F level as a saturated liquid. Since we deal with *differences* of enthalpy, this has no effect on the answers we seek. Substituting in (29·11) we get

$$H_m = 0.24 t_d + \overline{SH}_d(1060 + 0.45 t_d) \quad (29·12)$$

In this case H_{sd} is figured by an empirical formula, though we can take it from the *Steam tables* if we wish and be slightly more accurate.

The total heat for a saturated air-steam mixture is defined as

$$\overline{TH}_m = 0.24 t_w + \overline{SH}_w(1060 + 0.45 t_w) \quad (29·13)$$

Comparing with (29·12) we see that (29·13) is simply a special case of the former, being referred to the wet-bulb temperature.

29·7 Psychrometric chart. This chart, Fig. 29·5, summarizes the relations in Eqs. (29·2), (29·8), and (29·10) to (29·12). This skeleton chart usually has a counterpart with many more graduations than shown here; so it can be read to four significant figures. Ordinarily the chart is computed for standard atmospheric pressure of 29.92 in. Hg absolute (14.696 psia). In most work the chart can be used for pressures ranging from 29.0 to 31.0-in. Hg absolute without introducing significant error. Some charts incorporate correction factors to take care of these deviations.

The ordinates of the chart are the dry-bulb temperature and specific humidity of the air-steam mixture. With given total atmospheric pressure and specific humidities, Eq. (29·2) shows that the partial pressure of the water vapor has a simple relation to humidity, so we have a simple adjacent scale at the right of Fig. 29·5. Humidity is in grains of water vapor per pound of dry air, and 1 lb is equal to 7,000 grains.

Slightly tilted from the horizontal is a family of straight-line curves for wet-bulb temperature. These meet dry-bulb temperature lines with the same numbers at the saturation curve. The latter curve also gives 100%

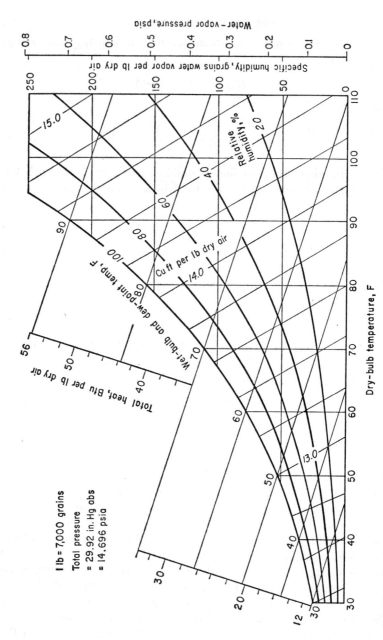

Fig. 29·5 Psychrometric chart shows eight properties of a steam-air mixture at any one state for a given pressure—in this case, standard atmospheric as the total pressure of the mixture. Process lines usually appear as simple straight lines when plotted on this chart.

relative humidity and the dew-point temperature. The family of relative-humidity curves flares up to the right.

Tilted sharply upward, another family of straight-line curves gives the specific volume of the dry air. Finally we have a scale for total heat at right angles to the wet-bulb curves. Strictly, these curves and scale apply only to the saturated mixture, but we can assume that they also apply to the enthalpy of the mixture if high accuracy is not essential.

So we have here nine factors tied together in a calculating chart: (1) dry-bulb temperature, (2) wet-bulb temperature, (3) specific humidity,

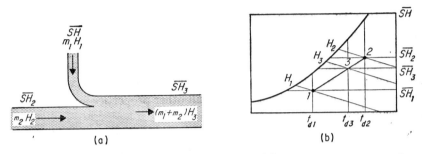

Fig. 29·6 (a) Two streams of air-steam mixtures with different properties merge to form a leaving stream with a third set of properties. (b) Mixing of two streams can be figured easily on a psychrometric chart to find the properties of the merged streams of mixtures.

(4) relative humidity, (5) water-vapor partial pressure, (6) dew-point temperature, (7) dry-air specific volume, (8) total heat, (9) total pressure of the mixture. The chart, of course, does not take any account of varying total pressure.

Knowing any two factors on this chart lets us fix six others with little or no auxiliary calculation. The chart proves to be an important tool for process calculations. Let us learn how to use it in air-conditioning problems.

29·8 Adiabatic mixing. This process, Fig. 29·6, considers two streams of air-steam mixtures merging into one to yield a given dry-bulb temperature and specific humidity. To solve this problem we apply a first-law energy balance and a conservation-of-mass balance. These tell us that

$$m_1 H_1 + m_2 H_2 = m_3 H_3 = (m_1 + m_2) H_3 \qquad (29·14)$$

where the m's refer to the masses of dry air in each stream and the H's to the enthalpy or total heat. Looking at the vapor in the streams we can write

$$m_1 \overline{SH}_1 + m_2 \overline{SH}_2 = m_3 \overline{SH}_3 = (m_1 + m_2) \overline{SH}_3 \qquad (29·15)$$

Rearranging both (29·14) and (29·15) and solving for the leaving factor, we find that

$$H_3 = \frac{m_1}{m_3} H_1 + \frac{m_2}{m_3} H_2 \qquad (29\cdot16)$$

$$\overline{SH}_3 = \frac{m_1}{m_3} \overline{SH}_1 + \frac{m_2}{m_3} \overline{SH}_2 \qquad (29\cdot17)$$

These equations show that the properties of the mixed stream are the weighted averages of the properties of the incoming streams.

Selecting any two points 1 and 2 on the psychrometric chart, Fig. 29·6b, we see that all weighted averages of any of the properties of the two states fall on a straight line connecting the two points. So to find the averaged state all we need do is locate both initial points on the psychrometric chart, draw a straight line between them, solve any one equation for a leaving property as in (29·16) or (29·17), and spot it on the connecting line as at 3. We then can read all other properties for the leaving air-steam flow.

Example: On a full-scale psychrometric chart find the final properties of mixing two streams of equal mass, one with $t_d = 93$ F and $\overline{TH} = 45.75$ Btu per lb dry air and the other with $t_d = 73$ F and $\overline{TH} = 23.3$ Btu per lb dry air.

Solution: We need to find the simple average of either property; so

$$\overline{TH}_3 = \frac{\overline{TH}_1 + \overline{TH}_2}{2} = \frac{45.75 + 23.3}{2} = 34.5 \text{ Btu}$$

$$t_{d3} = \frac{t_{d1} + t_{d2}}{2} = \frac{93 + 73}{2} = 83 \text{ F}$$

We find that both these averages lie on the straight connecting line between 1 and 2. Then we can read that $\overline{SH}_3 = 92$ grains per lb dry air, $t_{w3} = 70.5$ F, and $V_{a3} = 13.96$ cu ft per lb dry air.

29·9 Heating and cooling. Heating and cooling a stream of air-steam mixture, Fig. 29·7a to c, shows as horizontal process lines on the psychrometric chart, 29·7c. Figure 29·7a shows a stream $m_1 = m_2$ flowing through a heat exchanger that adds heat Q to the mixture. Since no vapor is added during this process, $\overline{SH}_1 = \overline{SH}_2$, but note that $\overline{RH}_1 > \overline{RH}_2$, $t_{d1} < t_{d2}$, and $t_{w1} < t_{w2}$.

Figure 29·7b shows a stream $m_3 = m_4$ flowing through a heat exchanger which extracts heat from the air-steam mixture. Since no vapor is added, $\overline{SH}_3 = \overline{SH}_4$, but other properties change in the reverse direction of heating.

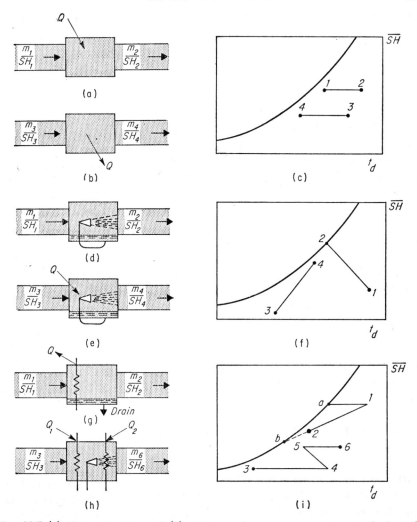

Fig. 29·7 (a) Mixture-heating and (b) mixture-cooling processes appear as horizontal lines on the psychrometric chart (c). (d) Adiabatic humidifying cools the mixture as shown in (f), but adding heat while humidifying as in (e) raises both temperature and humidity. (g) Combined cooling and dehumidifying is a two-part process with subsequent mixing as in (i). (h) Processes can be combined as needed to control final humidity and dry-bulb temperature of an air-steam mixture.

For the general case we can evaluate heat transferred as

$$Q = H_{m1} - H_{m2} \quad (29 \cdot 18)$$

for either heating or cooling the mixture stream.

29·10 Adiabatic humidifying. This process, Fig. 29·7d, adds water (in contact with the air-steam flow) to the stream. This arrangement cools the water to the saturation temperature t_w when steady flow has been established. We derived the energy balance for this process in Art. 29·5, formulating Eq. (29·9); in terms of Fig. 29·7f the balance is

$$m_1 H_1 + (\overline{SH}_2 - \overline{SH}_1) m_{a1} H_{fw} = m_2 H_2 \qquad (29\cdot 19)$$

where

$$m_{a1} + m_{s1} = m_1 \quad \text{and} \quad m_{a1} + m_{s2} = m_2$$

As Fig. 29·7f shows, this raises the specific humidity from 1 to 2 but drops the dry-bulb temperature from t_{d1} to t_{d2}. The process need not be carried to saturation as shown here. Point 2 can be controlled to fall short of the saturation curve and usually is so placed in practical systems. The process line can be assumed to follow the constant total-heat or wet-bulb-temperature line, where results need not be too accurate. For strict accuracy the process line follows the line of constant enthalpy, assuming the water injected is at some value other than t_w.

29·11 Heating and humidifying. Heating and humidifying, Fig. 29·7e, add the factor of heat addition Q to the adiabatic humidifying in Art. 29·5. For the energy balance we need to add Q to (29·19) to get

$$m_{a3} H_{a d} + m_{s3} H_{s d} + (\overline{SH}_4 - \overline{SH}_3) m_{a3} H_{f4} + Q = m_{a3} H_{a 4} + m_{s4} H_{s4}$$
$$(29\cdot 20)$$

Here we expanded the equation to show all the mass elements and individual enthalpies. Figure 29·7f shows the process line 3–4 indicating a rise in $\overline{SH}$, t_d and t_w as well as in V.

29·12 Dehumidifying and cooling. Dehumidifying and cooling an air-steam stream, Fig. 29·7g, pass the flow over cooling coils to lower its enthalpy by removing heat Q. The portion of the flow in direct contact with the coils follows the cooling process 1–a–b. In the first part 1–a, cooling takes place at constant specific humidity with that part of the flow becoming saturated at a. With further cooling below a, moisture starts condensing to some saturated state as at b. Then the directly cooled part at b mixes with the rest at state 1, and the final state of the total mixture will be at state 2 on a straight line connecting 1 and b.

The energy balance for the overall process is

$$m_1 H_1 = m_2 H_2 + (m_1 - m_2) H_f + Q \qquad (29\cdot 21)$$

For the entire mass of mixture to reach state b we would need a very large heat exchanger.

29·13 Combined processes. These, Fig. 29·7h, can use any of the principles we discussed to condition air. Here we see a heating process at constant humidity followed by an adiabatic humidifying process with a final constant-humidity heating process. Such an arrangement gives flexibility in conditioning large amounts of air to close tolerances.

Figure 29·7i shows the three processes on a psychrometric chart. Carrying on process 3–4 indefinitely would produce dry, hot air. If 4–5 were carried on to saturation, the mixture would have a higher humidity and be somewhat cooler. Process 5–6 brings the air to desired humidity and dry-bulb-temperature levels.

We have taken a detailed look at the relations of air and vapor in mixtures. Now let us apply some of the principles we learned in a few elementary processes.

29·14 Air conditioning. This process takes in atmospheric air and changes its properties to increase human comfort or for some processing need. In Fig. 29·8 we begin with air that has a high vapor load. The dry-bulb temperature is 90 F and wet-bulb 85 F; the total pressure equals 30.00 in. Hg absolute.

We pass the air through a cooler to reduce its dry-bulb and wet-bulb temperatures to 60 F by removing heat Q_1. Then a heater raises the dry-bulb to 75 F by adding heat Q_2.

First let us find, from Eq. (29·10), the specific humidity of the entering atmospheric air:

$$\overline{SH}_1 = \frac{0.622(P_{sw}/P_{aw})(H_{sw} - H_{fw}) + 0.24(t_w - t_d)}{H_{sd} - H_{fw}}$$

Substituting the given data in this equation,

$$\overline{SH}_1 = \frac{0.622 \times [1.2133/(30.00 - 1.2133)](1045.8) + 0.24(85 - 90)}{1100.9 - 53.0}$$

$$= 0.02514 \text{ lb vapor per lb dry air}$$

Now let us find the relative humidity of this air. First we must get partial pressure of steam in the mixture from

$$\overline{SH} = 0.622 \frac{P_s}{P_a} = 0.622 \frac{P_s}{P_t - P_s}$$

$$P_s = \frac{P_t \overline{SH}}{0.622 + \overline{SH}}$$

$$= \frac{30 \times 0.02514}{0.622 + 0.02514}$$

$$= 1.204 \text{ in. Hg abs}$$

$$\overline{RH}_1 = \frac{\overline{SH}_1}{0.622} \frac{P_t - P_s}{P_g}$$

$$= \frac{0.02514}{0.622} \frac{30.00 - 1.204}{1.4215}$$

$$= 0.819 \text{ or } 81.9\%$$

A study of the T-S diagram in Fig. 29·8b shows that the mixture is cooled below the initial saturation pressure of the steam at 60 F; so part of it condenses to water W_w.

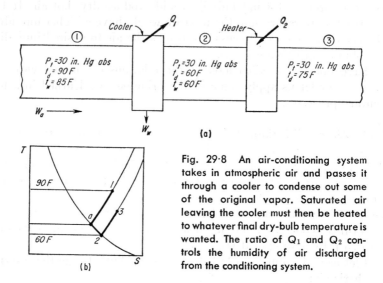

Fig. 29·8 An air-conditioning system takes in atmospheric air and passes it through a cooler to condense out some of the original vapor. Saturated air leaving the cooler must then be heated to whatever final dry-bulb temperature is wanted. The ratio of Q_1 and Q_2 controls the humidity of air discharged from the conditioning system.

Steam tables give us the partial pressure of the saturated steam; so we can figure the specific humidity of air leaving the cooler from

$$\overline{SH}_2 = \frac{0.622 P_s}{P_a}$$
$$= 0.622 \times \frac{0.5218}{30 - 0.5218}$$
$$= 0.01105 \text{ lb vapor per lb dry air}$$

Then the water condensed in the cooler is the difference of the specific humidities entering and leaving, or

$$W_w = \overline{SH}_1 - \overline{SH}_2 = 0.02514 - 0.01105$$
$$= 0.01409 \text{ lb water per lb dry air}$$

Now we are ready to figure the heat removed by the cooler. We shall do this by relating everything to 1 lb of dry air flowing through the system. First we take a first-law energy balance about the cooler:

$$w_a H_{a1} + \overline{SH}_1 H_{s1} = W_w H_f + Q_1 + w_a H_{a2} + \overline{SH}_2 H_{s2}$$

Substituting (remember that $w_a = 1.0$),

$$0.24 t_{d1} + \overline{SH}_1 (1060 + 0.45 t_{d1})$$
$$= W_w H_f + Q_1 + 0.24 t_{d2} + \overline{SH}_2 (1060 + 0.45 t_{d2})$$

Entering data from Fig. 29·8 and previous calculations,

0.24 × 90 + 0.02514(1060 + 0.45 × 90)
= 0.01409 × 28.06 + Q_1 + 0.24 × 60 + 0.01105(1060 + 0.45 × 60)

Solving,
$$Q_1 = 22.46 \text{ Btu per lb dry air}$$

Heater Process. Since no vapor leaves or enters the air-steam mixture after it leaves the cooler,

$$\overline{SH}_3 = \overline{SH}_2 = 0.01105 \text{ lb vapor per lb dry air}$$

As we can see from Fig. 29·8b, partial pressure of the steam does not change either; so
$$P_{s3} = P_{s2} = 0.5128 \text{ in. Hg abs}$$

Then the relative humidity of the mixture leaving the heater is

$$\overline{RH}_3 = \frac{P_s}{P_g} = \frac{0.5218}{0.8750} = 0.5962 \text{ or } 59.62\%$$

P_g is saturation pressure corresponding to 75 F. From the energy balance about the heater we get

$Q_2 = w_a(H_3 - H_2)$
$= 0.24 t_{d3} + \overline{SH}_3(1060 + 0.45 t_{d3}) - 0.24 t_{d2} + SH_2(1060 + 0.45 t_{d2})$

Substituting, $Q_2 = 0.24 \times 75 + 0.01105(1060 + 0.45 \times 75) - 0.24 \times 60 + 0.01105(1060 + 0.45 \times 60) = 3.68$ Btu per pound of dry air.

Practical application of Fig. 29·8 and the data we figured will let us size some of the equipment. For example, let us assume that the system takes in 1,000 cfm (cubic feet per minute) of air at the initial conditions. We want to find total water and heats that must be handled.

First we must realize that

1,000-cfm atmospheric air = 1,000 cfm dry air
= 1,000 cfm vapor

The weight of this 1,000 cfm of dry air can be figured from $w_a = PV/RT$, where P must be in psfa and V is in cfm. Then

$$w_a = \frac{144(30 - 1.204)0.4912 \times 1,000}{53.3 \times 550}$$

= 69.45 lb dry air per minute

Weight of the vapor in the atmospheric air is

$w_s = 69.45 \times \overline{SH}_1$
$= 69.45 \times 0.02514$
$= 1.75$ lb per min

Total mass flow of atmospheric air is

$$w = w_a + w_s$$
$$= 69.45 + 1.75$$
$$= 71.20 \text{ lb per min}$$

Heat abstracted by cooler is

$$Q_1 = 69.45 \times 22.46$$
$$= 1560 \text{ Btu per min}$$

Since 1 ton of refrigeration equals 200 Btu per min, we would need a refrigerator with a capacity of

$$1560/200 = 7.8 \text{ tons}$$

The refrigerator condenses $69.45 \times 0.01409 = 0.98$ lb per min of water out of the air. This equals $(0.98/8.33)60$ or 7.05 gal per hr of water.

29·15 Spray cooling of air. This, Fig. 29·9, is often used for processing work and occasionally for comfort conditioning. Here we use water at

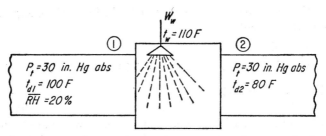

Fig. 29·9 Simple spray system can cool entering air at the expense of increasing the final humidity. The temperature of the spray water has little effect on the final dry-bulb temperature of the air.

110 F to cool air initially at dry-bulb temperature of 100 F and $\overline{RH}_1 = 20\%$ to a dry bulb of 80 F. How much spray water must be evaporated?

First let us find the entering $\overline{SH}_1$ by figuring partial pressure of the entering vapor from

$$P_{s1} = P_{g1}\overline{RH}_1$$
$$= 1.9325 \times 0.20$$
$$= 0.3865 \text{ in. Hg abs}$$

where 1.9325 is the saturation pressure at 100 F. Then

$$\overline{SH}_1 = \frac{0.622 P_{s1}}{P_t - P_{s1}}$$
$$= 0.622 \times \frac{0.3865}{30 - 0.3865}$$
$$= 0.00812 \text{ lb vapor per lb dry air}$$

From an energy balance about the cooler,

$$w_a H_{a1} + \overline{SH}_1 H_{s1} + W_w H_w = w_a H_{a2} + \overline{SH}_2 H_{s2}$$

But note that $W_w = \overline{SH}_2 - \overline{SH}_1$. Substituting, we get

$$0.24 \times 100 + 0.00812(1060 + 0.45 \times 100) + (\overline{SH}_2 - 0.00812)77.9$$
$$= 0.24 \times 80 + \overline{SH}_2(1060 + 0.45 \times 80)$$

Solving we find that

$$\overline{SH}_2 = 0.01291 \text{ lb vapor per lb dry air}$$

Then spray water added to reduce the temperature to 80 F is

$$W_w = \overline{SH}_2 - \overline{SH}_1 = 0.01291 - 0.00812$$
$$= 0.00479 \text{ lb of water per pound of dry air}$$

Now let us determine the relative humidity at the cooler outlet. First we must find P_{s2} from

$$P_s = \frac{P_t \overline{SH}}{0.622 + \overline{SH}}$$
$$= \frac{30 \times 0.01291}{0.622 + 0.01291}$$
$$= 0.6099 \text{ in. Hg}$$

Substituting in

$$\overline{RH}_2 = \frac{\overline{SH}_2}{0.622} \frac{P_a}{P_g}$$
$$= \frac{0.01291}{0.622} \frac{30 - 0.61}{1.0321} = 0.591$$

or 59.1% at the outlet.

29·16 Cooling tower. This, Fig. 29·10, acts as a device to transfer unavailable energy to the great natural sink—our atmosphere. More pointedly, the air is used to cool water that circulates through a plant to carry off unusable energy.

The water sprays into the top of the tower to fall in small droplets over "fill" material to the bottom of the tower. Atmospheric air flows upward through the tower, against the current of the water. The large aggregate surface of the water drops makes it easy to evaporate some water to mix with the air and so cool the remainder of the water.

Ideally the coolest and driest air meets the coolest water drops at the tower bottom and the water temperature equals the wet-bulb temperature of the entering air-vapor mixture; that is, $t_b = t_{w1}$. It would take a very large tower to do this; so the nearness to the ideal for an actual tower is measured by the *approach* $= t_b - t_{w1}$. The useful output of the tower is the amount of cooling done on the circulating water W_w and measured by *cooling range* $= t_a - t_b$. The ideal tower would have a cooling range of $t_a - t_{w1}$. Note that *approach* + *cooling range* $= t_a - t_{w1}$.

The energy balance of the cooling tower accounts for the dry air flowing through, the vapor entering with the air, the vapor leaving with the air,

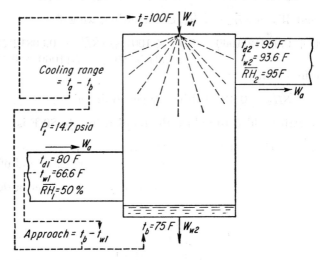

Fig. 29·10 Cooling tower takes heat out of the circulating water by evaporating a small part of it, raising the humidity of the air passing through, and so discharging energy to the atmosphere.

the cooling water entering, the cooling water leaving, and their respective enthalpies. We write the balance as

$$w_a H_{a1} + w_{v1} H_{v1} + W_{w1} H_{w1} = w_a H_{a2} + w_{v2} H_{v2} + W_{w2} H_{w2}$$

Dividing throughout by w_a, we can relate this equation to 1 lb of dry air

$$H_{a1} + \overline{SH}_1 H_{v1} + \frac{W_{w1}}{w_a} H_{w1} = H_{a2} + \overline{SH}_2 H_{v2} + \frac{W_{w2}}{w_a} H_{w2}$$

Taking a mass balance about the tower,

$$\frac{W_{w1}}{w_a} + \overline{SH}_1 = \frac{W_{w2}}{w_a} + \overline{SH}_2$$

$$\frac{W_{w1}}{w_a} = \overline{SH}_2 - \overline{SH}_1 + \frac{W_{w2}}{w_a}$$

Substituting this value of W_{w1}/w_a in the energy balance and clearing, we get

$$\frac{W_{w2}}{w_a} = \frac{H_{a1} - H_{a2} + \overline{SH}_1(H_{v1} - H_{w1}) - \overline{SH}_2(H_{v2} - H_{w1})}{H_{w2} - H_{w1}}$$

Example: An example, Fig. 29·10, lets us apply the above equations. Here we have air entering at 80 F dry bulb and 66.6 F wet bulb, which is a relative humidity of 50.0%. Cooling water enters the tower top at 100 F and leaves the bottom at 75 F. The air-vapor mixture leaves the tower at 95 F dry bulb and 93.6 F wet bulb, which is a relative humidity of 95% (the ideal tower would have $t_{d2} = t_a$ and $\overline{RH}_2 = 100\%$). How much cooled water leaves the tower per pound of dry-air flow? How much hot water enters the tower per pound of dry air?

Tower cooling range = $100 - 75 = 25$ F
Tower approach = $75 - 66.6 = 8.4$ F

$$P_{s1} = \overline{RH}_1 P_{g1}$$
$$= 0.5 \times 0.5069$$
$$= 0.2535 \text{ psia}$$
$$\overline{SH}_1 = \frac{0.622 P_{s1}}{P_t - P_{s1}}$$
$$= 0.622 \times \frac{0.2535}{14.7 - 0.2535}$$
$$= 0.01092 \text{ lb vapor per lb dry air}$$
$$P_{s2} = \overline{RH}_2 P_{g2}$$
$$= 0.95 \times 0.8513$$
$$= 0.8087 \text{ psia}$$
$$\overline{SH}_2 = \frac{0.622 P_{s2}}{P_t - P_{s2}}$$
$$= 0.622 \times \frac{0.8087}{14.7 - 0.8087}$$
$$= 0.03623 \text{ lb vapor per lb dry air}$$

With these factors figured, we are ready to calculate water cooled per pound of air:

$$\frac{0.24(80 - 95) + 0.01092(1060 + 0.45 \times 80 - 68) - 0.03623(1060 + 0.45 \times 95 - 68)}{75 - 100}$$
$$= \frac{W_{w2}}{w_a} = 1.196 \text{ lb cooled water per lb dry air}$$

Then it follows that

$$\frac{W_{w1}}{w_a} = \frac{W_{w2}}{w_a} + \overline{SH}_2 - \overline{SH}_1$$
$$= 1.196 + 0.0362 - 0.0109$$
$$= 1.2213 \text{ lb hot water per lb dry air}$$

Ideal-tower limits of performance are shown in Figs. 29·11 to 29·13. These are all based on an inlet-water temperature of 100 F, 100% outlet relative humidity, and an outlet dry-bulb temperature of 100 F. The boxes list fixed values.

In Fig. 29·11 we have additional constants: inlet dry-bulb of 80 F and zero approach. This graph shows what happens as the entering relative humidity varies from zero to 100%: The heat removed per pound of dry air drops inversely with rising humidity, which means that the cooled-water temperature rises. Water cooled per pound of dry air varies through only a narrow range.

Heat removal becomes more limited with rising humidity because of the reduced opportunity to evaporate water into the air-vapor mixture flowing through the tower. The heat absorbed by the dry air stays practically constant over the range of humidity. The cooled-water temperature

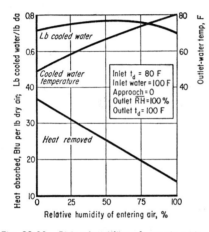

Fig. 29·11 Rising humidity of entering air raises the temperature of water leaving and cuts the heat removed from the water.

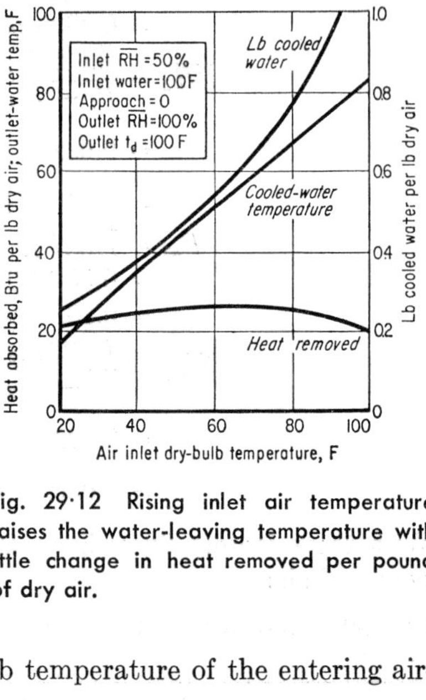

Fig. 29·12 Rising inlet air temperature raises the water-leaving temperature with little change in heat removed per pound of dry air.

for the ideal tower equals the wet-bulb temperature of the entering air-vapor mixture.

Figure 29·12 holds the entering relative humidity at 50% and varies the dry-bulb temperature. Here the heat-absorbing capacity of the dry air drops with rising inlet dry-bulb, but heat absorption involved in evaporating the water compensates; so the total heat removed per pound of dry air varies relatively little. Increasing wet-bulb temperature that goes along with rising dry-bulb makes the cooled-water temperature rise with the dry-bulb. Note that even though the dry-bulb rises to 100 F where the dry air cannot absorb heat, evaporation theoretically can cool the water to 83.2 F. The diminished cooling range with rising dry-bulb means that more pounds of water can be cooled per pound of dry air.

In Fig. 29·13 we study the effect of approach on the water-air ratio and the cooled-water temperature for 0, 50, and 100% entering relative humidity. As the approach rises, the cooled-water temperature also rises to shrink the cooling range at all humidity levels. Because of the diminishing cooling range, water cooled per pound of dry air rises with the approach, at varying rates for the different humidity levels.

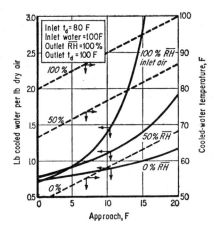

Fig. 29·13 Rising approach makes the water-leaving temperature go up but cools a greater amount of water flow.

A last word of caution: Do not confuse these curves with actual cooling-tower performance. These only define the limits within which cooling towers can work under ideal conditions.

REVIEW TOPICS

1. Why does atmospheric water vapor act like a perfect gas?
2. Show on a T-S chart the equal enthalpy of steam at the same temperature whether saturated or superheated.
3. Define specific humidity. Derive it in terms of partial pressures.
4. Define relative humidity. Derive it in terms of partial pressures, and express its relationship to specific humidity.
5. Define dew point. Explain how it can be calculated.
6. Write the general mixture equations for gas and vapor in regard to total volume, total temperature, total pressure, universal gas constant, total density, total mass, per cent mass, and per cent volume.
7. With the aid of a T-S graph explain the processes taking place in obtaining a wet-bulb reading.
8. Derive the relations in an adiabatic saturation process. Show how the entering specific humidity can be calculated.
9. Derive the equation for measuring entering specific humidity from readings of wet- and dry-bulb temperatures and the total pressure of the mixture.

10. Derive the equations for enthalpy and for total heat of an air-steam mixture.

11. Draw a rough sketch of a psychrometric chart, and describe its principal features.

12. Derive the equations for finding the enthalpy and specific humidity of two air-steam mixture streams after mixing. Show the process on a psychrometric chart.

13. What is the effect of heating or cooling an air-steam mixture. Show the processes on a psychrometric chart.

14. Show how adiabatic humidifying changes dry-bulb temperature on the psychrometric chart.

15. Show the effect of both humidifying and heating on a psychrometric chart.

16. On a psychrometric chart show the effects of dehumidifying by cooling the mixture of air and steam.

17. Indicate on a psychrometric chart the processes needed to give flexible control of air-steam mixtures.

18. How does a cooling tower function?

19. Derive the equation for determining the weight of cooled water produced per pound of dry air flowing through a cooling tower.

20. Define the *approach* and *cooling range* of a cooling tower. How are they related?

PROBLEMS

1. Find the specific humidity of a saturated air-steam mixture at 85 F dry-bulb temperature when atmospheric pressure is 30.5 in. Hg absolute.

2. An air-steam mixture has 3.2% of its volume in steam at a total pressure of 28.00 in. Hg absolute and 77 F temperature. Find its specific humidity and relative humidity.

3. An air-steam mixture has 1.9% of its volume in steam at a total pressure of 32 in. Hg absolute and 110 F temperature. Find its relative humidity.

4. An air-steam mixture with 72% relative humidity has a temperature of 92 F at a total pressure of 27.5 in. Hg absolute. Find the dew point.

5. A steam-air mixture at 15.2 psia total pressure and 120 F dry-bulb temperature has a wet-bulb temperature of 70 F. Find its (*a*) specific humidity, (*b*) relative humidity, (*c*) dew point, (*d*) specific volume, (*e*) enthalpy, (*f*) total heat.

6. Two streams of air-steam mixtures merge into a common stream. One with 1-lb-per-sec flow has a dry-bulb temperature of 60 F and wet-bulb of 55 F; the other with 2-lb-per-sec flow has a dry-bulb temperature of 98 F and wet-bulb of 81 F. Find the specific humidity and enthalpy of the merged streams when the total pressure of all streams is 29.92 in. Hg absolute.

GAS AND VAPOR MIXTURES

7. A steam-air mixture is cooled from 95 F dry bulb and 65 F wet bulb at constant pressure of 14.7 psia to 75 F dry bulb with no change in specific humidity. How much heat has been removed per pound of dry air.

8. An air-conditioning system has a cooler followed by a heater. Atmospheric air at 14.7 psia pressure, 95 F dry-bulb temperature, and 85 F wet-bulb temperature enters the system. The cooler lowers the dry-bulb temperature to 58 F, and the heater raises the dry-bulb temperature to 75 F. Find the final relative humidity of the air-steam mixture delivered by the system. If the system takes in 500 cfm of air-steam mixture, how much vapor is removed from the mixture? What must be the capacity of the refrigerator to remove this vapor?

9. A spray cooler uses water at 95 F to cool air initially at 100 F dry-bulb and 80-F wet-bulb temperatures to a dry-bulb temperature of 75 F. If the mixture is at 14.7 psia total pressure, how much spray water must be evaporated?

10. Air enters a counter flow cooling tower at 14.7 psia total pressure with 86 F dry-bulb and 70 F wet-bulb temperatures. Cooling water enters the tower top at 95 F and leaves at 78 F. The air leaves the tower at 91 F dry-bulb temperature and a relative humidity of 96%. (a) How much cooled water leaves the tower per pound of dry-air flow? (b) How much hot water enters the tower per pound of dry air? (c) What is the tower cooling range? (d) What is the tower approach?

APPENDIX A

WORK—CONSTANT TEMPERATURE

Work for a constant-temperature process, Fig. 6·6, is measured by the area under the process curve or

$$W = \int P\, dV$$

But since temperature is constant,

$$PV = C$$

Then,

$$W = \int \frac{C}{V} dV = C \int_{V_1}^{V_2} \frac{dV}{V} = C \log_e \frac{V_2}{V_1}$$
$$= P_1 V_1 \log_e \frac{V_2}{V_1}$$
$$= P_1 V_1 \log_e \frac{P_1}{P_2}$$

APPENDIX B

WORK—ADIABATIC PROCESS

During a reversible adiabatic process, Fig. 6·8, the energy equation can be written as

$$\Delta E + W = 0$$

or

$$c_v\, dT + P \frac{dV}{J} = 0$$

$$dT = -\frac{P\, dV}{c_v J}$$

Differentiating the perfect-gas equation $PV = RT$,

$$P\,dV + V\,dP = R\,dT$$

$$dT = \frac{P\,dV + V\,dP}{R}$$

Equating the two expressions for dT,

$$-\frac{P\,dV}{c_v J} = \frac{P\,dV + V\,dP}{R}$$

Multiplying both sides by $c_v R/PV$,

$$-\frac{R\,dV}{JV} = c_v \frac{dV}{V} + c_v \frac{dP}{P}$$

$$-\left(\frac{R}{J} + c_v\right)\frac{dV}{V} = c_v \frac{dP}{P}$$

$$-c_p \frac{dV}{V} = c_v \frac{dP}{P}$$

$$-k\frac{dV}{V} = \frac{dP}{P}$$

Integrating between states 1 and 2,

$$-k\int_{V_1}^{V_2} \frac{dV}{V} = \int_{P_1}^{P_2} \frac{dP}{P}$$

$$-k\log_e \frac{V_2}{V_1} = \log_e \left(\frac{V_1}{V_2}\right)^k = \log_e \frac{P_2}{P_1}$$

$$\left(\frac{V_1}{V_2}\right)^k = \frac{P_2}{P_1}$$

$$P_1 V_1^k = P_2 V_2^k = PV^k = C$$

Work for any reversible process is measured by the area under the process curve on P-V coordinates as in Fig. 6·8, or

$$W = \int P\,dV = \int \frac{C}{V^k}\,dV$$

Integrating between states 1 and 2,

$$W = C\int_{V_1}^{V_2} \frac{dV}{V^k} = \frac{CV^{-k+1}}{-k+1}\Big]_{V_1}^{V_2} = \frac{CV_2^{1-k} - CV_1^{1-k}}{1-k}$$

$$= \frac{P_2 V_2^k V_2^{1-k} - P_1 V_1^k V_1^{1-k}}{1-k}$$

$$= \frac{P_2 V_2 - P_1 V_1}{1-k} \qquad \text{ft-lb per lb gas}$$

APPENDIX C

ENTROPY—CONSTANT PRESSURE

In a reversible constant-pressure process,
$$dQ = c_p \, dT$$
and since
$$dS = \frac{dQ}{T}$$
$$\Delta S = c_p \int_{T_1}^{T_2} \frac{dT}{T} = c_p \log_e \frac{T_2}{T_1}$$

APPENDIX D

ENTROPY—CONSTANT VOLUME

In a reversible constant-volume process,
$$dQ = c_v \, dT$$
and since
$$dS = \frac{dQ}{T}$$
$$\Delta S = c_v \int_{T_1}^{T_2} \frac{dT}{T} = c_v \log_e \frac{T_2}{T_1}$$

APPENDIX E

ENTROPY—POLYTROPIC PROCESS

In a reversible polytropic process,
$$dQ = c_n\, dT$$
and since
$$dS = \frac{dQ}{T}$$
$$\Delta S = c_n \int_{T_1}^{T_2} \frac{dT}{T} = c_n \log_e \frac{T_2}{T_1}$$

APPENDIX F

Perfect-gas Formulas for One Pound of Gas

Process

Quantity	Constant volume	Constant pressure	Constant temperature	Constant entropy	Polytropic
Pressure, volume, and temperature	$\dfrac{T_1}{T_2} = \dfrac{P_1}{P_2}$	$\dfrac{T_1}{T_2} = \dfrac{V_1}{V_2}$	$P_1V_1 = P_2V_2$	$P_1V_1^k = P_2V_2^k$ $\dfrac{T_1}{T_2} = \left(\dfrac{V_2}{V_1}\right)^{k-1}$ $= \left(\dfrac{P_1}{P_2}\right)^{(k-1)/k}$	$P_1V_1^n = P_2V_2^n$ $\dfrac{T_1}{T_2} = \left(\dfrac{V_2}{V_1}\right)^{n-1}$ $= \left(\dfrac{P_1}{P_2}\right)^{(n-1)/n}$
Work	0	$P(V_2 - V_1)$	$P_1V_1 \log_e \dfrac{V_2}{V_1}$	$\dfrac{P_2V_2 - P_1V_1}{1 - k}$	$\dfrac{P_2V_2 - P_1V_1}{1 - n}$
Heat	$c_v(T_2 - T_1)$	$c_p(T_2 - T_1)$	$\dfrac{P_1V_1}{J} \log_e \dfrac{V_2}{V_1}$	0	$c_n(T_2 - T_1)$
Internal-energy change	$c_v(T_2 - T_1)$	$c_v(T_2 - T_1)$	0	$c_v(T_2 - T_1)$	$c_v(T_2 - T_1)$
Enthalpy change	$c_p(T_2 - T_1)$	$c_p(T_2 - T_1)$	0	$c_p(T_2 - T_1)$	$c_p(T_2 - T_1)$
Entropy change	$c_v \log_e \dfrac{T_2}{T_1}$	$c_p \log_e \dfrac{T_2}{T_1}$	$\dfrac{R}{J} \log_e \dfrac{V_2}{V_1}$	0	$c_n \log_e \dfrac{T_2}{T_1}$
Specific heat	c_v	c_p		0	$c_n = c_v \left(\dfrac{k - n}{1 - n}\right)$

APPENDIX G

OPTIMUM INTERMEDIATE PRESSURE

Considering Fig. 12·7b and $P_2V_2 = P_3V_5 = RT_a$ we can simplify Eq. (12·15) to

$$W = \frac{n}{1-n} RT_a \left[\left(\frac{P_3}{P_2}\right)^{(n-1)/n} + \left(\frac{P_6}{P_3}\right)^{(n-1)/n} - 2 \right]$$

Differentiating W with respect to P_3 and equating to zero gives us the relation needed to find the value of P_3 for minimum work input. In this equation T_a, R, n, P_2, and P_6 are constants.

$$\frac{dW}{dP_3} = 0 = \frac{(n-1/n)P_3^{-1/n}}{P_2^{(n-1)/n}} + \frac{n-1}{n} P_6^{(n-1)/n} P_3^{(1-2n)/n}$$

Solving for P_3 we find that

$$P_3 = \sqrt{P_2 P_6}$$

APPENDIX H

NOZZLE CRITICAL PRESSURE RATIO

Critical pressure ratio is that value where the ratio $w/A_2 = v_2/V_2$ becomes a maximum. From Eqs. (18·12) and (18·13)

$$\frac{v_2}{V_2} = \frac{223.8 \sqrt{c_p T_0 [1 - (P_2/P_0)^{(k-1)/k}]}}{V_0 (P_0/P_2)^{1/k}}$$

$$= \frac{223.8}{V_0} \sqrt{c_p T_0 \left[\left(\frac{P_2}{P_0}\right)^{2/k} - \left(\frac{P_2}{P_0}\right)^{(k+1)/k} \right]}$$

For a given nozzle with a given entering gas all the terms of this equation are constants except P_2, which may be varied for the conditions being studied. So v_2/V_2 becomes a maximum when the quantity in the brackets becomes a maximum. This becomes a maximum when its derivative in respect to varying P_2 becomes 0; so

$$\frac{d[(P_2/P_0)^{2/k} - (P_2/P_0)^{(k+1)/k}]}{dP_2}$$

$$= \frac{1}{P_0^{2/k}} \frac{2}{k} P_2^{(2-k)/k} - \frac{1}{P_0^{(k+1)/k}} \frac{k+1}{k} P_2^{1/k} = 0$$

$$\frac{P_2}{P_0} = P_c = \left(\frac{2}{k+1}\right)^{k/(k-1)}$$

APPENDIX I

Thermodynamic Properties of Sodium and Its Vapor

By E. L. Dunning, Associate Professor of Applied Science, Southern Illinois University. Reprinted from July, 1961, issue of Power. Superscript numbers refer to bibliography at end of this Appendix.

Sodium has become an important heat-transfer medium in power and process plants. Table I·1 gives data on saturated sodium liquid and vapor and Table I·2 gives data on superheated sodium vapor. A Mollier chart illustrates the variation of properties listed in the tables, Fig. I·1.

These properties are not complete, but their scope makes them useful in designing heat-transfer equipment.

Vapor-pressure data. Several empirical equations have been derived for relating vapor-pressure data of sodium. Published data on these have been inconsistent. Equation (I·1)[1] has been chosen as the most reliable:

$$\log P = \frac{-10{,}021}{T} - 0.5 \log T + 8.149 \qquad (I·1)$$

where P is in psia and T in degrees Rankine. This equation gives a normal boiling point of 1618 F. Another equation[2] in two parameters gives about the same boiling temperature:

$$\log P = \frac{-9396}{T} + 5.688 \qquad (I·2)$$

But equation (I·2) assumes that the latent heat of vaporization is independent of temperature. This is not so, as we shall show later.

APPENDIX I

Table I·1 Thermodynamic properties of sodium, saturated phases

Temp. F	Pressure, psia	Volume		Enthalpy			Entropy		
		v_f, ft³ per lb	v_g, ft³ per lb	h_f, Btu per lb	h_{fg}, Btu per lb	h_g, Btu per lb	s_f, Btu per lb, °R	s_{fg}, Btu per lb, °R	s_g, Btu per lb, °R
208	16.8×10⁻¹⁰	0.01726	18.53×10¹⁰	0	1964	1964	0	2.9407	2.9407
250	15.02×10⁻⁹	0.01736	21.98×10⁹	13.85	1962	1976	0.0201	2.7633	2.7834
300	1.07×10⁻⁷	0.01749	32.96×10⁸	30.20	1961	1991	0.0423	2.5802	2.6225
350	7.44×10⁻⁷	0.01762	50.45×10⁷	46.40	1952	1998	0.0629	2.4099	2.4728
400	37.52×10⁻⁷	0.01775	10.59×10⁷	62.45	1943	2005	0.0821	2.2593	2.3414
450	16.59×10⁻⁶	0.01788	25.36×10⁶	78.35	1933	2011	0.1000	2.1241	2.2242
500	51.50×10⁻⁶	0.01801	8.576×10⁶	94.10	1924	2018	0.1168	2.0041	2.1209
350	17.45×10⁻⁵	0.01818	26.52×10⁵	109.80	1913	2023	0.1327	1.8940	2.0267
600	52.00×10⁻⁵	0.01829	9.298×10⁵	125.40	1902	2027	0.1478	1.7944	1.9421
650	13.00×10⁻⁴	0.01843	38.78×10⁴	140.90	1890	2031	0.1621	1.7027	1.8648
700	3.05×10⁻³	0.01858	17.18×10⁴	156.30	1880	2036	0.1757	1.6207	1.7964
750	6.72×10⁻³	0.01871	8.072×10⁴	171.60	1864	2036	0.1886	1.5404	1.6683
800	14.30×10⁻³	0.01886	3.921×10⁴	186.85	1849	2036	0.2009	1.4674	1.6135
850	27.82×10⁻³	0.01901	2.081×10⁴	202.05	1835	2037	0.2127	1.4008	1.5629
900	53.40×10⁻³	0.01916	1.118×10⁴	217.20	1821	2038	0.2240	1.3389	1.5158
950	94.10×10⁻³	0.01931	6592	232.30	1806	2038	0.2349	1.2809	1.4728
1000	0.1662	0.01947	3801	247.35	1792	2039	0.2454	1.2274	1.4329
1050	0.2740	0.01963	2367	262.35	1778	2040	0.2555	1.1774	1.3961
1100	0.4280	0.01978	1555	277.35	1764	2041	0.2653	1.1308	1.3617
1150	0.6575	0.01995	1037	292.35	1750	2042	0.2748	1.0869	1.3298
1200	1.012	0.02011	689	307.35	1736	2043	0.2840	1.0458	1.2999
1250	1.489	0.02029	479	322.35	1722	2044	0.2929	1.0070	1.2714
1300	2.120	0.02045	344	337.35	1707	2045	0.3015	0.9699	1.2453
1350	3.060	0.02064	243	352.40	1693	2045	0.3099	0.9354	1.2203
1400	4.275	0.02081	178	367.45	1678	2045	0.3181	0.9022	1.1978
1450	5.855	0.02100	132	382.55	1665	2048	0.3261	0.8717	1.1777
1500	7.880	0.02118	100	397.70	1654	2052	0.3339	0.8438	1.1585
1550	10.18	0.02137	79.1	412.90	1642	2055	0.3416	0.8169	1.1408
1600	13.28	0.02156	62.0	428.20	1631	2059	0.3491	0.7917	1.1345
1618	14.70	0.02170	56.1	433.70	1626	2060	0.3519	0.7826	1.1233
1650	17.12	0.02181	48.8	443.55	1618	2062	0.3565	0.7668	1.1072
1700	22.00	0.02203	38.6	459.00	1606	2065	0.3637	0.7435	1.1072
1750	27.82	0.02225	31.14	474.55	1594	2069	0.3708	0.7212	1.0920
1800	34.42	0.02247	25.60	490.20	1584	2074	0.3778	0.7009	1.0787
1850	42.72	0.02271	20.99	505.95	1576	2082	0.3847	0.6823	1.0670
1900	52.05	0.02299	17.52	521.85	1567	2090	0.3915	0.6639	1.0554
1950	63.87	0.02317	14.52	537.80	1557	2095	0.3982	0.6461	1.0443
2000	75.75	0.02341	12.44	553.90	1546	2100	0.4048	0.6285	1.0333
2050	91.02	0.02372	10.31	570.15	1538	2108	0.4113	0.6128	1.0241
2100	107.23	0.02393	8.796	586.50	1530	2117	0.4177	0.5977	1.0154
2150	127.18	0.02919	7.457	603.00	1524	2127	0.4241	0.5839	1.0080
2200	147.29	0.02446	6.537	619.60	1518	2138	0.4304	0.5706	1.0010
2250	170.96	0.02474	5.719	636.30	1513	2149	0.4366	0.5583	0.9949
2300	198.45	0.02503	4.947	653.15	1507	2160	0.4428	0.5460	0.9888

Dimerization. Sodium vapor is monatomic at normal temperature and dimerizes to Na_2 as the temperature rises. The reaction is $Na + Na = Na_2$. Property values in the table and chart account for the dimer. The quantities were evaluated on a mole basis and then divided by the molecular weights.

Molecular weight is proportional to the amount of dimer in the mixture and can be figured as

$$M_m = X_1(M_1) + X_2(M_2) \tag{I·3}$$

Table I·2 Thermodynamic properties of sodium, superheated vapor

Temp, F	Pressure = 0.02 psia			Pressure = 0.05 psia			Pressure = 0.10 psia		
	v, ft³ per lb	h, Btu per lb	s, Btu per lb, °R	v, ft³ per lb	h, Btu per lb	s, Btu per lb, °R	v, ft³ per lb	h, Btu per lb	s, Btu per lb, °R
821	28,530	2036	1.6453	—	—	—	—	—	—
893	—	—	—	10,560	2038	1.5697	—	—	—
900	30,830	2036	1.6733	11,910	2044	1.5728	—	—	—
954	—	—	—	—	—	—	4019	2038	1.5124
1000	33,630	2117	1.7045	13,110	2089	1.6042	4985	2063	1.5293
1100	36,060	2144	1.7223	14,300	2134	1.6356	7085	2118	1.5661
1200	38,530	2172	1.7397	15,300	2163	1.6533	7611	2152	1.5868
1300	40,950	2196	1.7536	16,300	2193	1.6711	8138	2187	1.6076
1400	43,340	2221	1.7675	17,250	2216	1.6834	8622	2212	1.6213
1500	45,680	2242	1.7784	18,200	2239	1.6958	9106	2238	1.6350
1600	48,000	2264	1.7891	19,150	2262	1.7072	9575	2260	1.6457
1700	50,330	2286	1.7998	20,100	2286	1.7186	10,040	2282	1.6564
1800	52,670	2312	1.8106	21,030	2312	1.7294	10,520	2308	1.6672
1900	55,000	2335	1.8214	21,960	2335	1.7402	11,000	2335	1.6780
2000	57,320	2358	1.8310	22,890	2358	1.7497	11,470	2358	1.6885
2100	59,660	2382	1.8405	23,820	2382	1.7593	11,930	2382	1.6871
2200	61,990	2406	1.8495	24,750	2406	1.7683	12,400	2406	1.7061
2300	64,320	2430	1.8585	25,680	2430	1.7773	12,860	2430	1.7151

Temp, F	Pressure = 1.0 psia			Pressure = 5.0 psia			Pressure = 14.7 psia		
1198	702	2043	1.3309	—	—	—	—	—	—
1200	703	2044	1.3315	—	—	—	—	—	—
1300	769	2148	1.3906	—	—	—	—	—	—
1400	825	2173	1.4032	—	—	—	—	—	—
1423	—	—	—	157	2049	1.2097	—	—	—
1500	882	2198	1.4158	164	2092	1.2320	—	—	—
1600	938	2233	1.4335	177	2151	1.2606	—	—	—
1618	—	—	—	—	—	—	56.1	2060	1.1345
1700	995	2268	1.4512	191	2210	1.2892	60.2	2111	1.1285
1800	1043	2296	1.4635	202	2251	1.3075	64.8	2169	1.1893
1900	1092	2324	1.4759	214	2293	1.3259	69.4	2228	1.2102
2000	1140	2350	1.4864	224	2323	1.3382	72.9	2269	1.2269
2100	1188	2376	1.4970	234	2354	1.3506	76.5	2310	1.2436
2200	1236	2403	1.5072	245	2386	1.3626	80.9	2352	1.2596
2300	1285	2430	1.5173	255	2919	1.3745	85.4	2395	1.2756

Temp, F	Pressure = 30 psia			Pressure = 75 psia			Pressure = 150 psia		
1767	29.3	2071	1.0876	—	—	—	—	—	—
1800	30.0	2092	1.0967	—	—	—	—	—	—
1900	32.2	2155	1.1242	—	—	—	—	—	—
1998	—	—	—	12.5	2099	1.0337	—	—	—
2000	34.2	2210	1.1465	12.6	2100	1.0347	—	—	—
2100	36.2	2265	1.1689	13.0	2167	1.0604	—	—	—
2200	38.4	2315	1.1877	14.1	2230	1.0841	—	—	—
2206	—	—	—	—	—	—	6.44	2139	1.0003
2250	—	—	—	—	—	—	6.70	2172	1.0142
2300	40.5	2365	1.2064	15.2	2293	1.1078	6.96	2215	1.0282

where M_m = molecular weight of mixture
M_1 = molecular weight of component 1
M_2 = molecular weight of component 2
X_1 = mole fraction of component 1
X_2 = mole fraction of component 2

In these calculations, the molecular weights of Na and Na₂ were taken as 23 and 46, respectively. The mole fractions were calculated by using these equations:[3]

$$Kp = \frac{P\text{Na}_2}{(P\text{Na})^2} \tag{I·4}$$

APPENDIX I

where Kp = equilibrium constant
PNa_2 = partial pressure of Na_2 vapor
PNa = partial pressure of Na vapor

and
$$N_1 = \frac{1}{1 + 0.272 P_t K_p} \tag{I.5}$$

where N_1 = moles of Na vapor at equilibrium
P_t = total pressure psia

also,
$$X_1 = \frac{2N_1}{1 + N_1} \quad \text{and} \quad X_1 + X_2 = 1$$

Values of PNa and PNa_2 were found in the literature[4] for the range of 208 to 2050 F. Beyond this range, extrapolated values for K_p were used.

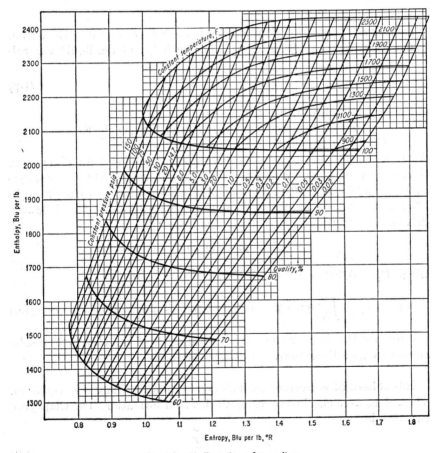

Fig. I·1 Mollier chart for sodium.

Specific volume. This was determined from the density-temperature relationship. Equation (I·6)[4] below was chosen as being sufficient for density evaluation of liquid sodium:

$$d = 59.47 - 0.0077t - 3.48 \times 10^{-7}t^2 \qquad (\text{I·6})$$

where d = density, lb per cu ft
t = temperature, F

Specific volume of the vapor was determined by using the perfect-gas equation and applying the compressibility factor z:

$$pv = zRT$$

Factor z was found from the generalized compressibility charts using 4091°R and 5,050 psia as the critical temperature[3] and pressure, respectively.

Enthalpy. The melting point of sodium, 208 F, was used as the datum plane for evaluating all enthalpies. The enthalpy of the liquid was calculated as

$$\Delta h = C_p \Delta t \qquad (\text{I·7})$$

where h = enthalpy, Btu per lb
C_p = specific heat at constant pressure, Btu per lb, F
t = temperature, F

Average specific heats were used over small temperature differences. Specific-heat data are available in the literature[4] for the temperature range from 208 to 2050 F. Extrapolated values were used from 2050 to 2300 F. As a further check on liquid-enthalpy values, the following differential and algebraic equations were used:

$$dh = C_p \, dt$$

where $C_p = 0.389 - 1.106 \times 10^{-4}T + 0.3410 \times 10^{-7}T^2$

$$h_1 - h_2 = \int_{T_1}^{T_2} 0.389T - 0.553 \times 10^{-4}T^2 + 0.1137 \times 10^{-7}T^3 \qquad (\text{I·8})$$

T is temperature in degrees Rankine. Evaluation of Eq. (I·8) gives results consistent with those derived by direct methods.

Latent heat of vaporization. This comes from the Clapeyron equation. The values were checked by using the heats of formation. The Clapeyron equation is

$$\frac{dp}{dT} = \frac{h_{fg}}{T(V_g - V_f)} \qquad (\text{I·9})$$

where h_{fg} = latent heat of vaporization, Btu per lb
V_g = specific volume of gas, cu ft per lb
V_f = specific volume of liquid, cu ft per lb

If V_f is neglected and V_g obtained by $V_g = RT/p$, Eq. (I·9) can be written

$$h_{fg} = \frac{dp}{dT}\frac{RT^2}{p} \tag{I·10}$$

Substituting Eq. (I·10) into Eq. (I·1) and integrating give

$$h_{fg} = \frac{45{,}860 - 0.993T}{M_m} \tag{I·11}$$

where M_m is the molecular weight of the gaseous mixture at saturation. The latent heat of vaporization can also be determined by using the heats of formation:

$$h_{fg} = \frac{X_1(\Delta H_{f1}) + X_2(\Delta H_{f2})}{M_m} \tag{I·12}$$

where ΔH_{f1} = heat of formation of component 1, Btu per mole
ΔH_{f2} = heat of formation of component 2, Btu per mole

Reliable values of heats of formation were found in item 5 of the bibliography.

The enthalpy of saturated vapor was found by

$$h_g = h_f + h_{fg} \tag{I·13}$$

The enthalpy of superheated vapor was found by using heats of formation as follows:

$$h = h_{fg} + \frac{X_1(\Delta H_{f1}) + X_2(\Delta H_{f2})}{M_m} \tag{I·14}$$

Entropy. The melting point of sodium, 208 F, was used as the datum plane for evaluating all entropy quantities. The entropy of saturated liquid was figured by

$$\Delta s = \frac{\Delta h}{T}$$

where s is the entropy in Btu per pound, degree Rankine. This method was checked by evaluating

$$s_1 - s_2 = \int_{T_2}^{T_1} C_p \frac{dT}{T} \tag{I·15}$$

where $C_p = 0.389 - 1.106 \times 10^{-4}T + 0.341 \times 10^{-7}T^2$

The entropy of vaporization was figured as

$$s_{fg} = \frac{h_{fg}}{T}$$

The entropy of the saturated vapor is $s_g = s_f + s_{fg}$ and the entropy of the superheated vapor is $\Delta s = \Delta h/T$.

BIBLIOGRAPHY

1. Ditchburn, R. W., and J. C. Gilmour: "Vapor pressures of monatomic vapors," *Review of Modern Physics* 13, pp. 310–327 (1941).
2. Makansi, M. M., C. H. Muendel and W. A. Selke: "Determination of the vapor pressure of sodium," *Journal of Physical Chemistry*, 59, 40 (1955).
3. Inatomi, T. H., and W. C. Parrish: *Thermodynamic diagrams for sodium*, North American Aviation, NAA-SR-62 (1950).
4. Sittig, Marshall: *Sodium, its manufacture, properties and uses*, Reinhold Publishing Corporation, New York, 1956.
5. Evans, W. H., et al, "Thermodynamic properties of the alkali metals," *Journal of Research* of the National Bureau of Standards.

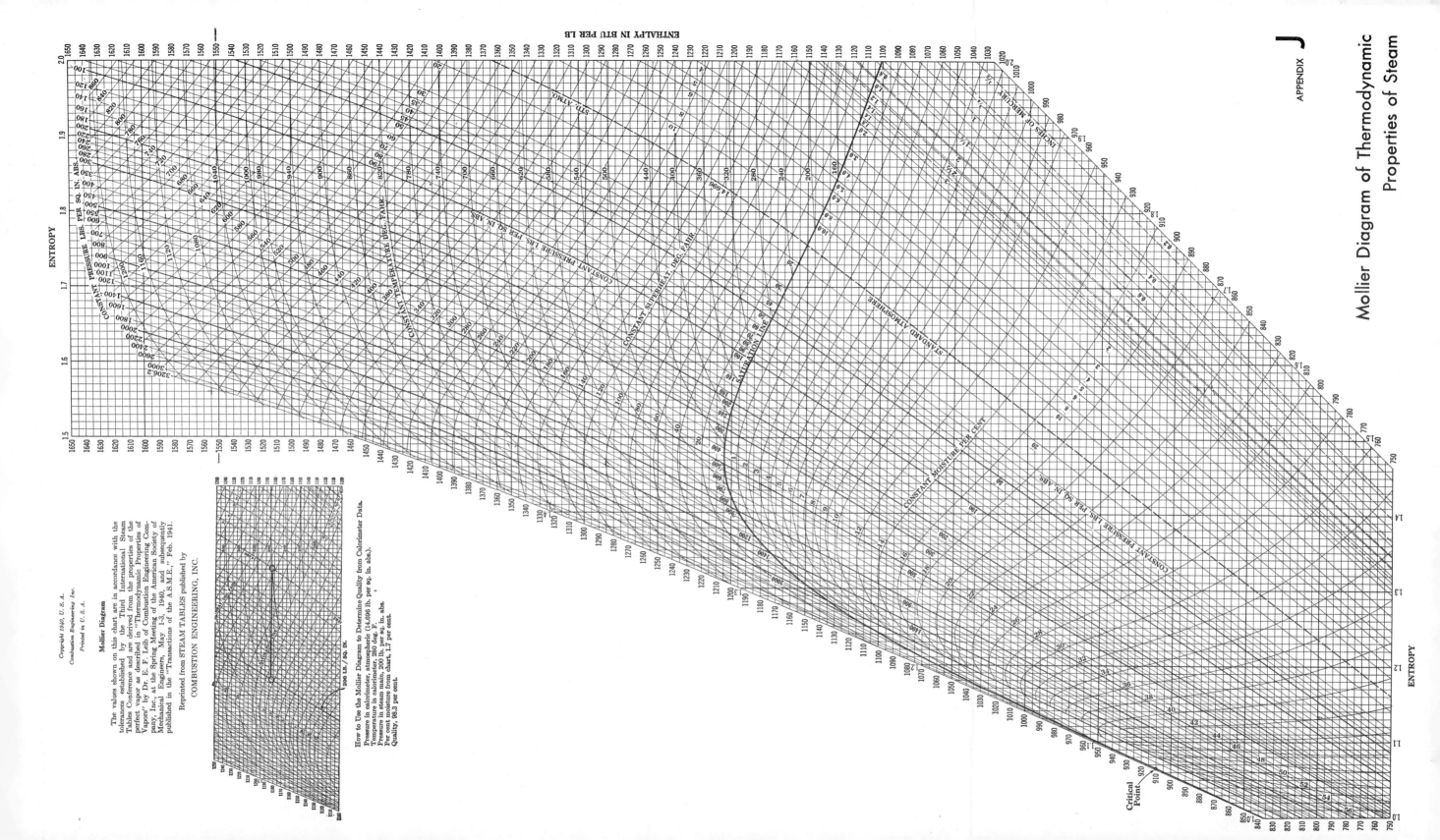

Index

Adiabatic entropy change, 106
Adiabatic humidifying, 469–470
Adiabatic mixing, air-steam, 467–468
Adiabatic processes, 80–81
    Carnot cycle, 95, 102–104
        irreversible, 129–130
    comparison of, 85–87
    irreversible, steam, 365–367
        compression, 375–376
        expansion, 373–375
    reversible, 81, 102–104
        constant-S, 145, 159
        energy equation, 483–484
        steam, 364–365
Admission, engine, 155, 430
Aftercooler, 157
Air, compressed (see Compressed air)
    perfect-gas equation, 31, 32
    and steam, mixtures, 456–481
    thermodynamic properties, 303–305
Air compressors (see Compressors)
Air conditioning, 464–465, 471–474
    adiabatic humidifying, 469–470
    dehumidifying and cooling, 470
    heating and cooling, 468–471
    heating and humidifying, 470
    spray cooling, 474–475
Air engine, 155–156
Air induction heating, 243–245
Air supply, combustion, 324–325
Ammonia refrigerant, 441, 443–446
Atmospheric humidity, 456–457
Atmospheric pressure, 16–18

Atoms, 6–8
    gas and vapor, 16
    motion of, 8–9
Availability loss, 393
Available energy, 101
Avogadro's law, 36–38, 309–310, 457

Back pressure, Rankine cycle, 386
Barometer, 16
    standard, 18
Biphenyl, thermodynamic properties, 358
Bleeding regenerative cycle, 401–413
    energy flow, 405–406
    multiple, 408–409
    one-heater, 406–408
    practical, 409–410
    T-S chart, 404
Blowdown cooling, 241–243
Boiling, 10
Boltzmann, Ludwig, 67
Bomb calorimeter, 43–44
    constant-volume, 334–336
Bounce cylinders, 207
Bourdon tube, 19
Boyle's law, 20–22
    perfect-gas equation, 32, 33
Brayton, George, 137
Brayton cycle (Joule cycle), 137–140
British thermal unit (Btu), 40, 43
Buckets, gas turbine, 283–297
    efficiency, 287–288, 290
    energy transfer, 268–269

Buckets, gas turbine, force developed, 286–287
   pressure compounding, 288–289
   velocity compounding, 289–290
   velocity relations, 285–286

Calorimetry, 43–45
   constant-pressure, 335–336
   constant-volume, 334–336
Carnot, Sadi, 91
Carnot cycle, 90–95, 102–103
   compared, with Otto cycle, 175
     with Rankine cycle, 389–390
     with Stirling and Ericsson cycles, 133–134
   efficiency, 95–96, 99
   irreversibility, 120–122, 130–131
   temperature influence, 96–99
Carnot engine, 89–100, 225
   design factors, 112–115
Carnot heat pump, 118
Charles' law, 26–27
   perfect-gas equation, 30–33
Chart, psychrometric, 465
Chemoelectric production, 322
Clearance, compressor, 150
Coefficient of performance, 117, 440
Cold room, 116
Combustion, 318, 323–324
   air supply, 324–325
   conditions, 332–333
Combustion products, 326, 329–332
Compounding, steam engine, 435
Compressed-air cycles, 142–170
   constant-S, 145, 159
   constant-T, 148–149, 158
   polytropic, 149, 158–159
Compressed-air systems, 142, 156–168
   analysis, 159–162
   energy flow, 158, 162–164
   performance, 161–162, 166–168
Compressed liquid, 348, 357
Compressibility factor, 306–307
Compression ratio, diesel cycle, 194–195
   Otto cycle, 178–179
Compressors (air and gas), 142–170
   air output, varying, 151–152
   axial-flow, 298
   centrifugal, 299
   clearance, 150–151
   constant-S, 145, 159
   constant-T, 148–149, 158
   discharge stroke, 146
   efficiency, 257

Compressors (air and gas), intake stroke, 146
   multistage, 152–154
   performance, 166
   polytropic, 149, 158
   rotary, 300
   steady-flow, 298–301
   three-stage, 154–155
Condensation, cylinder, 434
Condensing, 342
Condition line, turbine, 293
Conservation, of energy, law of, 45
   of momentum, 60
Constant-entropy process, 302–305
   steam, 364–365, 372–373
Constant-pressure process, 26–31, 52–54, 76
   comparison of, 84–85, 87
   energy conversion, 72
   entropy change, 107–108
   reversible, 73, 485
   steam, 361–362, 368–369
   volume relations, 53–54
Constant-pressure specific heat, 50–52
Constant-temperature process, 77–79
   Carnot cycle, 102
   comparison, 85, 87
   compressed air, 148–149, 158
   entropy calculation, 103–106
   steam, 363–364, 371–372
   work, 103–106, 483
Constant-volume process, 26–31, 77
   comparison, 83
   entropy change, 108–109
   reversible, 485
   steam, 363, 369–370
Constant-volume specific heat, 47, 50–52
Conventional volumetric efficiency, 167
Conversion of energy (*see* Energy conversion)
Cooling, blowdown, 241–243
Cooling tower, 475–479
Corresponding states, 307
Critical point, 344–345
Critical pressure ratio, nozzle, 274
Cutoff, engine, 155, 432
Cycle efficiency, Carnot, 95–96, 99
   gas turbine, 252, 254, 257
   Rankine, 383, 386, 391
   steam reheat, 396
   (*See also* Thermal efficiency)

Cycles, actual, 215–227
  basic, Brayton (Joule), 137–140
  Carnot (*see* Carnot cycle)
  Ericsson, 133, 134, 136
  Stirling, 133–136
  compressed-air, 142–170
    (*See also* Compressors)
  gas-turbine, 249–267
  ideal, 204, 223
  internal-combustion-engine, 171–206, 215–227
    (*See also* Diesel cycle; Otto cycle)
  steam, 382–422

Dalton's law, 310–311, 457
Dehumidifying and cooling, 470
Density, 37
Detonation, diesel engine, 199–200
  Otto engine, 197–199
Dew point, 461
Diesel cycle, 193–197
  compared with free-piston gas generator, 214
  load control, 196
  supercharging, 197
  thermal efficiency, 194–196
  two-stroke, 219–221
Diesel engine, 193–197
  detonation, 199–200
  with free-piston gas generator, 207
  performance, 226
  preignition, 197
  supercharging, 197
Diffusers, 428–430
Diffusing action, 281
Dissociation, 333–334

Efficiency, compressor, 257
  conventional volumetric, 167
  cycle (*see* Cycle efficiency)
  machine, 257–258
  thermal (*see* Thermal efficiency)
  turbine, 257–258
Einstein's relation, 1
Electrical energy, 321
  measurement of, 42–43
Energy, changing form of, 4, 15, 49
  conservation of, 45
  convertibility of, 2, 49–52
  definition of, 1
  entropy of, 126–128
  forms, 3
  internal, 2, 5, 13, 40
  kinetic, 3, 60, 229

Energy, measurement of, 40–56
  units, 43
  molecular, 9–12
  potential, 3, 229–230
  release of, 11–12
  steady-flow, 228, 230
Energy conversion, 49–52, 71–72, 101
  basic processes, 319–321
  constant-pressure, 72
  Otto cycle, 186, 187
Energy equation, 1, 71–88, 483–486
  fan, 232
  gas turbine, 233
  heat exchanger, 234
  nonflow (simple), 75
  nozzle, 233, 269
  steady-flow (*see* Steady-flow energy equation)
  throttling, 235
Energy flow, compressed-air system, 158, 162–164
  free-piston cycle, 214
Energy sources, 318–319
Energy transfer, turbine, 268–269
Engine (*see* specific engine)
Engine cutoff, 432
Engine cycle (*see* Cycles; specific cycle)
Enthalpy, 76, 231
  fusion, 341
  latent, 348
  stagnation, 271
  steady-flow equation, 231–232
  vaporization, 341
Entropy, 101, 103–106
  of energy, 126–128
Entropy change, adiabatic, 106
  constant-pressure, 107–108, 485
  constant-volume, 108–109, 485
  polytropic, 109, 486
Ericsson, John, 136
Ericsson cycle, 133, 134, 136
Evaporation, 341
Excess air, 325
Expander, air, 165

Fahrenheit temperature, 26, 28, 32
Fan energy equation, 232
First law of thermodynamics, 45, 318
Flow, mass, 230
  nozzle, 236–237, 268–282
  molecular, 238–239
  vortex, 239–240
  (*See also* Steady-flow energy equation)

Flow work, 228–229
Foot pound, 42
Forces, molecular, 7–8
Free air, 143
Free-piston gas generator, 207–214
Friction, 125–126
Fuel, heating value, 43–44
   properties, 326
Fuel-cutoff rates, diesel cycle, 194–195

Gas (gases), actual, 305
   analysis, molal, 314–317
      volumetric, 313–314
      weight, 314
   molecular action, 10–11
   molecular weights, 36
   properties, 302–308
      mixed gases, 309–317
      and vapor, mixtures, 456–481
Gas constant, R, 31
   universal, 37
Gas flow, nozzle, 268–282
Gas tables, 302–305
Gas-turbine cycles, 249–267
   actual, 256–257
   analysis, 250–252
   efficiency, 252, 254, 257–258
   intercooling, 254
   regenerative, 252–254, 258–261
      intercooled, 261
      irreversible, 263–264
   reheat, 254
   reheat-regenerative, 254–256, 261–263
      intercooled, 262
   simple, 259–260
      irreversible, 263
Gas-turbine engines, 249–250
   compounding, 288
   condition line, 293–294
   energy equation, 233, 293
   energy transfer, 268–269
   examples, 259–264
   forces, 286
   impulse, 284–285
   locked wheel torque, 284
   pressure lines, 294–295
   reaction, 291–292
      blading, 291
   shaft work, 283–284
   stage efficiency, 294
   temperatures, 259
   velocity relations, 285–286
   (*See also* Buckets; Nozzles)

Gas-turbine–steam-turbine cycle, 417–420
General gas equation, 73
   (*See also* Perfect-gas equation)
Gibbs-Dalton law, 310–313, 457
Gibbs paradox, 317
Gravitation, Newton's law of, 8
Gravitational acceleration, 58
Gray, Paul N., 221

Heat, definition of, 13, 15
   electrical equivalent of, 42–43
   mechanical equivalent of, 41–42
   specific (*see* Specific heat)
Heat exchanger, energy equation, 234–235
Heat pump, Carnot, 118–119
Heat rate, bleeding cycle, 413
   gas-turbine–steam-turbine cycle, 420
   Rankine cycle, 388–389
   reheat-regenerative cycle, 416
Heat transfer, 13–15
   constant-temperature process, 103–105
   gas turbine, 254–256
Heaters in regenerative cycle, 406
Heating, and cooling, air-steam mixture, 468–471
   liquid and vapor, 340–346
   low-pressure, 344
   regenerative, 136–137
Heating value, 43, 334
   calculated, 337
   higher, 335
   lower, 336
Hilsch, Rudolph, 240
Horsepower, indicated engine, 433–434
Humidifying, adiabatic, 469–470
   heating and, 470
Humidity, atmospheric, 456–457
   relative, 460–461
   specific, 457–460
   wet-bulb thermometer measurement, 462–464

Ice, saturated, 340
   subcooled, 340
Impulse, 59
Impulse turbine, 284
Indicated horsepower, steam engine, 433–434
Input, station, 392
Intercooler, 152

Internal-combustion engine cycles, 171–206
  actual, 215–227
  ideal, 204, 223
  (*See also* Diesel cycle; Otto cycle)
Internal-combustion engines, pressure-volume changes, 215–216
  processes, 201–203
  rotary, 200–201
  shape, 203–204
  work output, 216–217
Internal energy, 2, 5, 13, 40
Irreversibility, Carnot cycle, 120–122, 129–131
  external mechanical, 124, 126
  external thermal, 123
  gas-turbine cycle, 263–264
  internal mechanical, 124, 125
  internal thermal, 123, 124
  Rankine cycle, 391
  reheat steam cycle, 397–398
  thermal, 120
Irreversible adiabatic processes (*see* Adiabatic processes)

Jet velocity, 272
Joule, James, 41, 64, 137
Joule cycle, 137–140
  regenerative, 139–140
  thermal efficiency, 139–140
Joule's experiment, 41–42, 74–75

Kilowatt-hour, 42–43
Kinetic energy, 3, 60, 229
Kinetic molecular theory, 56–70, 237–239
Knocking, diesel engine, 199
  Otto engine, 197

Laws of motion, Newton's, 56–59
Laws of thermodynamics, first, 45
  second, 101–102
Liquids, 9–10, 340–360
  compressed, 357–358
  molecular action, 9–10
Logarithms of pressure ratio, 79

Mach number, 278
Machine efficiency, 257–258
Magnetohydrodynamics, 322
Manometer, 19
Mass flow, 230
Matter-energy equation, 1
Maxwell, James Clerk, 67

Maxwell velocity distribution, 69
Mean effective pressure, 112
Mechanical equivalent of heat, 41–42
Mechanical reversibility, 89
Mechanical work, 1–2, 49, 52–53
Melting, 10, 340
Mercury–steam cycle, 416–417
Metastable state, 424
Mixing, adiabatic, 467
Mixtures, gas and vapor, 456
Moisture, steam, 356
Molal analysis, gas, 315
Mole volume, 37, 310
Molecular action, 7–12, 22–23, 47
  in steam cycle, 425–426
Molecular energy, 9–12
  kinetic, 60–63
Molecular flow, nozzle, 238–239
Molecular kinetic theory, 56–70
Molecular weights, 35–36
Molecules, 6–11
  collision, 11
  in straight path, 61–63
  with walls, 48, 49, 63–64
  forces, 7–8
  in heat transfer, 13–15
  motion, 8–9
  velocity, 66–68, 237–238
    temperature and, 68–70
Mollier diagram, 349–350, *following* 496
Momentum, 59
  conservation of, 60

Newton's laws, of gravitation, 8
  of motion, 56–57
    first, 57–58
    second, 58–59
    third, 59
Nonflow energy equation, 75
Nozzles, buckets and, 283–297
  coefficient of discharge, 427
  converging, 270
  converging-diverging, 270
  critical pressure ratio, 274, 277
  efficiency, 294, 427
  energy equation, 233–234, 236–237, 269, 427
  flow, 236–237, 271
  gas flow, 268–282
  molecular flow, 238–239
  performance, 273, 276, 427
  pressure, 270–277
  sonic velocity, 278
  static pressures, 279–281

Nozzles, steam, 426–428
  diffusers, 428–430
  throat, 274, 278
  turbine, 283
  variable back pressure, 277
  variable inlet pressure, 278–279
  velocity coefficient, 427
Nuclear fission, 320
Nuclear fusion, 320

Organic vapors, 358
Otto cycle, 172, 174–193
  analysis, actual, 215
    theoretical, 179–182
  compared with diesel, 193–197
  performance, 175, 183–184, 189–190
  polytropic, 184
  pressure, temperature, and compression ratios, 178–179
  supercharged, 186–193
  temperature effects, 175–177
  thermal efficiency, 175–177
  water-cooled, 182–186
  work variation, 177–182
Otto engine, 171–174
  detonation, 197–199
  pressure-volume changes, 215, 216
Oxygen supply, combustion, 324–326

Perfect-gas equation, 26, 30–35, 73, 457, 487–488
Polytropic process, 81–82, 149, 158
  entropy change, 109
  Otto cycle, 182–186
  reversible, 486
Porous-plug experiment, 75
Potential energy, 3, 229–230
Preignition, diesel engine, 199
  Otto engine, 197
Pressure, 16–18, 49
  absolute, 18
  atmospheric, 16–18
  gage, 18
  gas, kinetic theory, 64–66
  mean effective, 112
  optimum intermediate, 488
  stagnation, 270
  static, 270, 279
  steam reheat, 396
  variations, irreversibility, 123–125
Pressure compounding, turbine, 288
Pressure drop, steam, 391–392
Pressure gages, 19–20

Pressure ratio, 115
  critical, 274, 489
  logarithms, 79
  Otto cycle, 179
Pressure-temperature relations, 28–29, 343–344
Pressure-time diagram, internal-combustion engine, 219
Pressure-volume chart, 61
Pressure-volume relations, 23–24, 53–54
  Boyle's law, 20–22
  internal-combustion engine, 215–216
  P-V graph, 61, 215, 216, 345
  steam engine, 345–346
Processes, air, 144–146
  comparison of, 82–87
  energy equations for, 76–82
  irreversible, 129–131
  steam, 361–381
  (See also specific processes)
Psychrometric chart, 465–467
Pump work, 383

Rankine cycle, 382–383
  availability loss, 393
  back pressure, 386
  compared with Carnot, 389–390
  efficiency, actual, 391–393
    thermal, 383–386
  heat rate, 388–389
  ideal, 389
  practical, 388, 389
  steam rate, 387–388
  superheating, 385
  superpressure, 385–386
  throttle pressure and temperature, 387
Rankine temperature, 28, 32
Ranque, Georges, 240
Ranque-Hilsch tube, 240–241
Ratio of specific heats k, 52
Reaction turbine, 291
Receiver, 91
  compressed air, 156
  refrigerated, 119–120
Reciprocating compressors, 143
Reduced properties, 307
Refrigerating cycles, absorption, 450–451
  Carnot, reversed, 438–439
  cascaded, 449–450
  compressed air, closed system, 165–166

INDEX

Refrigerating cycles, intercooled, 446–447
  performance, 439–440
  regenerative intercooled, 447–449
  vacuum, 451–452
  vapor-compression, 441–443
Refrigerating fluids, 441–446
Refrigeration, tons of, 440
  vapor-compression, 438–455
Refrigerator, Carnot, 115–117, 438
Regenerative cycle, gas turbine, 252–254
  Joule, 139–140
  steam, 399–401
    bleeding, 401–413
Regenerative heating, 136–137
Regenerator, gas turbine, 252
  Stirling cycle, 135
Reheat cycle, steam, 394–399
  irreversibility, 397–398
  performance, 398–399
Reheat factor, turbine, 295
Reheat pressure, 396
Reheat-regenerative cycle, steam, 413–416
Reheating, multiple, 396–397
Reversed Carnot cycle, 112–122, 438–439
Reversibility, mechanical, 72–73, 89
  Stirling cycle, 135
  thermal, 89–90
Root-mean-square velocity, 68, 70
Rotary engine, 200–201

Saturated vapor, 342
Second law of thermodynamics, 101–102
Shaft work, 94
Shock wave, 280
  condensation, 423
Simple energy equation, 75
Sodium, thermodynamic properties, 490–496
Solar radiation, 322–323
Solids, 6
Sonic velocity, 278
Source of energy, 91, 92
Specific heat, 45–47
  common substances, 46
  constant-pressure, 50–52
  constant-volume, 47, 50–52
  gas, 47–52
  pressure-volume relations, 53–54
  steam, 348–350
Stagnation enthalpy, 271

Stagnation pressure, 270
Stagnation temperature, 271
Standard commercial ton, 440
Steady-flow energy equation, 228–248
  air induction, 243–245
  blowdown cooling, 241–243
  fan, 232–233
  gas turbine, 233
  heat exchanger, 234–235
  nozzle, 233–234, 236–237
  Ranque-Hilsch tube, 240–241
  summary, 230–231
  throttling, 235–236
  vortex flow, 239–240
Steam, Mollier chart, 349, *following* 496
  P-T graph, 344
  P-V graph, 345
  quality, 356
  T-H graph, 348
  T-S graph, 346
  thermodynamic properties (*see* Steam tables)
  wet mixtures, 356–357
Steam cycles, 382–422
  bleeding regenerative, 401–413
  Carnot, 90, 134, 389
    reversed, 112
  components, 423–437
  engine, 430–432
  gas-turbine–steam-turbine, 417–420
  irreversibility, 397–398
  mercury–steam, 416–417
  Rankine, 382–390
  regenerative, 399–401
  reheat, 394–399
  reheat-regenerative, 413–416
Steam engines, 430–436
  compound, 435
  reciprocating, 430
  uniflow, 435
Steam processes, adiabatic, irreversible, 365–367, 373–376
  reversible, 364–365
  constant-entropy, 364–365, 372–373
  constant-pressure, 361–362, 367–369
  constant-temperature, 363–364, 371–372
  constant-volume, 363, 369–370
  reversible heating, 378–380
  throttling, 367–368, 376–378
Steam rate, 387–388
Steam tables, 350–356
  high-pressure, 352, 355, 356
  saturation–temperature, 350–352

Steam tables, superheated steam, 352–354
Stirling, Robert, 221
Stirling cycle, 133–136, 223, 224
Stirling engine, 221–226
Subcooled liquid, 342
Sublimation, 342
Supercharging, diesel engine, 197
　Otto cycle, 186–193
Superheated vapor, 342
Superheating, Rankine cycle, 385
　steam engine, 434–435
Supersaturated state, 424
Supersaturation, 423–425

Temperature, absolute, 27–28
　in Carnot cycle, 96–99, 102
　dry-bulb, 462
　in energy conversion, 101–102
　molecular effects, 8–10
　pressure relations, 28–29, 343–344
　scales, 26, 28
　stagnation, 271
　velocity and, 68–70
　volume relations, 29–30
　wet-bulb, 462–464
Temperature-entropy graph, 109, 347–348
Temperature ratio, Otto cycle, 178
$T\text{-}H$ graph, 348–349
Thermal efficiency, Carnot cycle, 95–96, 139
　diesel cycle, 194–196
　free-piston cycle, 213
　gas turbine, 257–258
　Joule, regenerative, 139
　　simple, 139
　Otto cycle, 175–177
　Rankine cycle, 383–386
　reheat cycle, 396
Thermal irreversibility, Carnot cycle, 120–122
　external, 123
　internal, 123, 124
Thermal reversibility, 89–90
Thermionic generator, 321
Thermoelectric generator, 321
Thermometer, wet-bulb, 462–464
Throat, nozzle, 274
　properties, 278
Throttle $P\text{-}T$, Rankine cycle, 387
Throttling, energy equation, 235–236
　steam process, 367–368, 376–378
Triple point, steam, 343

$T\text{-}S$ graph, 109, 347–348, 384
Tube, Ranque-Hilsch, 240
Turbine (see Buckets; Gas-turbine engines; Nozzles)
Two-phase flow, 425
Two-stroke-cycle engines, 219

Unavailable energy, 101
Universal gas constant, 37

Valve timing, 218–219
van der Waals' equation, 305–306
Vapor, 10, 340–360
　and gas, mixtures, 456–481
　metallic, 490–496
　molecular action, 10–11
　organic, 358
Vapor-compression refrigeration, 438–455
　(See also Refrigerating cycles)
Vaporizing, 341–342
Velocity, distribution, 66–68
　jet, 272
　molecular, 66–70, 237
　sonic, 278
　temperature and, 68–70
Velocity compounding, turbine, 289
Volume, specific, 30
Volume-pressure relations (see Pressure-volume relations)
Volume-temperature relations, 29–30
Volumetric analysis, gases, 313–314
Volumetric efficiency, conventional, 167
Vortex flow, 239–240

Wankel, Felix, 200
Wankel engine, 200
Water, phase characteristics, 340
　specific, 46
Water-cooled (polytropic) Otto cycle, 182–186
Water vapor, combustion, 332
　(See also Vapor)
Watt, James, 200
Weight analysis, gas mixture, 314
Wet-bulb thermometer, 462–464
Wet mixtures, steam, 356–357
Wilson line, 424
Work, 1, 15
　adiabatic, 483
　constant-temperature, 103–106, 483
　flow, 228
　mechanical, 1–2, 49, 52–53
　pump, 383
　shaft, 283